W0018583

RADIOPHARMACEUTICALS

RADIOPHARMACEUTICALS
Introduction to Drug Evaluation and Dose Estimation

Lawrence E. Williams

City of Hope National Medical Center
Duarte, California, USA

CRC Press
Taylor & Francis Group
Boca Raton   London   New York

CRC Press is an imprint of the
Taylor & Francis Group, an **informa** business

Cover image from Proffitt, R. T., L. E. Williams et al. *Science*, 220, 4596, 1983. With permission.

CRC Press
Taylor & Francis Group
6000 Broken Sound Parkway NW, Suite 300
Boca Raton, FL 33487-2742

© 2011 by Taylor and Francis Group, LLC
CRC Press is an imprint of Taylor & Francis Group, an Informa business

No claim to original U.S. Government works

International Standard Book Number: 978-1-4398-1067-5 (Hardback)

This book contains information obtained from authentic and highly regarded sources. Reasonable efforts have been made to publish reliable data and information, but the author and publisher cannot assume responsibility for the validity of all materials or the consequences of their use. The authors and publishers have attempted to trace the copyright holders of all material reproduced in this publication and apologize to copyright holders if permission to publish in this form has not been obtained. If any copyright material has not been acknowledged please write and let us know so we may rectify in any future reprint.

Except as permitted under U.S. Copyright Law, no part of this book may be reprinted, reproduced, transmitted, or utilized in any form by any electronic, mechanical, or other means, now known or hereafter invented, including photocopying, microfilming, and recording, or in any information storage or retrieval system, without written permission from the publishers.

For permission to photocopy or use material electronically from this work, please access www.copyright.com (http://www.copyright.com/) or contact the Copyright Clearance Center, Inc. (CCC), 222 Rosewood Drive, Danvers, MA 01923, 978-750-8400. CCC is a not-for-profit organization that provides licenses and registration for a variety of users. For organizations that have been granted a photocopy license by the CCC, a separate system of payment has been arranged.

Trademark Notice: Product or corporate names may be trademarks or registered trademarks, and are used only for identification and explanation without intent to infringe.

Library of Congress Cataloging-in-Publication Data

Williams, Lawrence E., author.
    Radiopharmaceuticals : introduction to drug evaluation and dose estimation / Lawrence E. Williams.
       p. ; cm.
    Includes bibliographical references and index.
    ISBN 978-1-4398-1067-5 (hardcover : alkaline paper)
    1. Radiopharmaceuticals. I. Title.
    [DNLM: 1. Radiopharmaceuticals. 2. Clinical Trials as Topic. 3. Radiopharmaceuticals--administration & dosage. WN 415]
    RS431.R34W55 2011
    615.8'424--dc22   2010044822

Visit the Taylor & Francis Web site at
http://www.taylorandfrancis.com

and the CRC Press Web site at
http://www.crcpress.com

I would like to dedicate this volume to my wife, Sonia Bell Williams, and our two children, Erica Helen and Beverley Ann. I could not have achieved these results without their continuing support and love. They have been the wonderful lights in my life.

Contents

Foreword

The topical material and the purposes encompassed in this text on drug targeting are timely and relevant to current drug development. The text has been edited by an illustrious medical physicist whose career has been "in the trenches" of a leading program involved in the investigation and clinical uses of radiation targeted by drugs.

During his and my careers, the term *targeted* has become adulterated. The term was originally intended to refer to drugs that achieved a substantially greater concentration in the target, most commonly cancer cells, than in normal nontarget cells made possible because of specific and tight binding to a moiety characteristic of and accessible in the target cells. Targeted was intended to distinguish new classes of drugs, notably monoclonal antibodies (mAbs), from the chemotherapeutic drugs used during the previous one half century. The latter rarely achieve target to nontarget concentration relationships greater than 1 1/2, whereas mAb-based drugs commonly achieve concentration ratios greater than 10. Unfortunately, newer cancer chemotherapeutics, such as the protein kinase inhibitors, are often referred to as *targeting drugs*. Like most earlier chemotherapeutic drugs, they have a specific target but do not achieve favorable cancer to nontarget concentration ratios. It should not be surprising that these drugs have undesirable nontarget effects.

Whereas mAbs recognize cancer cells, size limits their value: as size decreases, tissue penetration increases. Although smaller than an intact mAb, mAb fragments are appreciably larger than chemotherapeutic drugs. In addition to newer chemotherapeutics and to smaller drugs derived from mAbs, there is an almost infinite list of targeting drugs, such as peptides, aptamers, affibodies, and selective, high-affinity ligands (SHALs) in development. Peptides are very small. Notable groups bind to somatostatin and related growth-factor receptors; they have been translated to the clinic. Aptamers, affibodies, and SHALs are promising classes of molecules having small size, high affinity, and specificity.

A good targeting drug is often referred to as a *magic bullet*. However, all drugs move throughout the body and thus can have many effects within the patient. Effects may occur because of the targeting drug or because of the drug attached to the targeting drug. If the drug is radiolabeled, its movement in the patient may be followed using external detection equipment. Cancer chemotherapeutics and other drugs can be followed to assess their pharmacokinetic behavior and their targeting in an individual in this manner. If a diagnostic radionuclide (radioisotope) is attached to the drug, its movement in vivo can be followed noninvasively over time and as a function of manipulations. In addition to initial detection of a cancer, drug-based imaging to determine its *phenotype*, to measure its target levels, to select patients more likely to respond, to determine dose, and to assess treatment response plays important roles in the development, evaluation, and implementation of a therapeutic. Additional purposes for pharmacokinetics (and radiation dosimetry) in drug development and use include (1) determining drug amount needed for targeting; (2) characterizing drug pharmacokinetics (and radiation dosimetry); (3) comparing competing drugs; (4) assessing pharmacokinetic variability; (5) decisions regarding sequence and timing of drugs; and (6) providing data for patient-specific dosing and planning.

The screening, development, and implementation of drugs can be improved by using the approaches described in this text for characterizing their tissue distributions (concentrations) and pharmacokinetics, when given alone, when attached to another drug, or when given along with other drugs. In recognition of the importance of this information for drug development and implementation, pharmaceutical companies have developed "in house" these capabilities even though they are not trivial.

Most patients with locoregional cancer are cured by surgery, radiotherapy, chemotherapy, and combinations thereof; despite systemic chemotherapy, those with distant metastases often are not. Cancers become chemoresistant yet remain responsive to radiotherapy. Although these patients respond to local radiotherapy, they require systemic therapy. Molecular-targeted radiotherapy (MTRT) is a strategy for the treatment of multifocal and radiosensitive cancers.

Drug effects can occur because of the bioactivity of the targeting drug or the drug attached to the targeting drug or both. The cytocidal potency of a targeting drug can be greatly augmented by attaching a therapeutic radionuclide as a radiation source for MTR. Radionuclide emissions can destroy cells to which the drug is attached and surrounding cells by a *bystander effect*. Fixed, population-based, or individualized radionuclide (radiation) dosing has been used. A fixed radionuclide dose assumes little pharmacokinetic variability between patients. Individualized dosing requires estimated radiation dosimetry analogous to that known critical for other forms of radiation therapy. In an era of "personalized" medicine, individualized dosing is attractive because few would suggest that a drug has the same effects in all patients.

If a therapeutic radionuclide is attached to the targeting drug to provide systemically delivered radiotherapy, then the radiation distributions that logically follow provide essential information. Evidence is abundant for the radiation dose and tissue response relationship for radiotherapy, including for radionuclide therapy. However, the radiation effect on a tissue reflects both its radiosensitivity and the radiation absorbed by the tissue.

This text amplifies issues for drug development and describes methods for estimation of radiation doses. The latter process, *radiation dosimetry*, is complicated for internal emitting radionuclide therapy. In the past decade, the concept of a patient-specific radiation dose has become accepted.

Gerald L. DeNardo, M.D.
Professor Emeritus of Internal Medicine, Radiology, and Pathology

Preface

Having a physicist write on the topic of radioactive drugs may seem usual. Yet there are important issues that appear to be sparsely described in the development of such agents as well as in their entry into clinical trials. In a therapy context, the estimation of radiation dose must be computed with a particular patient rather than a phantom for therapy applications of such materials. In this second context, a physicist's exposition can also be helpful. I feel that the text would be useful to pharmacists, physicists, radiation dosimetrists, nuclear medicine practitioners, medical oncologists, and radiation oncologists. As will be seen clearly in the following, many disciplines are involved in the generation and testing of novel agents. We must all work together to evolve the best radiopharmaceuticals (RPs) that we can make for the use that is proposed. To make that evolution practical, improved pharmaceutical criteria and absorbed dose estimates are essential.

My interest in this field is due to an approximately 30-year involvement with cancer patients at the City of Hope National Medical Center. Of particular interest is the location and possible treatment of multiple sites of metastatic disease. This situation is one whereby patients and their families feel utterly lost—without having any real solution at hand. Chemical therapies may simply not target, and local strategies such as surgery or external beam cannot be applied to the multiple sites that are present.

Probably the fundamental aspect of medicine continues to be an ongoing desire to have specific drug treatment for a given set of symptoms. This goal is presumably prehistoric and was equally true for primitive humans woefully looking upon their ill fellows. An ideal result would be a one-to-one mapping of pharmaceutical to disease. Such a perfect agent is often called a *magic bullet*. However, this concept is a logical paradox as all drugs move, in principle, throughout the body and thus can have effects at many sites within the patient. Additionally, today, multiple possible drugs may be applied, at least theoretically, to any clinical situation. For a particular individual, the correctness of the nonradioactive drug selection cannot be clearly demonstrated since the targeting to the tissues of interest is not overtly demonstrable. In other words, the patient is a "black box" into which the drug is given with hope of eventual resolution of the symptoms. Possible toxic effects as well as wasted time and prolongation of the illness are downsides of selecting an incorrect drug in the first analysis of treatment. I intentionally do not use the term *side effects*. By definition, a drug can have only effects. If some of these outcomes are of a morbid nature, it is particularly absurd to refer to them as side effects.

Because of recombinant technology and the desire to find specific targeting agents, we live in an era in which drug development is going on at a geometrically increasing pace. The bioengineering world in which these manipulations are occurring has spatial dimensions on the order of nanometers. In the text, I call this the problem of plenty. Many new designs are being proposed, almost daily, and multiple variations on each theme are possible by changing atomic constituents, electric charge, molecular weight, and other parameters. Additionally, the various agents may be given in time sequence to achieve the desired targeting. A variety of different types of agents are being constructed in nanotechnology starting from classical entities such as liposomes and going through to antibodies and their engineered descendants. Segments of amino acids, DNA, or RNA are also being proposed for specific targeting applications. In these research efforts, the investigators have often attempted to design their constructs to attach to a given molecular target. A primary application of many of these novel structures is locating a tumor. Following success in that process, there may be the additional aspect of therapy. In the latter case, treatment may be effected by the radioactivity associated with the targeting agent or by the agent itself. Normal tissues—and their associated molecules—are other possible sites of interest to the pharmacologist. However, in these cases, there is generally not going to be an intended therapy by means of ionizing radiation. Instead, the nanostructure can be engineered to carry a therapeutic entity such as a missing gene or protein into the tissue.

One person often mentioned in the conceptual development of engineered structures at nanometer scale is Richard Feynman. He is reported to have said in 1965 that there is "plenty of room at the bottom" when referring to designing entities that could function at molecular sizes. In this regard, I would like to relate a story told me by my colleague Richard Proffitt. Dick remembers that in the late 1950s or early 1960s a speaker from CalTech came to his junior high school in Pasadena, California. Upon taking the stage, Dr. Richard X (Dick remembers that he and the speaker had the same first name but does not recall the lecturer's last name) produced several pieces of rubber inner tube, some mercury, and a container of liquid nitrogen. A reference to the low temperature of space and possible lunar travel was then given. After placing the mercury into a spoon and inserting an ice cream stick vertically into the metallic liquid, Dr. X lowered the combination into the nitrogen. This action was replicated with the inner-tube segments. Keeping his insulated gloves on and using the stick as a handle, the demonstrator then proceeded to use the frozen mercury as a hammer head to drive the solidified rubber into a piece of wood. The now solid mercury flashed in the stage lighting as he repeatedly lifted the hammer while driving in the solid rubber nails. Students, generally inattentive at such talks, went wild.

During the time of immersion and hammering, there was an ongoing patter indicating that reducing heat levels may make liquids or soft materials into hard objects by fixing atomic bonds. Dr. X also mentioned that there were large spaces between atoms inside the molecules and that future scientists would use robots, perhaps iteratively, to produce smaller and smaller devices that, in turn, could extend engineering down to the molecular level. Dick recalls that he did not believe that such entities could ever be built and that molecular robots were beyond imagination—at least in 1960. At that time, the only robots looked like people and shuffled around on the set of B-grade films. (It is not far from Pasadena to Hollywood.) Moreover, everyone knew that there was a B-grade actor inside those metallic suits so that robots were creatures of the stage and not of the real world.

My guess is that Dr. X was Richard Feynman and that his discussion at Marshall School was a precursor to his recorded comments to the American Physical Society in

1965. He might have hoped that someone in his audience that day would carry out his schematic research project. Indeed, that is what happened approximately 20 years later. In fact, the earliest work at City of Hope involved the chemistry department at CalTech as our manufacturing source of such small robots—phospholipid vesicles. These entities were engineered to be stable in blood—unlike naturally occurring liposomes. In addition, their size was controlled so they could pass through the capillary walls of solid tumors. Liposomes became a trafficking method to carry materials inside the lesion space for both imaging and therapy. To follow these engineered structures in the body of mice and patients, we used a radioactive label (^{111}In).

When we radiolabel the biological construct, we may track its movement using external detection equipment. This striking aspect is missing in conventional drug development, although there is the possibility of labeling a given material and then attempting to follow its physiological processing. In the text, we point out that chemotherapeutics against cancer should be labeled and followed in this fashion to test—in an explicit way—targeting to disease sites in an individual. It is also possible that targeting may change as a function of time following initial treatment. To some degree, such variation is expected due to the tumor's attempt to replicate in a hostile environment (i.e., the Wallace–Darwin argument for species survival).

This text is written to amplify two basic issues in our era of multiple agent development. First on the agenda is the historical lack of a good method of differentiation of one radiopharmaceutical (RP) from another. I point out early in the discussion that traditional figures of merit used in imaging and therapy are not sufficient. In the former case, the ratio of tumor to blood uptake (r) cannot generally be used for an imaging indicator for a number of reasons. Five similar engineered antibodies are used in the comparison of r and the novel imaging figure of merit (IFOM) derived at City of Hope. In both direct animal imaging comparisons and simulations done with Monte Carlo images, the r indicator is not appropriate, and IFOM is a distinctly improved measure of the improvement obtainable by changing proteins or radiolabels. A similar argument is made for the therapy figure of merit (TFOM) compared with the traditional ratio (R) of the area under the curve for tumor divided by the same integral for the blood.

Our second extensive area of exposition is improving the methods for estimation of radiation doses. This process, unfortunately termed *dosimetry* in most of the nuclear medicine discussions, is complicated in internal emitter therapy of cancer. Complexities occur due to the several analytic stages necessary in going from an image distribution of relative activities to the actual amount of activity in a given organ or voxel. Only in the past decade has the concept of patient-specific radiation dose become well established in the RP literature. Yet, even today, correlations between these estimated absorbed doses and clinical outcomes are rare. Reasons for this discrepancy reflect at least two areas: difficulty in prediction of absorbed dose as well as the question of exactly what is the dose parameter. Various correction factors are now used to moderate the analytic dose estimate to better predict the experimental or clinical outcome. Such factors reflect the type of radiation and its spatial as well as its temporal variation.

A third area that I describe in some detail is the correlation, if any, between animal and human biodistributions and kinetics for a given RP or other agent. It has often been assumed that the mouse is a good representation of the patient when testing new RPs. What should be attempted in the allometry is a direct comparison of one mammal with another with regard to specific engineered nanostructures. I recommend that the U.S. Food and Drug Administration (FDA) require applicants for clinical trials to submit a direct comparison of their eventual clinical outcome with the mouse (or other animal) results originally submitted for clinical approval. Only in this explicit way can we, as a scientific body, come to recognize how to correlate results in the various species. To paraphrase George Orwell, it may turn out that some agents are more equal, across species, than others.

Acknowledgments

I would like to salute the technical and medical staff members at City of Hope who have made so much of this work possible. Particular thanks go to Richard Proffitt, Cary Presant, John Shively, Anna Wu, David and Barbara Beatty, Jeffrey Wong, and Andrew Raubitschek. The support of my chairs, Hyman Gildenhorn and J. Martin Hogan, has been very important. I want to acknowledge Dave Yamauchi, the head of nuclear imaging, who has suffered through many hours of reading clinical images, and also An Liu, the implementer of much of the programming that was needed to set up the dose estimation system. Finally, I wish to remember my late Ph.D. advisor, John H. Williams. He would admonish both students and staff alike that part of our work must be "putting some chips back in the pot." Here are a few chips in John's memory.

About the Author

Lawrence E. Williams, Ph.D., is a professor of radiology and an imaging physicist at City of Hope National Medical Center in Duarte, California. He is also an adjunct professor of radiology at University of California–Los Angeles (UCLA). While in high school, he was one of 40 national winners of the Westinghouse (now Intel) Science Talent Search. Dr. Williams obtained his B.S. from Carnegie Mellon University and his M.S. and Ph.D. degrees (both in physics) from the University of Minnesota, where he was a National Science Foundation (NSF) fellow. His initial graduate training was in nuclear reactions at Minnesota, where he demonstrated excited states of the mass-4 system (^{4}He*). He later extended this work by finding excited levels of mass-3 nuclides while working at the Rutherford High Energy Laboratory in England. Since obtaining the National Institutes of Health (NIH) support to become a medical physicist, Dr. Williams has devoted most of his research to tumor detection and treatment and has written approximately 250 total publications as well as a number of patents in nuclear imaging and radionuclide therapy. He is a coauthor of *Biophysical Science* (Prentice Hall, 1979) and editor of *Nuclear Medicine Physics* (CRC Press, 1987). He has been a grant and site reviewer for NIH since the mid-1990s. Dr. Williams is associate editor of *Medical Physics* and a reviewer for several other journals. He is a member of the American Association of Physicists in Medicine (AAPM), the Society of Nuclear Medicine, the New York Academy of Sciences, Sigma Xi, Society of Imaging Informatics in Medicine (SIIM), and the Society of Breast Imaging. Dr. Williams has received a lifetime service award from the American Board of Radiology.

Among Dr. Williams' most significant biophysical discoveries is the mass-law for tumor uptake as a function of tumor size. He was also codiscoverer (with Richard Proffitt) of tumor targeting with liposomes. This work involved one of the first applications of normal organ blockage by use of an unlabeled agent—that is, a two-step process. Dr. Williams has developed a pair of indices for quantifying the ability of a radiopharmaceutical to permit imaging or therapy of lesions in animals or patients. He has also demonstrated that radioactive decay must be considered inherently as one possible exit route in modeling analysis of radioactive drugs. With his colleagues at City of Hope, Dr. Williams measured and calculated the brake radiation dose result for a source of ^{90}Y in a humanoid phantom. This study remains as one of the few examples of a comparison of dose estimates and measurement in the nuclear medicine literature.

Tumor Targeting and a Problem of Plenty

1

This text deals with issues relating to the selection and use of radiopharmaceuticals (RPs), which are designed to target disease sites in the patient. Ideally, specific molecular targets may be found in this fashion. Nuclear medicine imaging has thereby been recently redefined as an aspect of the more general topic: molecular imaging. One could generalize these considerations to include all pharmaceuticals—if the latter were labeled so that they could be tracked inside the animal or patient. We will also propose that such labeling be increased in particular areas such as chemotherapeutics. One eventual objective would be a patient-specific imaging and, perhaps, therapy.

Of greatest interest in molecular imaging today are the possibly multiple anatomic locations associated with malignant disease. Treating disseminated tumor cells has been an ongoing problem in oncology since its inception and has had no realistic solution. This failure has two obvious aspects that occur in a logical sequence: finding these locations and then treating disease *in situ*. The medical problem is essentially one of hunting down malignant cells and then rendering them incapable of dividing at each location. A first step in any such strategy is to image at least some of the sites and to determine the existence of disease prior to beginning any possible therapy or surgical intervention.

A problem of plenty arises because of the advent of extensive engineering at the nanometer scale. As Richard Feynman reportedly said at an American Physical Society meeting at Pasadena in 1965, "There is lots of room at the bottom." Thus, many different types of agents are in the process of being invented or, if already invented, somehow perfected in a clinical context. As we will see, more than six different types of tracer are already in at least the development stage. Yet each of these has many possible variants due to changes in the construction parameters. For example, physical size, molecular weight (MW), amino acid sequence, RNA sequence, DNA sequence, electric charge, and other variables can be adjusted. But how is the clinical investigator going to select one type or variant over another for a possible trial? Any trial is time-consuming and expensive; if a wrong agent is chosen, the entire project of molecular detection is set back since a limited number of local patients are being wasted in a vain pursuit. This aspect is described more completely in the next two chapters. There, we consider strategies to select an optimal agent based on animal data sets and certain figures of merit.

It may be possible that an engineered creation is itself designed to be toxic or to carry a chemical toxin to the tumor. For example, an antibody-based targeting molecule could, in principle, trigger the patient's immune system to recognize one or more proteins at the malignant cell's surface. Short lengths of RNA or DNA, if achieving cell entry, could interfere with the production of proteins or the replication of tumor cells. Ultimately, the designer of the targeting moiety may simply include one or more standard chemotherapy drugs within the agent. Upon targeting, these may be released on site using the tumor cell's metabolism to "open the drug package" and release the toxin. These strategies will also be described in the following chapter.

Radiopharmaceuticals have another therapeutic aspect that is inherent in their use. In the early history of nuclear medicine, this was exemplified by the treatment of thyroid cancer using radioactive ^{131}I given as Na^{131}I. Radiation doses are administered via the RPs in a tissue-specific and individual way to each patient. By this we mean that the amount of radiation damage would vary greatly from one organ to another in a given person while that person's results could be very different from a second individual receiving the same amount of RP. While probably not very large in a diagnostic context, these variable dose values must be greatly increased and controlled to be useful in radiation therapy. It is this objective that will be covered extensively in several of the following segments such as Chapters 4, 8, and 9.

We emphasize at the beginning, as will be shown in Chapter 4, that such absorbed dose values must be estimated. There are practical and ethical reasons doses cannot be measured in living tissues. Because of the associated uncertainty of these estimates, positive correlations between absorbed dose and clinical effects are often difficult to observe. In Chapter 9, we discuss the very limited history of such correlations.

1.2 THE EXTENT OF DISEASE

Figure 1.1 shows a medullary thyroid cancer patient's antibody scan with multiple lesions appearing in the pelvis and femurs. Chemotherapy has been tried and has not been able to reduce these lesions in size or number. External beam therapy would be limited to "sharp shooting" of a few particular sites that cause pain. While site-directed surgery is generally not an option in bone lesions, it may be a possible intervention for soft-tissue metastases such as liver sites in the case of a breast or colon primary cancer. Even in these situations, the surgeon will generally tell the radiologist that the number of lesions seen *a priori* on the nuclear medicine (or computed tomography [CT] or magnetic resonance imaging [MRI]) scan is probably only a fraction of the total number seen via exploratory laparotomy in the operating room. Likewise, the pathologist will point out to the surgeon that among the tissue samples taken (some of which are presumed to be benign), there are often surprising discoveries of unknown tumors or

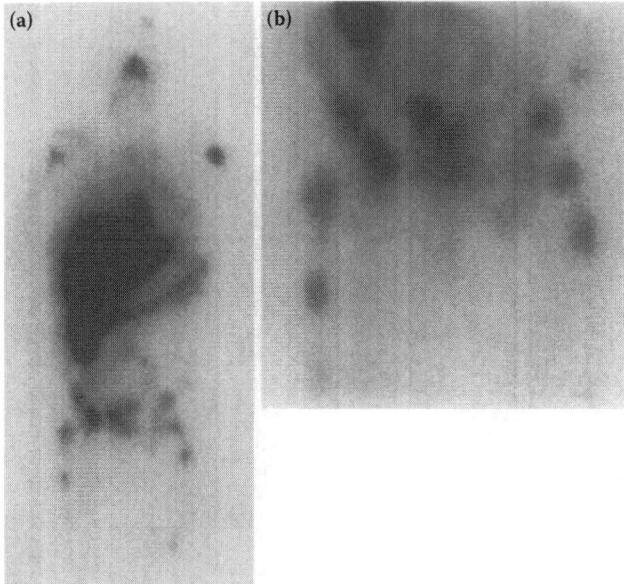

FIGURE 1.1 Medullary thyroid cancer patient imaged at 48 h postinjection of intact anti-carcinoembryonic antigen (anti-CEA) antibody labeled with ^{111}In. An anterior image, obtained with a gamma camera, shows multiple metastatic lesions in pelvis, femurs, and shoulders. (From Wong J. Y. C. et al., *J. Nucl. Med.,* 38, 12, 1997. Reprinted by permission of the Society of Nuclear Medicine.)

even single malignant cells. These sites can be seen only upon microscopic examination in the laboratory. Thus, through direct investigation, the number of disease locations is often found to be greater than that estimated using any single radiological examination—or even a combination of such exams. Patients and their oncologists are therefore presented with the problem of treating those tumors that are seen as well as those that are unseen. This text will describe both of these objectives.

Disparity between the number of visualized disease sites and the actual value determined upon direct investigation is important. It implies that a systemic process must be used in the treatment phases. Usually, this is done via the blood circulatory system with an intravenous (IV) injection into the patient. Alternatively, an intraperitoneal (IP) injection may be employed in ovarian Ca or even a direct injection into the tumor bed, such as in a breast cancer patient after partial mastectomy.

The reader will notice an inherent dependency on the blood circulatory system for the eventual success of almost any targeting agent. A living body is a busy enterprise where there is a long evolutionary history of multiple and interacting organs. Important among these are the liver, kidneys, and marrow. Any one of the tissues may interrupt the smooth passage of the tumor-seeker due to biologically preordained clearance mechanisms. For example, the liver may sequester colloidal materials or liposomes of certain sizes due to the clearance function of its Kupffer cells. Hepatocytes become involved in the ultimate sequestration of antibodies. Kidneys excrete small molecules such as short lengths of proteins or nucleic acids. It is therefore unlikely that the "magic bullet" so beloved by clinicians can ever be engineered. Any constructed moiety must pass through a number of organs, via their circulation, before getting to all tumor sites. The various organs, in an evolutionary sense, have "seen it all" and are able to respond to almost any RP design that fits into their sequestration or excretion programming.

One feature of targeting research debated by all RP investigators is an interest in the kinetics of an agent's blood curve. This function is generally monotonically decreasing and is often measured in chemotherapy trials. Yet the ideal form of the curve remains elusive, and very different strategies have emerged. Some investigators prefer that the radioactive agent's blood circulation time in man be short—on the order of minutes. This is to eliminate the moiety from the patient before toxicity may occur in the red marrow, for example. Others argue for a much long circulation time so that the agent may continue to accumulate and deliver its cargo to the (possible) multiple tumor locations. In this case, blood circulation lifetimes on the order of multiple hours to days are desired. Clearly some sort of compromise is in order whereby a balance between toxicity and therapy is possible. A common name for this eventual clinical strategy is treatment planning; it will be described more completely in Chapters 8 and 9 for RPs.

To find a specific molecular or other entity requires that the targeting moiety have access to the sites of that target. When these molecules are on cell surfaces or interstitially among the cells, the process is more efficient than if the targets are inside a cell or its nucleus. Implicit in the strategy is that blood or lymphatic circulatory systems will bring the radiotracer into proximity of the target species. One difficulty with some molecular targets is that they may be shed from the original cells into one or both circulatory systems. If this occurs, the agent may complex with them in sufficient numbers so that the fraction localized at their cellular origins may be greatly reduced.

Disseminated disease has long been a justification for chemotherapy. Yet, even today, little is known about the actual distribution of such toxic agents as they are rarely radiolabeled. Beyond that, the question arises as to how effective a given chemotherapeutic is in a particular patient at any one time. It may be that the requisite targeting is lacking in an individual such that the intervention causes only normal tissue toxicity and little benefit. There are also issues of tumor cells becoming resistant over the course of treatment to any previously effective therapeutic due to Darwin–Wallace survival selection. Tumors are not a set of cellular clones whereby each cell is identical to all others. Instead, there is probably variable expression of any molecular marker that may be used in the targeting phase. Such variability may be one reason the malignancy has escaped destruction by the body's own immune system in the first place. In other words, systemic treatments must be evaluated, *a priori*, each time they are to be performed. This issue occurs in any therapy protocol, and we will return to it again in our discussion of monoclonal antibody treatments.

A proof of targeting of the purported therapy agent is needed prior to beginning treatment. Radiolabeling of the agent is one of the best ways to demonstrate such localization, although, as already noted, resultant images cannot give a complete description of the extent of the disease. An obvious question is how to produce chemical moieties that target to cancer sites—or even to cancer cells in vivo. Before we describe production of possible novel RP agents, we will need to discuss various radiolabels and the several radiolabeling techniques commonly used in clinical practice.

1.2.1 Radioactive Decay

Radiopharmaceuticals require radioactive atoms to be attached to them to be followed from outside the living body. Nuclei of such atoms spontaneously decay because there is a lower-energy nucleus (which may in turn also be radioactive) available. This is an example of Albert Einstein's energy–mass relationship whereby the daughter nucleus has the lower mass value. Decays go on independently of whether the parent nucleus is attached to any specific chemical form and are defined by a half-life. This interval is the time for one-half of the original parent radionuclide to decay. Decay is measured in the International System of Units (SI) unit of becquerel (Bq), where one Bq is one decay per second. An older and still common American unit is the curie, defined as 3.7×10^{10} decays per second. Both units occur in the literature, and neither is particularly convenient. The becquerel is too small, and the curie is too large. In practice, MBq (10^6 decays/s) and mCi (3.7×10^7 decays/s) are most commonly used.

Generally, investigators prefer radioactive species that resemble naturally occurring atoms in the moieties of interest. However, these may not be convenient as to either their half-life or their chemical properties. RP development is, in essence, a field lying between chemistry and physics with heavy overtones of physiology. Some discussion of the possible labels is now in order. Most labels are appropriate for either imaging or therapy. Only a very limited number may be used for simultaneous detection and treatment.

1.2.2 Radionuclide Labels

A large number (on the order of 30 or more) of radioactive atoms are generally available for labeling of tumor-targeting and other nuclear medicine RP agents. Table 1.1 lists these as well as their half-lives and important emissions. Three types of emitted particles are used in clinical nuclear medicine: photons, beta particles (both e+ and e–), and alpha particles. The most important of these entities for imaging are the gamma and x-rays. Included in this category are photons from the annihilation of e+ (positron) particles in the subject. Therapy requires use of ordinary electrons (e–) and alpha particles. The e– emission is called a beta ray and is the same physical entity as an orbital electron. We will also consider alpha rays ($^4He^{2+}$), which have important applications in therapy of relatively small targets (i.e., those of cellular size). It is important to recognize that the physical description of emitted particles is very different for charged versus uncharged emissions. The former lose energy continuously while passing through an absorbing medium due to interaction with the ambient electrons. Uncharged particles lose energy by sporadic processes that are best described statistically.

1.3 RADIONUCLIDE EMISSIONS

1.3.1 Charged Particles

Particles that have an electric charge are described by a range parameter in soft tissue or other absorbing medium. Charged particles of a given energy and mass can penetrate a distance only equal to or less than their range into the surrounding material. The inequality occurs for electrons and positrons and follows from the fact that an e^+ or e^- may collide with ambient atomic electrons while going through the absorbing substance. This multiscattering process essentially expends the initial kinetic energy over a smaller linear distance than the range value. For electrons and positrons emitted in a typical radiodecay, the range lies between 1 to 12 mm in soft tissue. This result implies that radiation-based treatments may still be possible if the target cell lies within the range of the electron from the source location. With poorly vascularized or very large lesions, therapy using electrons may not be possible. Corresponding range values for alpha particles are only on the order of 30 to 50 μm; that is, they are restricted to several cell diameters. Alphas are much more massive than any electron scatterer and thus do not undergo multiscattering processes in a medium. Thus, to affect therapy, the alpha sources must be quite close to malignant cells. Such striking differences from the beta radiation lead to very different applications for these two emissions in radiation therapy using nuclear decays.

An additional detail is important in understanding beta decay and electron ranges. Nuclear emission of an e^- or e^+ particle, termed electron or positron decay, respectively, is a three-body process. For example, ^{18}F decays via

$$^{18}F \rightarrow ^{18}O + e^+ + \upsilon_e \tag{1.1}$$

Note the presence of the ordinary electronic neutrino (υ_e) on the right side of the equation. This particle is always present in beta decays of either type (e^+ or e^-) and has been one of the great enigmas of modern physics. The neutrino has recently been shown to have a finite rest mass. Neutrinos can interact with matter only by inverse beta decay and are very difficult to detect using absorbers much smaller than a planet. They are uncharged and describable by a half-value layer (HVL) as defined following. For typical neutrinos, that length is on the order of light-years of water so that their direct application in clinical imaging is nil. The indirect importance of the existence of neutrinos on beta decay is very significant as it leads to the production of electrons with a continuum of energies rather than a discrete set of energies as in the case of gamma emission.

Since neutrinos are produced along with the electron or positron, three bodies are generated via beta decay. Yet there are only the two restrictions on momentum and energy conservation, so that each body ends up sharing a variable amount of the total available kinetic energy. This means that the kinetic energy of the e^+ varies between essentially zero and a maximum value whereby the neutrino has no kinetic energy. It is this maximum positron energy that defines the range for a given positron (or electron) emitter. Because of this variation and the resultant variable distance the positron may go in tissue, there is an additional parameter used for beta decays: the average range. This distance parameter includes the probability of each emission kinetic energy in its computation. An additional effect of this statistical variation in emitted positron energies leads to an inherent reduction of the sharpness of an image using a positron emitter as a radiolabel.

Positron labels are usually not detected directly by nuclear technology. Instead, they are observed using characteristic annihilation radiation that occurs near the end of the path of the e^+:

$$e^+ + e^- \rightarrow \gamma_1 + \gamma_2 \tag{1.2}$$

where two 511 keV photons (symbolized by γ) are produced. This is an annihilation process between matter (e^-) and antimatter (e^+). By conservation of momentum, these photons move 180° apart and produce the characteristic decay signal that is recorded by the positron emission tomography (PET) instrument. Such detectors are described more completely in the following chapter.

Another very important aspect of e^+ emission is that the radioactive nucleus has an inherent alternative mode of decay. By quantum mechanical arguments, a decay such as given in Equation 1.1 can be rewritten as

$$^{18}F + e^- \rightarrow {}^{18}O + \nu_e \tag{1.3}$$

Here we have taken the e^+ particle from the right-hand side (RHS) of the equation and transposed it to the left-hand side (LHS) while simultaneously changing it into its antiparticle, the electron. In other words, particle and antiparticle transform into each other by going backward in time. This is equivalent to going from the right side to the left side in any decay equality. The process demonstrated in Equation 1.3 is, in fact, not a theoretical conjecture but a physical reality. It was discovered by Luis Alvarez to occur in competition with e^+ emission in the case of any positron emitter.

The competitive process shown in Equation 1.3 is called K-capture. Its terminology indicates that an orbital electron (from the K-shell of ^{18}F in this case) can be caught by the radioactive nucleus and thereby permit its decay to ^{18}O. When this occurs, the decay cannot be detected by standard gamma-counting instruments since no e^+ and hence no high-energy photons are produced. In our discussion of possible radiolabels (Table 1.1), we list the fraction of each positron emitter that decays by K-capture. While small for ^{18}F (around 3%), this fraction may approach 80% or more for some potentially interesting positron-emitters such as ^{124}I. The amount of K-capture is one of the most restricting aspects of positron-emitter applications in nuclear imaging. It effectively shuts off the positron emission route to reduce the amount of externally detectable activity at the site of positron decay. It is a stringent restriction on selection of a potential positron emitter as a label for an RP.

Investigation of K-capture has led to situations where the decay rate can be altered by putting external pressure on a physical sample containing the radionucleus. The K electron shell is a diffuse sphere (from the Gr. Kugel) that encloses and touches the nucleus.

TABLE 1.1   Abbreviated List of Radiolabels for Targeted Agents

RADIONUCLIDE	$T_{1/2}$ (PHYSICAL HALF-LIFE)	GAMMA ENERGY (keV)	MEAN BETA ENERGY (keV)	MEAN POSITRON ENERGY (keV)
Significant Gamma/Beta Emitters				
^{67}Ga	3.26 d	93 (39%)		
		185 (21%)		
		300 (17%)		
^{99m}Tc	6.01 h	140		
^{131}I	8.05 d	364 (82%)	192 (90%)	
		637 (7%)	97 (7.3%)	
^{123}I	13.3 h	159 (83%)		
^{111}In	67.3 h	171 (90%)		
		245 (94%)		
^{153}Sm	46.3 h	103 (30%)	200 (32%)	
			226 (50%)	
			265 (18%)	
^{177}Lu	6.73 d	113 (6.4%)	48 (12%)	
		208 (11%)	111 (9.1%)	
			149 (79%)	
^{105}Rh	35.4 h	319 (19%)	70 (20%)	
			180 (75%)	
^{186}Re	3.72 d	137 (9.5%)	306 (22%)	
			359 (93%)	
^{188}Re	17.0 h	155 (15.6%)	728 (70%)	
			795 (26%)	
Important Beta Emitters				
^{32}P	14.3 d		649 (100%)	
^{89}Sr	50.5 d	0.909 (0.01%)	585 (100%)	
^{90}Y	64 h		937 (100%)	
^{67}Cu	61.8 h	93 (16%)	121 (57%)	
		185 (49%)	154 (22%)	
			189 (20%)	
^{166}Ho	26.8 h	81 (6.7%)	651 (49%)	
			694 (50%)	
Important Positron Emitters				
^{18}F	109 m			250 (97%)
^{68}Ga	68 m	1,080 (3%)		836 (88%)
^{64}Cu	12.7 h	1,350 (0.47%)	190 (39%)	278 (17%)
^{76}Br	16.2 h	559 (74%)		375 (6%)
		657 (16%)		427 (5%)
		1,850 (15%)		1,530 (26%)

TABLE 1.1 (continued)   Abbreviated List of Radiolabels for Targeted Agents

RADIONUCLIDE	$T_{1/2}$ (PHYSICAL HALF-LIFE)	GAMMA ENERGY (keV)	MEAN BETA ENERGY (keV)	MEAN POSITRON ENERGY (keV)
86Y	14.7 h	627 (33%) 703 (15%) 777 (22%) 1,080 (83%)		535 (12%) 883 (4.6%)
89Zr	78.4 h	909 (100%)		396 (23%)
124I	100.2 h	603 (63%) 1,690 (11%)		686 (12%) 974 (11%)

ALPHA EMITTERS		GAMMA ENERGY	ALPHA ENERGY	MEAN BETA ENERGY
211At	7.2 h	687 (0.2%)	5,870 (42%)	
213Bi	46 m	440 (26%)	5,550 (0.15%) 5,870 (1.94%)	320 (31%) 490 (66%)

Source: http://www.doseinfo-radar.com

It is not an elliptical orbit around the nucleus at some great distance as shown in many textbooks and popular literature. Thus, the pair of K-shell electrons moves over the parent nuclear surface continuously. If the shell can be made smaller by external compression, this contact is increased and the decay rate enhanced. It is interesting to think of squeezing a radioactive specimen and causing it to decay more readily.

1.3.2 Uncharged Particles

Uncharged emissions, such as photons (or neutrinos), cannot be described by a range parameter since they lose energy only in discrete events and not continuously. They inherently follow a circuitous route through materials and may interact with local atoms in several different ways before ending their existence. Photons are described by statistical considerations when they interact with the electrons in the medium. These interactions, at the energies used in RP research, include the photoelectric effect, Compton scatter, and pair production. Mathematically, this is equivalent to saying that photons have a so-called HVL or, equivalently, a mean-free path in a medium. Mathematically, a beam of parallel photons of initial intensity I_0 decreases with distance x in the medium via

$$I \equiv I_0 \exp(-\mu x) \tag{1.4}$$

Here, μ, in units of inverse distance, is termed the linear attenuation coefficient (1/mean-free path). It is a decreasing function of photon energy until $E\gamma$ approaches 1.02 MeV where the photon can produce e^-, e^+ pairs. As the gamma energy goes above 10 MeV, other processes become energetically possible including the production of single nucleons (protons or neutrons) from atomic nuclei in the medium. This is termed a photonuclear process and is generally not important in RP research.

1.4 METHODS OF LABELING

There are several options available to attach a radioactive atom to a chemical agent. Originally, a radioactive ion was used directly instead of the commonly available stable isotope such as ^{123}I in lieu of stable ^{127}I to track the movement of iodine ions from the stomach into the thyroid. This method is still used for following other elements, such as iron, copper, or lead, in the various metabolic processes of the body.

A similar technique is the insertion of the radioactive isotope into a molecule to replace a stable isotope of the same element. One may use ^{11}C as a label in the glucose molecule while doing PET imaging of a diabetic patient. Here, the assumption is that the kinetics of the labeled sugar are not significantly different from that of the native molecule even though the molecular weight has been shifted slightly downward due to ^{11}C replacing ^{12}C. If we had substituted ^{14}C for ^{12}C and used a beta counter, the isotopic effects, if seen, would be in the opposite direction since the MW has now increased.

Effects due to shifting molecular weight also occur in the use of a single ion as described already for following the iodine pathway in the body. Clearly, this replacement method, like the use of the pure iodine ion, is intellectually satisfying as no demonstrable atomic change has occurred to the atom or molecule of interest. Unfortunately, such elegant strategies are not easily done in general, and other, less appealing, techniques are common in the production of RPs.

A typical labeling is the *ad hoc* coupling of a radioactive atom to a molecule or other entity using one of several attachment methods. Here, the label is simply carried by the agent as long as the attachment remains intact. Two types of labeling are in general use: iodination and chelation. They are not mutually exclusive, and both can be used on the same moiety. We discuss these in turn.

Iodination is probably the more common of these techniques whereby radioactive iodine ions are attached to a tyrosine or lysine amino acid residue in a protein. One important feature of iodination is that the method requires relatively small amounts of the biomolecule—typically on the order of 50 to 100 µg for antibody proteins. This is significant in engineering practice since the rate of production of a new agent may be very slow and expensive. By being able to label small samples with little loss, the bioengineer can test novel entities sooner in the development process. This biological testing is described more completely in Chapters 2 and 3. There we will study novel agents that have been iodinated because of this reduction in the amount of material required.

While simple to apply as a labeling process, iodination has a unique problem. Because of the necessity of iodine in the production of thyroid hormones, mammals have developed extensive enzyme systems to liberate iodine atoms from almost any molecules or other constructs within various normal tissues. Since tumors develop from such tissues, they also possess one or more of the enzymes. This process is termed dehalogenation. Freed iodine atoms are then sent to the thyroid via the circulation. Other organs in the body are involved in the trafficking. Patients receiving radioiodinated compounds often display subsequent uptake of the label in the stomach, salivary glands, kidneys, bladder, and other tissues. This confounds the imaging process and adds to the radiation dose to these organs, as we will see in Chapter 8. It also leaves uncertain

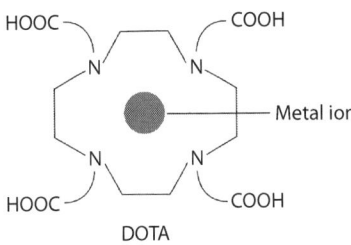

HOOC   COOH

N   N

Metal ion

N   N

HOOC   COOH

DOTA

FIGURE 1.2 Structural model of the DOTA molecule. Notice the four carboxyl radical COOH arms, which eventually close over the chelated radiometal ion.

the quantitative question of how much of the original chemical moiety has gone to the various locations inside the animal or patient. Instead, the label has become the object being observed, and the tumor agent has been at least partially lost from the imaging process. This topic is further discussed in Chapter 6 on modeling biodistributions.

Chelation (from the Greek word for claw) is a more general concept than iodination. In this case, a special molecule (the chelator) is initially engineered to tightly hold a particular radioactive metal ion. In turn, the chelator is attached to the construct or molecular structure of interest. Chelators are often referred to as bifunctional; that is, they both bind the radiometal as well as attach covalently to another ligand. Among the best-known examples of such structures are diethylene triamine pentaacetic acid (DTPA), ethylenediaminetetraacetic acid (EDTA), and 1,4,7,10-tetraazacyclododecane-1,4,7,10-tetraacetic acid (DOTA). The molecular structure of DOTA is shown in Figure 1.2. Chelators generally have several molecular "arms" that wrap over the metallic ion and capture it for extended periods of time—even in biological milieus. These are the acetic acid linkages in the case of DTPA and EDTA. Given a chelator, labeling is generally a two-step process: attachment of the chelator to the moiety of interest and then the labeling of the resultant new entity or molecule by incubation with the desired radiometal. In some cases, this process can be reversed such that the radiolabeling is done first and then the ion-containing chelator is attached to the possible tumor-targeting agent.

Because of the two-step nature of this metal-labeling process, larger amounts of original material are generally needed for chelation than for iodination. Typically, the additional amount is between five-fold and approximately one order of magnitude such that a large fraction of one mg of protein might be required for chelation labeling of an antibody, for example. Alternatively, the obvious advantage of this approach is that multiple radioactive metals (cf. Table 1.1) may be used as labels for engineered moieties. Thus, unlike the iodination case where only a few radioactive isotopes are available, there is a much larger set of possible chelation-attached radiolabels. Of course, a user may be forced to design an appropriate chelator for a given metal ion. This process can be quite daunting and often is not completely satisfying due to release of the label in vivo (i.e., the so-called off rate of the captured metallic ion).

Chelators have their own special literature, and a number of chemists have devoted a substantial portion of their careers to designing and then producing novel molecules designed to capture and hold a specific radiometal. The insertion of the radioion into the space is quite precise; for example, ^{111}In is not as good a fit as is ^{90}Y for DOTA even though both metals have the same valence and are in the same column of the periodic table. In fact, this insertion process is a quantum-mechanical problem that depends on ionic size and the electronic orbitals in the chelate as well as in the metal used as radiolabel.

With chelators, several issues occur during biological testing. Primary is the fractional uptake of the label by the chelator during a given period of incubation time. Various strategies may be employed to more effectively load the radiometal: changing solution pH, increasing ambient temperature, or using cold (nonradioactive) metal as a first step to saturate other binding sites on the moiety of interest. After loading comes the question as to whether the inserted metal ion remains within the molecular "claw" while the labeled moiety is in the blood of an animal or patient. This is tested using incubation with plasma at 37°C. No chelator is perfect, and some loss of the label occurs with time. Finally, with animal biodistribution studies (cf. Chapter 2), the investigators determine where any liberated metal ions go within the animal's body. Since these free radioactive ions may contribute to radiation dose at one or more undesirable locations, such testing in living mammals is essential.

If an unattached radiometal ion is present in circulation, one method of capture and excretion is use of a second chelator to take up the liberated ion. In the case of ^{90}Y-DOTA-cT84.66 monoclonal antibody used at City of Hope, the second chelator has been DTPA injected intravenously into the patient at 48 h after the original injection of the therapy dose. In this case, excretion is via the renal system and bladder to eliminate the ^{90}Y ions prior to their trafficking to the bone marrow. Since the therapeutic radioactive species is the pure beta-emitter ^{90}Y in this case, there is no external imaging possible to track this elimination from outside the body of the patient.

While labeling is a necessity for the imaging and therapeutic processes, an investigator must have associated targeting moieties to carry that label (or set of labels) to the selected sites within the body. Several standard agents have been engineered to target to tumors or their associated lymph nodes. The following list is not complete but does include common agent types as well as several novel molecular forms that are becoming more useful in oncology practice.

1.5 NANOENGINEERING

During the last three decades, enormous effort has gone into the generation of novel agents that may potentially target malignant sites in the body and follow the draining nodes. Generally, these entities are manufactured in the 50 to 200 nm size range, so the term *nanoengineering* has been introduced into the literature. Such small structures are efficient at passing from the circulation into the tumor milieu due to the fenestrations found in the vascular walls at the site of many malignancies. Some of these moieties, such as antibodies, can also have direct therapeutic effects on cancer cells. Usually, however, a radiolabel is needed for radiation therapy treatment. In most cases, this label is a beta emitter (e$^-$), although alpha emissions have also been used.

The following section describes a sequence of nanoengineered agents beginning with small colloids and liposomes and ending with short segments of nucleic acids. This list covers the dominant agents in present usage. Members of the group are not mutually exclusive. Combinations of agents will therefore also be described such as liposomes with antibodies or short segments of RNA encapsulated within their lipid phase.

1.6 COLLOIDAL DESIGNS

Small colloidal particles have been of continuing interest to the bioengineer since the 1960s. Originally, these were micron sized and made of heated sulfur. Their modern descendants are in the 100 nm range and are still of interest to the lymph-node investigator. While not directly targeting to tumors in the node, these primitive examples of nanoengineering are valuable in tracking the lymphatic path from the tumor bed to the important (sentinel) nodes. Labels include both ^{99m}Tc and ^{111}In. The latter offers a much-increased half-life as well as dual photons to expedite detection at longer times.

Colloidal RPs may be imaged using a gamma camera and also detected in the operating room by the surgeons using probes. Both of these technologies are described in the next chapter. In the breast patient, injections are made around the tumor bed—often at the four points of the compass (i.e., in breast surgeon notation at the 9, 12, 3, and 6 o'clock positions). Some controversy arises concerning the optimal size of the colloidal particles. Certain surgeons prefer the smaller entities, whereas others favor use of a simple mixture of variously sized particles as produced by the heating process.

1.7 LIPOSOMES

Among the most important tumor-targeting agents has been the liposome or phospholipid vesicle (PV). Beginning in the late 1960s with the work of Bangham (Bangham, Standish, et al. 1965) and continuing into the 1990s, small unilamellar (single-layer) phospholipid structures were constructed in early experiments at the nanometer scale. Figure 1.3 gives a schematic picture of a liposome. Liposomes form naturally within the animal and human bloodstream, particularly after consuming a fatty meal. This has led some biologists to believe that cellular life originated within the relative safety of PVs that had formed earlier. In this conjecture, the prototypic cell wall was originally a liposomal bilayer.

If the liposome is stable in biological solutions such as blood plasma, the phospholipid bilayer wall provides a membrane that protects the inner (aqueous) phase material from the external environment. To some developers, this is the single greatest feature of the liposome and one of continuing interest to many other nanotechnologies. It is known that other possible engineered agents, if directly injected into the mammalian blood or lymphatic supply, would suffer sequestration and metabolism if left unprotected. Bioengineers then often see the liposome as offering possible shelter to their novel agent just as it may have provided for primitive life forms. Both hydrophilic and

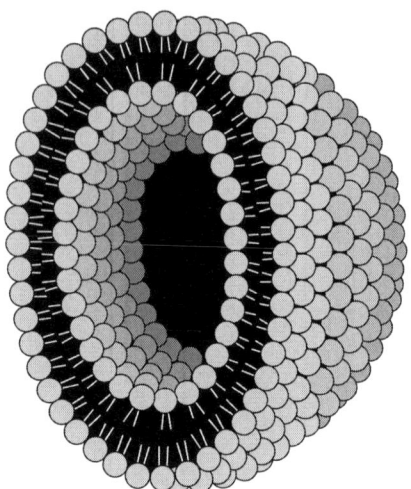

FIGURE 1.3 Schematic drawing of a single-layer liposome. Wall thickness is given by the size of the bilayer of phospholipid and is fixed. Under quality control, the overall diameter may be specified by the manufacturing process. Courtesy of Professor Frederick Hawthorne, International Institute of Nano and Molecular Medicine, Columbia, Missouri.

lipophilic molecules may be inserted into the inner space and bilayer, respectively, of the PV. Thus, the liposome is inherently important as a cargo carrier in the circulation.

There is little likelihood of immune system responses to the natural PVs that have typical biomolecules in their outer wall. Such natural structures, however, are usually short-lived in the bloodstream—persisting for periods of seconds to minutes. Likewise, these liposomes have a wide range of sizes and electric charges in vivo. Generally, the natural PV would consist of a number of concentric phospholipid bilayers built around the original single sphere. Man-made PVs were engineered to get around all these restrictions. Stability in vivo was achieved using a rigorous chemical mixture to produce a very stable PV wall. A typical combination would be distearoyl phosphatidyl-choline (DSPC), cholesterol (CH) in 2:1 molar ratio to achieve neutral PVs.

The two other important parameters, size and surface electric charge, are also manip-ulated to form liposomes for clinical trials. Chemists have limited PV size ranges by using ultrasonic oscillators having a short wavelength or extrusion plates. These devices are used to limit the diameter to lie in a predetermined range between 50 and 200 nm. This size is probably the most effective to allow passage from within the bloodstream and into the tumor environment. Resultant sizes of the product liposomes are measured using light-scattering devices. By changing the chemical composition of the vesicle wall, all three possible values of charge (positive, negative, or neutral) can be achieved. It has been found that certain tissues prefer to take up liposomes of a given electric charge. Liver, for example, shows enhance accumulation of positively charged PVs.

Labeling of the PV can be done with various strategies. In the early work at Caltech, the investigators fabricated an ionophore into the liposomal wall to allow the movement of radioactive ions into the center of the preformed structure. Mauk and Gamble (1979) employed the ionophore A23187 to allow ^{111}In 3^+ ions into the aqueous phase at the center of unilamellar liposomes. To keep those ions within the structure, a chelator (nitrilotriacetic acid or NTA) was placed inside the PV during production. Thus, radio-active indium ions, once in aqueous phase, were kept there by the NTA molecules so that reverse transitions through the ionophore were not possible.

A second labeling locus is the outermost wall of the PV. Iodination may be performed, for example, by placing lipophilic molecules containing tyrosine within the liposomal wall during formation and then using the previously outlined method. Alternatively, some investigators have followed the technique used to label red blood cells and using a Sn ion ("tinning") to coat the surface of the PV and then follow with incubation inside a solution of ^{99m}TcO$_4^-$ ions. The technetium labels the tinned sites through a reduction process, and the PV becomes radiolabeled with a nuclide that is ideal for imaging in the standard gamma cameras used in nuclear medicine.

Phillips and coworkers (Phillips, Rudolph, et al. 1992) established a novel labeling method for preformed PV using a two-step process. First, hexamethylpropyleneamine oxime (HMPAO) is incubated with ^{99m}TcO$_4$ for a short period, on the order of several minutes. Then the chelated and lipophilic result is added to the preformed liposomes that contain aqueous glutathione. As a result, the ^{99m}Tc is transferred into the aqueous phase where it is unable to pass back through the bilayer. This strategy is similar to that using the ionophore and was based on an analogy with previous brain imaging studies using HMPAO.

Restrictions on the in vivo usage of liposomes have centered on issues of stability in the blood and potential sequestration by the reticuloendothelial (RES) system. If a PV product quickly falls apart when immersed in plasma, there can be little hope of its eventual targeting to the tumor sites in animal or patient. Likewise, if rapidly taken up by the liver, spleen, and red marrow, the liposome may not be able to get to the possibly multiple tumor sites in large enough numbers.

Stability of liposomes in vivo is a fundamental issue and must be measured. Several methods may be used to test the labeled product; one of the more interesting is use of perturbed angular correlation method (PAC) involving the ^{111}In radionuclide. The two gamma rays given off in the decay of ^{111}In are actually in cascade (i.e., occur shortly after one another in a direct temporal sequence). Note that this photon-producing process is in distinction to the two simultaneous 511 keV photons produced in positron decay. If the radioactive indium is within the PV, the attached NTA chelate has a relatively low MW and consequently rotates rapidly at room temperature. If, however, the liposomal wall has broken, the indium can move away from the intentionally weak chelator and label larger proteins within the plasma medium. In this case, the indium-labeled molecule is now larger and rotates correspondingly more slowly. By counting the two gammas in sequence by a compass-array of four photon-detecting probes arranged around the sample at 90° intervals from each other, the observer can measure the probability that the indium is attached to NTA versus attached to a larger plasma protein. This observation

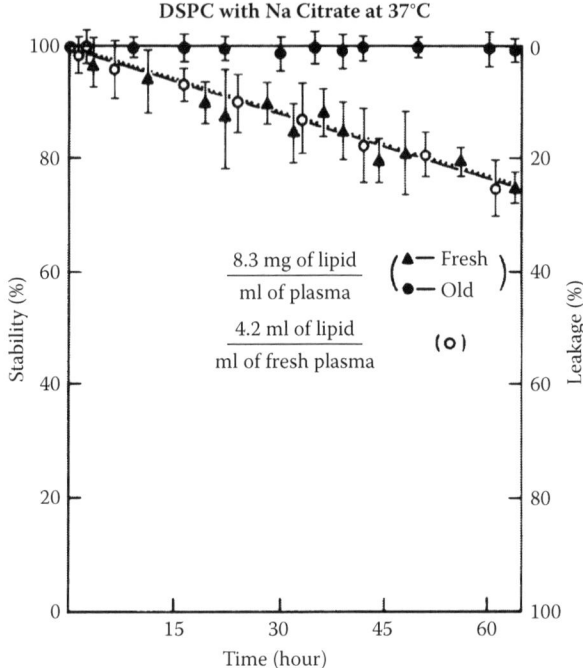

FIGURE 1.4   Decay of DSPC:CH = 2:1 liposomal integrity as a function of time in human plasma. Note the importance of having fresh human plasma as the medium. (From Wallingford, R. H. & Williams, L. E., *J. Nucl. Med.*, 26, 10, 1985. Reprinted by permission of the Society of Nuclear Medicine.)

may be done over a period of time (Wallingford and Williams 1985) to document the breakdown of the vesicle wall in plasma or other fluids. Figure 1.4 gives results obtained in human plasma for unilamellar liposomes tested at the City of Hope.

One issue that arose with liposomes was the avid targeting of some formulations to the liver and spleen. This phenomenon can also be observed with other engineered agents, as will be described in the following section. It was found that this uptake could be reduced by partially saturating the RES receptors at those sites with an unlabeled ("cold") liposome that was particularly attractive to the liver. One of the best of these saturating agents was a PV formulated with aminomannose (AM) in the bilayer with the molar ratio of 8:3:1 for DSPC, CH, and AM. This pretreatment led to an approximately 50% increase in tumor uptake compared to nonpretreated mice (Proffitt, Williams, et al. 1983). These results are described more completely in the next chapter.

A second method to reduce RES uptake of liposomes is the addition of a coating of polyethylene glycol (PEG) to the exterior membrane. Such so-called stealth liposomes (Allen, Hansen, et al. 1989) were able to exhibit much extended times in the plasma of animals compared with the standard form. The explanation is that the liver and spleen are less able to recognize the presence of a PV when it has been coated with PEG. This strategy has been expanded to include a number of other engineered agents—particularly those with low molecular weight. Stealth has now become a general technique that has been applied to other tumor-targeting agents such as antibodies.

After extensive murine testing, liposomes were applied to patient imaging in a variety of clinical cancers. Using ^{111}In as label, the rate of detection of 97 known tumor sites was reported as 85%—independent of the disease type (Presant, Proffitt, et al. 1988). Of particular interest was the targeting to Kaposi's Sarcoma in AIDS patients (Presant, Blayney, et al. 1990) Here, small (45 nm) vesicles had been loaded with daunorubicin for an effective therapy against the disseminated form of this disease (Gill, Wernz, et al. 1996). In another example of chemotherapy with PVs, pegylated liposomes containing doxorubicin (Park 2002) have been shown to be effective against metastatic breast cancer—particularly sites in the skeleton. In both examples, cardiac toxicity was reduced compared with giving the same amount of drug without liposomal encapsulation. Note that there was no specific targeting to a molecule in these PV applications; the liposome was apparently degraded at the tumor site with its label remaining as the tumor marker.

1.8 ANTIBODIES

While manufactured liposomes represent a generic agent that can carry materials to lesion sites, they are inherently not sensitive to any tumor molecular marker or even a given tumor type. For specific targeting, the researcher must find an agent that recognizes the predefined molecule and binds appreciably to it in a reasonable time. Currently, the standard such agent is either an intact antibody (Ab) or one of its engineered cognates. Originally, antibodies were derived from the challenging of an animal by a human tumor mixture including adjuvant material. Resultant animal plasmas were then searched for a high-affinity Ab that could be labeled using previously outlined techniques. It was found, unfortunately, that the reinjected patient would eventually respond to the alien proteins by producing their own human antibodies to those derived from the animal. When this happens, injection of the alien Abs causes antibody–antibody complexes to form in the plasma; these are taken up extensively by the liver and spleen to prevent the animal antibody from reaching the malignant targets. Order, Stillwagon, et al. (1985), at Johns Hopkins University, attempted to ameliorate these immunity effects by rotating the animal sources, such as using a goat-derived, followed by a rabbit-derived, followed by a mouse-derived antiferritin antibody. In this way, a given patient's immune system was exposed to a sequence of various animal-derived Abs to mitigate the immune response. Because of the complicated nature of this plan as well as the necessity of keeping large numbers of unique animals alive, other methods were needed to simplify patient tumor therapy.

Kohler and Milstein (1975) developed a cell-fusion technology to eliminate the difficulty of raising titers of antibodies in a specific animal model. In their method, murine myeloma cells were fused with murine spleen cells to produce an implantable structure called a hybridoma. If the mouse had been challenged earlier with a human tumor extract, this long-lived fusion structure would produce murine antibodies to the human tumor via its spleen cell component. The myeloma component would assure longevity. Hybridomas could be generated, in principle, for most human tumor extracts and the binding constant measured for the resultant murine antibodies. With hybridomas, the cell fusion construct was the entity that was kept alive—generally in the abdomen of the same mouse species as all cells were of murine origin. This replaced the necessity of keeping immunized large animals alive as in the earlier strategy of Order et al. (1985).

As in the original work at Johns Hopkins University, it was found that hybridoma-derived antibodies, termed monoclonal antibodies (Mabs), also led to immune responses by the patient. The response would occur after several injections of the same mouse-derived Mab into a given individual and would lead to the sequestration of the resultant complexes as mentioned. This is referred to as a HAMA (human antimouse antibody) response. In some cases, this reaction is not just to the original murine protein but to other murine hybridoma-raised Mabs as well since these entities all share a common protein backbone. Because of such cross-reactivity, an entire avenue of possible therapy could be cut off for a particular patient.

Since the early 1990s, there has been extensive engineering effort expended to reduce this response by the patient. Several different concepts are in use and are listed in Table 1.2. These strategies have centered on two topics: making the antibody more human-like and making it smaller to reduce the probability of patient immune response. By comparing the amino acid sequences of humans and mice, sequences that are more human-like could be substituted to the purely murine form from the hybridoma. This result is a chimeric antibody since it does not occur in nature and combines two species' amino acid sequences. Note, in Figure 1.5, that the recognition segment of the antibody is kept in the original (murine) form since it is specific to the antigen. Changes to make the Mab more acceptable to the patient must occur further down the sequence.

Unfortunately, chimeric antibodies also led to the generation of human antichimeric antibody (HACA). This result may not be surprising. Reducing the MW of the Mab by gene engineering was also found to be limiting due to the reduced blood circulation times and enhanced renal accumulations of these novel constructs. It was found that

TABLE 1.2   Methods to Reduce Immunogenicity of Animal-Derived Antibodies

METHOD	MOLECULAR RESULT	CLINICAL RESULT
Make intact antibody more human	Chimeric (C) antibodies with both human and murine components	HACA is found in patients
Make antibody completely human	Restructure the Mab so that it has a human (H) framework and no obvious murine amino acid sequences; may be difficult to achieve	HAHA is observed
Reduce the size of the antibody	Structural alterations including F(ab')$_2$, minibodies, and diabodies.	Targeting is reduced due to shorter circulation times in patient or animal
Cover the antibody with polyethylene glycol (PEG)	Pegylation; sometimes called stealth technology	Longer circulation times and reduced patient immune response

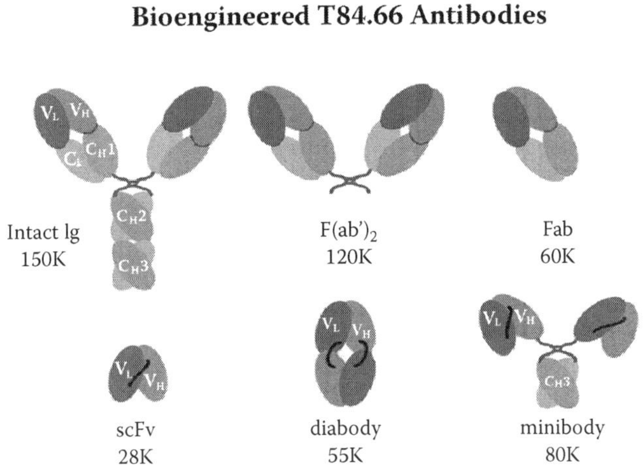

Bioengineered T84.66 Antibodies

FIGURE 1.5 Schematic design of five cognate anti-CEA antibodies. Only the intact form (upper left corner) occurs in nature. Molecular weights are included with the respective drawings. (From Williams, L. E. et al., *Cancer Biother. Radiopharm.*, 16, 2001. With permission.) The CH$_2$ and CH$_3$ segments constitute the Fc portion of the antibody.

smaller molecules, particularly down to 25 kDa in the case of the single-chain antibody, had significantly shortened times in the blood of both mice and patients. This clearance reduced the amount available at the tumor sites—a topic that is more completely described in the next two chapters. Figure 1.5 gives a graphic display of several of these engineered antibody candidates that were designed, coded for in yeast or other cells, and then generated in a cell culture material. Notice that genetic engineering has now obviated the need for hybridomas. But the price paid is an enormous increase in the number of possible agents for targeting tumor cells in vivo.

Genetic engineering can produce a very large set of possible protein agents based on the original antibody framework. There are approximately 1,500 amino acids (AA) in the intact mammalian Ab. To change any single one of these leads to approximately 20 alternatives due to the number of amino acids as possible candidates for substitution at that given site. Yet which AA do we select for our revised form? There might be 20^{1500} results if one accepted the entire protein as being open to manipulation. Clearly, some limits must be put into place by the genetic engineers prior to intervention.

Typically, those seeking to swap out a murine AA for a more human-like candidate AA at any point along the antibody protein need to refer to sequences from both species. This was done for the chimeric and humanized forms with approximately 10 to 20 important amino acids being changed in the process. That such novel constructs still led to patient immune response has been one of the more limiting results in the application of antibodies to tumor imaging. Notice that, as might not be expected, there have been immune responses seen in patients even with supposedly totally human-type antibodies. Tabular indication of the presence of HAHA indicates human antihuman antibodies being observed clinically.

Among the more recent antibody redesigns has been coating the exterior of the moiety with PEG. This surface rendering has had earlier successful use in liposomes (i.e., stealth liposomes) and was likewise seen to prolong the circulation times for Mabs. Immune response was also reduced—presumably since the patient's immune system could not get a "clear look" at the offending species in the blood. Stealth has now become a standard addendum to a number of possible tumor-targeting agents.

Two applications of intact murine antibodies to treatment of non-Hodgkin's B-cell lymphoma (NHL) are currently approved by the U.S. Food and Drug Administration (FDA). In a Phase III trial of a ^{90}Y-labeled anti-CD20 Mab (Zevalin) versus unlabeled anti-CD20 Mab (rituximab), the clinical response rates were found to be 80% and 44%, respectively (Wiseman, White, et al. 2001). The CD20 marker is a protein found on the surface of the malignant, as well as the normal, B cell. Use of labeled antibodies to this marker is the most notable achievement in clinical application of Mabs and has become, essentially, the treatment of choice for NHL of the B-cell type. Lack of HAMA response in the NHL patients was attributed to the reduced effectiveness of their own immune systems. These clinical studies are described in more detail in Chapters 8 and 10.

1.9 SMALL PROTEINS

Besides using an intact antibody or one of its lower-mass, engineered cognates, investigators have looked at short sequences of amino acids that bind to a predefined molecular target. This binding is analogous to that between an antibody and its antigen. For small proteins, particular sequences of amino acids in the target may have nanomolar affinities for a complementary sequence of amino acids in the designed agent. This process can be described as a "key that fits the lock." Fundamental to the small protein strategy is the concept that a lower MW will be less likely to trigger an immune response in the patient—even after multiple injections. Probably the best known of these proteins is octreotate, an eight amino acid moiety (octamer) that targets somatostatin receptors (Kwekkeboom, Krenning, et al. 2000). The latter are found in a number of neuroendocrine tumors such as pheochromocytoma. Labeling has been done with a various labels including ^{111}In, ^{90}Y, and ^{177}Lu to allow imaging and therapy in a clinical context.

Because of their low MW, excretion of octreotate and its analogs is primarily via the renal system. During neuroendocrine tumor therapies, this has led to issues of elevated kidney absorbed radiation dose resulting in eventual toxicity for some patients. Effects are typically not seen acutely but occur some months to years after the therapy is completed. Estimation of renal absorbed dose prior to treatment is essential (Lambert, Cybulla, et al. 2004) and will be described more extensively in Chapter 9.

Another variant of the small protein concept is the use of selective high-affinity ligands (SHALs). Here, a molecular target associated with the tumor is selected by one or more of its epitopes, much as in an antibody study. In initial work done at the University of California–Davis, the DeNardos have generated a mimic of the well-known Lym-1 intact antibody that targets to the HLA-DR10 antigen on malignant B cells (DeNardo, Natarajan, et al. 2007). Molecular weight is on the order of 2 kDa for these artificial constructs. A prototype SHAL consisted of two of six possible simple proteins separated by a variable length spacer. Resultant binding affinities were in the nanomolar range provided that the lysine-PEG spacer was adjusted to the separation between two partial epitopes on the B-cell antigen. By having two binding sites, the bidentate SHALs are less likely to come off the epitope due to thermodynamic motion since both sites must be separated simultaneously from the targeting molecule. Similarly, attachment to other irrelevant molecules in vivo is reduced since both parts of the bidentate form must couple simultaneously to the target molecule.

Blood clearance of the SHALs in mice is via the kidneys and rapid; times on the order of several hours are measured using a ^{111}In label held within a DOTA chelator attached to the protein structure. Future work is indicated in the addition of other reactive proteins such that the total structure becomes tridentate or higher order so that the affinities are improved. As with antibodies, there is also a possibility of uniting one SHAL with another to provide a dimer similar to the diabody or minibody construct in antibodies. Again, this design change should effect a higher affinity and greater specificity in vivo. Future engineering seems assured and clinical trials are being planned.

1.10 OLIGONUCLEOTIDES

1.10.1 Aptamers

An area of growing interest in tumor targeting is the use of short chain RNA or DNA molecules that recognize specific proteins at the sites of disease. Affinities from the nanomolar to the picomolar range can be achieved by engineering these single-chain constructs against a given purified protein target. Aptamers (Hicke, Stephens, et al. 2006) are generated using combinatory libraries of DNA or RNA. By testing the various possible constructs (perhaps 10^{15} combinations) against the purified target protein, the inventor may select for binding. Amplification using polymerase chain reaction (PCR) is done to enhance the appropriate combination of DNA. This has been termed systematic evolution of ligands by exponential enrichment (SELEX). A similar method holds for RNA aptamers. After some 10 to 20 rounds of selection and amplification, the agent is available for animal testing. Automated techniques now make this selection relatively rapid; new agents can be developed within a month or two.

Protein blocking was one of the original applications as in the unlabeled treatment of macular degeneration using an aptamer that targeted vascular endothelial growth factor (VEGF). The size of a typical aptamer is between 10 and 20 kDa. Thus, they are more massive than a small protein but considerably smaller than an intact antibody (150 kDa). Circulation times are relatively brief: usually measured in minutes rather than hours or days. Thus, a short-lived radiolabel is appropriate, and ^{99m}Tc has become a favorite for gamma camera imaging. Because of their nucleic acid makeup, aptamers do not cause an immune response in the subject even after repeated injections.

Two blood clearance mechanisms are observed in animals injected with a typical aptamer. Renal filtration and biliary elimination are comparable and amount to approximately 50% each. It should be mentioned that aptamers, because of their size and content, have the ability to cross the cell membrane and enter the cytosol. This transition (internalization) is unlikely for antibodies—even the low-mass cognates previously described. Thus, intracellular target proteins are within the striking range of this agent.

While immune response is lacking, nuclease activity can act on the circulating agent—particularly if it is RNA based. Such nucleases are less prevalent at tumor sites so that substantial degradation does not occur after movement out of the bloodstream. Aptamer protection in the blood can be effected via pegylation as in the liposomal example. Additionally, also as in the case of liposomes, there is the possibility of saturating clearance mechanisms by injecting initially an irrelevant, unlabeled aptamer before injecting the eventual RP. By this method, the amount of radiotracer accumulating in the animal tumors may be significantly increased since clearance mechanisms have been saturated by the unlabeled injection. Note that the nonspecific aptamer would not contain a radionuclide marker so that the only external signal would arise from the desired agent finding its protein.

Another strategic aspect of the aptamer is the possibility of turning the protein blocking off at a given time. It may occur, for example, that a therapy has gone too far in a given direction. Rusconi, Roberts, et al. (2004) have been able to reverse prior blocking of an anticoagulant aptamer to allow blood clotting in the patient. Such techniques are termed *antidote aptamers* and are applicable if circulating aptamers are still present in the bloodstream.

1.10.2 RNA Interference

Double-stranded RNA (dsRNA) has been discovered to have an important role in the silencing of genes (Downward 2004). Originally discovered in invertebrates, this mechanism has now been observed in mammalian species. For long, double-stranded moieties appearing in the cytosol of fruit flies or worms, a cellular enzyme (dicer) binds to the double strand and reduces it to short segments of approximately 20 base pairs each. These are called small interfering RNA (siRNA). In turn, these segments bind to an enzyme complex called RISC (RNA induced silencing complex) that takes one of the strands to bind to complementary mRNA single-strands in the cell's interior. This RISC complex is then degraded to silence the gene producing that mRNA. As viewed by biologists, this mechanism provides a technique for the invertebrate to reduce the toxicity of an invading virus that carries dsRNA as its genome.

Mammals also possess a dicer–RISC combination that becomes manifest if the number of base pairs in the dsRNA is approximately 30 or less. Often termed a *knock-down* strategy, RNA interference is found to operate in combination with interferon production upon viral attack on a mammalian cell. In the case of interferon response, however, the cell is destroyed by apoptosis. Thus, there is the possibility of using long-chain dsRNA molecules to trigger the apoptotic response in tumor cells.

One limitation of the method is that the dsRNA must move into the cytosol for this process to be initiated. This can be done artificially with an engineered virus that carries the desired dsRNA into the cell interior; essentially, the investigator uses the viral approach that probably gave rise to the RNA interference process in the first place.

The malignant cell's target molecule must be a gene that produces the mRNA of interest. A dsRNA targeting would then inherently reduce this protein's production and also provide a mechanism to bring radioactivity into the cancer cell's interior. Attachment of chemotherapeutics is another possibility. Notice that the oncologist will probably want to kill the cell or at least render it unable to divide. Initial results have been promising in cancer cell culture, but clinical results are difficult to achieve due to degradation of the double-stranded RNA in the bloodstream.

Bioengineers have a history of similar nucleic acid agents. For example, there has been testing of so-called antisense single-chain DNA molecules that were designed to complex with mRNA of the target type. Degradation of the complex with consequential loss of the protein is observed. These scDNA moieties do not survive well in vivo and also are not particularly effective once inside the tumor cell. It is felt that the dsRNA will prove more robust in circulation and more effective in destroying the mRNA of the cancer protein.

1.10.3 Morpholino Adaptations

Nuclease activity on circulating nucleic acids has led to a number of innovative strategies. One of these, the morpholino (MORF), is engineered with a replacement of the natural pentose-based backbone with a six-member ring. In addition, the phosphorodiester intersubunit bonds are replaced with phosphorodiamidate linkers (Amantana and Iversen 2005).

Besides avoiding lysis by nucleases in the blood, morpholinos have other advantages in the gene-silencing operation. They do not activate the interferon system, and their lack of degradation products makes them less toxic at the cell environment. Nude mice studies with [188]Re label in a two-step process have shown the agent to be effective in tumor size reduction (Liu, Dou, et al. 2006). In this experiment by the groups at the University of Massachusetts–Worcester and the University of Oklahoma, three control groups of mice were compared with the one group with a targeted morpholino agent carrying the [188]Re label. It was found that statistically significant ($p < .05$) reduction in LS174T colon cancer growth occurred in the therapy group compared with control animals. The latter included untreated, MORF alone, and [188]Re complementary MORF alone.

1.11 SUMMARY

Metastatic cancer is an ongoing clinical problem. After attempting chemotherapy, there is no obvious treatment strategy using standard technology. To first discover and then possibly treat such disseminated disease, specific tumor-targeting agents must be developed. These must be radiolabeled to observe the sites from outside the patient and to provide a possible therapy mechanism using attached beta or alpha emitters. It is best if these agents are targeting to specific molecules at the tumor cell sites or nearby. Because of engineering at the nanometer scale, multiple technologies are available to attempt this attack. Currently, our problem is finding the best agent—or perhaps combinations of agents and labels—that permit this process to proceed optimally in vivo.

A number of generic agents are currently available to the clinician. Among the nanometer-sized structures are colloids, liposomes, antibodies, small proteins, and oligonucleotides based on RNA and DNA. A listing is given in Table 1.3. More and more exotic entities are being produced as a result of other nanoengineering. It is difficult to compare one of these moieties with another for a specific disease. We have referred to this situation as a problem of plenty. There is the confounding matter of which attached radionuclides are optimal for detection and treatment of a specific disease. Issues of both the agent as well as its label must be resolved for an optimal course to be planned for a specific disease. Tumor clone evolution during this or prior therapy can further complicate the process so that the therapy may have to change over time. Thus, the methodology is inherently time dependent and patient specific.

Chapters 2 and 3 describe how to select one agent over another based on results from appropriate in vitro and animal studies. A regulatory reviewer may object in that the animal species chosen need not turn out to be a good representation of the eventual patient population. Ethically, of course, evaluation must begin with nonhuman mammalian data presumed to be representative of clinical results. These laboratory data are shown to the regulatory and other official bodies to obtain approval for—and perhaps funding of—a clinical trial. Comparison of animal versus human biological functions and, in particular, their relative pharmacokinetics is an aspect of the more general topic of allometry. This subject and the current data on such animal–human comparisons are discussed in Chapter 11.

TABLE 1.3   Table of Engineered Agents

AGENT	METHODS OF PRODUCTION	SIZE OR MW	POSSIBLY EFFECTIVE UNLABELED
Colloids	Making sulfur colloids; sizing	50–2,000 nm	No
Liposomes	Phospholipids plus water; sizing	50 nm	No
Antibodies	Animals; hybridomas, engineering	160 kDa	Yes
Small proteins	Engineering	1–30 kDa	Yes
SHALs	Protein engineering	2 kDa	No
miRNA	RNA engineering		Yes
Aptamers (DNA)	DNA engineering	10–20 kDa	Yes
Morpholinos	Change DNA backbone	6.5 kDa	Yes

Note: Not a complete list.

Finally, there are issues of cost. Assuming encouraging animal results, taking an agent to clinical trial is expensive and possibly counterproductive. As we will see in the next chapter, total funding for a 20-patient imaging–therapy study may be several millions of dollars. If spent on the wrong agent, the money is wasted. It is also important to emphasize that only a limited number of cancer patients of a particular type are available at a given research institution for the experimental group. By beginning a trial with a less than optimal moiety, patients in the test group are being put into the wrong trial while a more useful agent lies in the lab, untested. In part, it is this problem of plenty that this text attempts to alleviate.

REFERENCES

Allen, T. M., C. Hansen, et al. 1989. Liposomes with prolonged circulation times: factors affecting uptake by reticuloendothelial and other tissues. *Biochim Biophys Acta* **981**(1): 27–35.

Amantana, A. and P. L. Iversen. 2005. Pharmacokinetics and biodistribution of phosphorodiamidate morpholino antisense oligomers. *Curr Opin Pharmacol* **5**(5): 550–5.

Bangham, A. D., M. M. Standish, et al. 1965. Diffusion of univalent ions across the lamellae of swollen phospholipids. *J Mol Biol* **13**(1): 238–52.

DeNardo, G. L., A. Natarajan, et al. 2007. Pharmacokinetic characterization in xenografted mice of a series of first-generation mimics for HLA-DR antibody, Lym-1, as carrier molecules to image and treat lymphoma. *J Nucl Med* **48**(8): 133–47.

Downward, J. 2004. RNA interference. *BMJ* **328**(7450): 1245–8.

Gill, P. S., J. Wernz, et al. 1996. Randomized phase III trial of liposomal daunorubicin versus doxorubicin, bleomycin, and vincristine in AIDS-related Kaposi's sarcoma. *J Clin Oncol* **14**(8): 2353–64.

Hicke, B. J., A. W. Stephens, et al. 2006. Tumor targeting by an aptamer. *J Nucl Med* **47**(4): 668–78.

Kohler, G. and C. Milstein. 1975. Continuous cultures of fused cells secreting antibody of predefined specificity. *Nature* **256**(5517): 495–7.

Kwekkeboom, D., E. P. Krenning, et al. 2000. Peptide receptor imaging and therapy. *J Nucl Med* **41**(10): 1704–13.

Lambert, B., M. Cybulla, et al. 2004. Renal toxicity after radionuclide therapy. *Radiat Res* **161**(5): 607–11.

Liu, G., S. Dou, et al. 2006. Successful radiotherapy of tumor in pretargeted mice by 188Re-radiolabeled phosphorodiamidate morpholino oligomer, a synthetic DNA analogue. *Clin Cancer Res* **12**(16): 4958–64.

Mauk, M. R. and R. C. Gamble. 1979. Preparation of lipid vesicles containing high levels of entrapped radioactive cations. *Anal Biochem* **94**(2): 302–7.

Order, S. E., G. B. Stillwagon, et al. 1985. Iodine 131 antiferritin, a new treatment modality in hepatoma: a Radiation Therapy Oncology Group study. *J Clin Oncol* **3**(12): 1573–82.

Park, J. W. 2002. Liposome-based drug delivery in breast cancer treatment. *Breast Cancer Res* **4**(3): 95–9.

Phillips, W. T., A. S. Rudolph, et al. 1992. A simple method for producing a technetium-99m-labeled liposome which is stable in vivo. *Int J Rad Appl Instrum B* **19**(5): 539–47.

Presant, C. A., D. Blayney, et al. 1990. Preliminary report: imaging of Kaposi sarcoma and lymphoma in AIDS with indium-111-labelled liposomes. *Lancet* **335**(8701): 1307–9.

Presant, C. A., R. T. Proffitt, et al. 1988. Successful imaging of human cancer with indium-111-labeled phospholipid vesicles. *Cancer* **62**(5): 905–11.

Proffitt, R. T., L. E. Williams, et al. 1983. Liposomal blockade of the reticuloendothelial system: improved tumor imaging with small unilamellar vesicles. *Science* **220**(4596): 502–5.

Rusconi, C. P., J. D. Roberts, et al. 2004. Antidote-mediated control of an anticoagulant aptamer in vivo. *Nat Biotechnol* **22**(11): 1423–8.

Wallingford, R. H. and L. E. Williams. 1985. Is stability a key parameter in the accumulation of phospholipid vesicles in tumors? *J Nucl Med* **26**(10): 1180–5.

Wiseman, G. A., C. A. White, et al. 2001. Biodistribution and dosimetry results from a phase III prospectively randomized controlled trial of Zevalin radioimmunotherapy for low-grade, follicular, or transformed B-cell non-Hodgkin's lymphoma. *Crit Rev Oncol Hematol* **39**(1–2): 181–94.

Preclinical Development of Radiopharmaceuticals and Planning of Clinical Trials

2

2.1 INTRODUCTION: NUCLEAR MEDICINE

Nuclear medicine personnel employ a variety of devices to detect radioactive decay. Generally, the result of this effort is a set of images at various levels of organization. At the largest scale, entire systems are imaged using agents that target to those tissues. For example, using sulfur colloid labeled with ^{99m}Tc, the radiologist is able to image the reticuloendothelial system (RES). Here, one is involved with cells found in liver, spleen, and bone marrow. At the next level of organization, specific cell types within an organ are made visible. Galactosylated chitosan, tagged with ^{99m}Tc, may be used to image hepatocytes within the patient's liver (Kim, Jeong, et al. 2005). The smallest scale level of targeting involves finding molecules within the patient's body. For example, the surgeon may wish to find possible metastatic sites in the liver due to a primary colon cancer. In this last case, an antibody to carcinoembryonic antigen (CEA) labeled with ^{111}In may be a useful tracer. Note that such cancer-associated targets can, in principle, be in any tissue so that the first and second levels of targeting are now not relevant as the observer is looking for a specific molecule. The last strategy is of primary interest in contemporary nuclear medicine and is the featured topic in this text.

Molecular imaging has taken the field into some remarkable adventures of discovery. It has become almost conventional to look for tumor-associated molecular markers in the cancer patient. Included in these have been CEA, CA-125 (a marker for ovarian Ca), epidermal growth factor receptor (EGFR), prostate specific antigen (PSA), HER2 neu (a breast cancer marker), and multiple other molecules. It is important to recall that no molecule uniquely assignable to a cancer has ever been found. Thus, there is usually some production of these various species in normal tissue throughout adult life. The biological function of the molecule of interest may not even be clear (e.g., as in the case of CEA). This molecule has its highest production in the fetus and then decreases with age. Its particular function within or on the cell remains unknown, although it appears to have some association with cell-to-cell attachment—hence its fetal importance.

2.2 THE TOOLS OF IGNORANCE: PHOTON DETECTION AND IMAGING DEVICES

Nuclear medical photon recording relies on relatively primitive technology that is little changed over the past half century. In baseball, the catcher's protective equipment and glove are referred to as the "tools of ignorance." Nuclear medicine can be categorized in a similar way insofar as its various photon detection devices are concerned. The radiologist has to suffer through an ongoing period of uncertainty upon viewing the generally hazy output of the imaging device at hand. Referring physicians also appreciate the extent of this ignorance but do allow that the methodology does yield some idea of what is going on inside the patient. In the following, the various instruments are described, and their advantages and disadvantages are listed. It is important to realize that these devices are "all we have" and that there are no other ways presently available to obtain radionuclide data—at depth—within a living animal or patient.

2.2.1 Single Probes

One of the oldest instruments for gamma detection is the NaI (Tl) probe. The NaI (Tl) scintillation crystal is generally on the order of 6 mm to several cm thick to have high efficiency for detecting gamma photons up to hundreds of keV in energy. Scintillation is a process whereby the high-energy particle is converted into visible light of relatively short pulse length. This light flash is, in turn, converted to an electronic signal using a photomultiplier (PM) tube. Around the crystal, a Pb collimator is arranged in the form of a cylinder to allow some degree of shielding and hence directionality to the device. Lead is also fixed at the posterior end so that the detector is open only at the one side that faces the source of radiation. Like all gamma ray instruments, an associated electronic circuit allows counting of impinging radiation only if its energy lies between a preordained lower and upper set of bounds. For example, one might operate a ^{99m}Tc probe system with an energy window of 120 to 160 keV. The lower limit is designed to eliminate low energy photons that may have resulted from Compton scatter in the emitting patient, the collimation, or even walls of the room. In other words, only direct (unscattered) gamma rays should be recorded. An upper-level discriminator is usually not as important unless one has other radionuclides in the patient or in the general vicinity. The latter may be a problem when using the probe in a clinical setting where patients in nearby rooms may be contributing to incidental radiation. If we consider a gamma camera tuned to ^{99m}Tc, a thyroid therapy patient walking past the imaging suite would be an example of an outside, high-energy (360 keV) source of photons due to their having ^{131}I onboard.

Handheld detectors have a long and distinguished history in the nuclear and endocrine clinics for use in thyroid counting. These devices are also used to record the activity in the entire animal or patient. Here, the observer moves the detector sufficiently far away from the living emitter so that the entire body of the subject is within the view of the collimator. When repeated in this fixed geometry over an extended period of time, the observer has a record of the total body accumulation of activity over that interval. Some caution is advised as the activity may move internally during that period so that attenuation effects may vary and produce a somewhat uncertain result. This is discussed further in Chapter 5.

In oncology practice, probes are quite commonly used in the operating room to look at so-called sentinel nodes in breast cancer, melanoma, and other patients. If a molecular target, such as a cancer-associated antigen, is to be detected, the observer uses a radiolabeled molecular probe that combines with the target and remains at the nodal site for extended periods. Most of the time, the surgeon is not interested in such particulars and wants only to follow the lymphatic system from the tumor bed back to the one or more sentinel nodes. In this case, a colloid is generally employed to simply track the path of lymphatic flow from the (former) tumor site to the vena cava. Most colloids have sizes that are large compared with the agents described in Chapter 1: typically diameters up to 1 μm or even larger are used.

Upon discovery via the probe counts, these nodes are excised and passed to the pathologist for examination. Staging of the disease is done with the negative or positive outcome of this assay done microscopically. Because of the simultaneous application of unlabeled antibodies by the pathologist, it is now not unusual for single malignant cells to be discovered in some excised lymphatic specimens. Clinical significance of this outcome is currently unclear, although such results are technically metastatic sites. Several hundred unlabeled antibodies are in current use for pathological specimen assay.

Not all ionizing radiation detectors are based on NaI(Tl). Over the past 20 years, solid-state devices have been developed in an attempt to miniaturize the detection system—particularly for operating room usage. Such probes are also less sensitive to thermal, physical, and other shocks. Sodium iodide is notorious for its thermal sensitivity. It is not that high (or low) temperatures are detrimental to its operation. Instead, the rate of temperature change is of primary importance. If the janitor leaves the window slightly open during room cleaning, the crystal may very well crack on a cold winter night in Minnesota. A similar problem occurs with other uses of this scintillator material, including the gamma camera described next.

Commercial solid-state probes have been made using CdTe and CsI(Tl) as the sensitive crystals. In the CsI(Tl) example, a photodiode is used to carry the light signal back to the discriminator system. Operating room use requires sterilization with appropriate gas immersion prior to use; in addition, a plastic sleeve is put over the device before moving it into the surgical field.

Any single detector gives rise to a radioactive decay signal that may come from any point in space that can directly impact the detection crystal. Use of one probe with such a limited field of view may not be adequate in covering an extended object wherein regions may be of interest. Arrays of detectors have been designed to look over a 3-D source by having each single probe within the array individually aimed at one specific source region. This technology was extensively applied to measuring the blood flow to the brain during clinical evaluations. Here, the neurologist or psychologist would have the patient perform various mental tasks. Processing might include listening to commands, reading sentences, or viewing images. It was discovered, to the amazement of many physiologists, that brain blood flow varied regionally with the actual task. Areas of the brain associated with speech would show greatly elevated flow rates while the subject was speaking a given line of text. Generally, ^{133}Xe gas was the radiotracer of choice in these studies; it was able to leave the body via the lungs after only a few circulations. Such an array system is still clinically usable, but most studies of this type now use 3-D imaging via a gamma camera or positron emission tomography (PET) system.

2.2.2 Well Counters

Closely analogous to the probe is the concept of a well counter. Here, a radioactive injection, tissue sample, or unknown specimen is loaded into the well, which has a hollow crystal or set of detectors arrayed around the underside and outside of the counting volume. Unlike the single probe, the geometric efficiency is markedly higher since a large fraction of the 4π solid angle is covered by a sensitive detection volume. There are two such devices in common usage: the dose calibrator system and specimen well counter. For dose calibrators, argon gas is the detection medium, which is arrayed around the volume of the detector. Data taken from the dose calibrator are used to assay the total activity injected into the animal or patient prior to beginning an imaging or therapy protocol. As such, the calibrator reading is very important and provides one of the fundamental inputs into the total data mix. Periodic activity assay (usually done quarterly) is important in making quantitative images in the case of gamma cameras and PET scanners as described in Chapter 5.

Specimen counters are built on the same technology as the single probe. With NaI(Tl) crystals, they provide data on the activity of a gamma emitter in the sample. If the sample is a beta emitter, the well is filled with a phosphor-laden medium such that the electrons generate visible light upon their movement through the solution. This light is recorded by the PM technology as previously described.

One unusual feature may arise in beta assays using a well counter. If the beta (or positron) energy is sufficiently high, another type of emission, termed Cerenkov radiation, may be produced in the medium. This is the bluish glow seen in the water surrounding the nuclear reactor rods and represents a situation where the charged particle is moving faster than the speed of light in H_2O. Cerenkov radiation is analogous to the sonic boom of an aircraft moving faster than the speed of sound in air. Note that this does not contradict Einstein's speed limit of c since we are not considering movement in a vacuum. Instead, the beta particle is moving faster than only the speed of light in the local material. Cerenkov radiation is continuous out to a maximum at the kinetic energy of the beta. Several well-known therapy radionuclides can provide Cerenkov output, including ^{32}P and ^{90}Y. Detection of this radiation provides another method to quantify source strength given a set of activity standards.

For animal studies, the well counter is the dominant instrument that permits direct measurement of tissue accumulations of radioactivity over time. Here, after sacrificing the animal, various organs are taken to the counter, and the counts are recorded along with an activity standard for the radiolabel of interest. Tissues taken typically include the dominant ones in the anatomy: liver, lung, spleen, kidneys, heart, blood, and tumor. Blood may be sampled from the heart or taken directly from the tail vein in the case of a mouse or rat. There is some controversy as to whether the animal is bled or not bled prior to sacrifice. At City of Hope, the tissues are taken without prior bleeding. This is to make their activities comparable to those seen by the imaging studies that are often concurrent with these bioassays. Imaging of animals is, in fact, a superior method to determine activity levels as a function of time in the various tissues. These results, however, may have to be normalized to the bioassays done with the well counter. Animal imaging results are further described herein.

In clinical protocols, well counting is generally limited to patient blood samples taken during an imaging or therapy phase of the study. A nurse is typically involved in drawing blood at the same time as the imaging is being done. Unless surgeons are associated with the protocol, well counting of organ or tumor samples is uncommon in clinical activity assays. Even if the surgeon can obtain such samples, a caveat must be associated with the operating room tissue specimens. Any sample, almost by definition, would be only a small portion of the tumor or organ being investigated. A majority of the specimen must be delivered to the pathologist for tissue assay. Thus, assay results on such samples may be criticized as being nonrepresentative of the entire organ or tumor.

In the early 1950s, Cassen (1957) and coworkers at the University of California–Los Angeles (UCLA) developed a moving probe device for animal and clinical imaging. It is termed a rectilinear scanner since the probe and its collimator are moved in raster motion over selected portions of the subject's anatomy. A motor-driven gantry permitted data to be taken without human attention. It was necessary that the patient be kept in a fixed position for these studies. Today, the rectilinear scanner is rarely used for imaging of animals or patients due to the subsequent development of the gamma camera. The camera's advantage is that it then became possible to image simultaneously a relatively large extent of the patient anatomy.

2.2.3 Gamma Cameras

The conventional gamma camera was invented by Anger (1952) at the Lawrence Radiation Lab in the late 1950s. Anger's insight was to place a set of PM tubes on the backside of a relatively large single NaI(Tl) crystal to localize the impact point (scintillation) of each detected gamma or x-ray photon. In prior work with the probes, this information had been lost since no localization within the crystal had been attempted. Generally arranged in a close-packed hexagonal array, the tubes give signals that are proportional to the closeness of the scintillation to a given PM. Tubes closest to the impact point will have the highest signal, whereas those that are one lattice spacing away will be correspondingly reduced. Signal strength is essentially a solid-angle effect assuming that all the PMs are closely matched as to magnitude of their electrical output. Originally, the system was controlled by analog circuitry to determine the most likely location of the photon's scintillation point. Today, computers are installed in the camera head to provide localization of the impact as well as the PM tube signal normalization. Modern designs are generally based on rectangular crystals with their greater dimension being 50 cm or larger to cover the entire width of a patient in a single pass of the camera over the patient's body.

Temporal formatting of data taken via gamma cameras may be done in one of two ways. In principle, the memory associated with the camera can record each count by storing three parameters: (t, x, y). Here, time (t) is the elapsed time since the start of the imaging process, and x and y refer to the two spatial coordinates of the scintillation on the camera face. This type of recording is called list mode and is very useful if the kinetics of the radiopharmaceutical are not known prior to the start of the tracer experiment. Alternatively, the camera memory can be programmed to take data for a fixed length of time (Δt) and add all counts at each x and y location in what is termed frame mode. Frame acquisitions are essentially recording a snapshot of the emitter over a fixed time interval—much like an ordinary visible-light camera. Commercial gamma cameras manufactured at the present time generally do not allow list acquisitions. Memory size limitation is the primary reason for this restriction as list mode requires one additional word of memory for each recorded photon. When millions of counts come in during an imaging session, the use of list mode becomes problematic.

The most important result of the Anger design is that it became possible to make images by recording the locations of the registered photons serially in time. A set of static pictures or a motion picture of the movement of the radioactivity can be made. These pictures are a direct record of the motion of the radioactive pharmaceutical (RP) through the animal or patient. Also, as in the case of the probe, the total electrical signal of the camera is passed into a discriminator circuit to make certain that the recorded photon was of the appropriate energy to have come directly from the patient without scattering. While the original design had an output display that was shown on a slow-phosphor cathode-ray tube, computer memories are now used to immediately store the camera image. Display is via flat panel monitor under software control to establish a window and level much as in a computed tomography (CT) or magnetic resonance imaging (MRI) display. Long-term filing is in a picture archiving and communication system (PACS). This storage also involves other modalities that may be associated with the nuclear image, such as CT or MRI imaging results from the same patient. There arises the possibility of combining such images ("fusion") to produce both physiologic and anatomic representations of the individual. Other methods to combine such images more readily are given next.

If radioactive objects are two-dimensional, no Pb collimation is needed on the camera surface. For example, if one were to measure regional activity in thin pathology slices from the operating room, specimens could be placed directly on the gamma camera face. Suitable plastic wrapping would be required to prevent contamination of the detector. For 3-D objects, a directional feature has to be added to the Anger system. Just as in the case of the single probe, the required collimator defines the finite angular range over which a photon is allowed into a given segment of the crystal face. Walls (septae) of the collimator are generally five or more half-value layers (HVLs) at the appropriate photon energy being used for imaging. The collimator essentially produces a projection image of the radioactive distribution upon the NaI(Tl) crystal face of the camera. Unfortunately, the lead septae absorb a large fraction of the emitted photons so that the collimated camera has relatively poor sensitivity. As a general rule, a parallel-hole collimator system has sensitivity on the order of 1.8×10^{-4} if both the collimator and crystal (photon detection) efficiencies are included. This means that of 10,000 emitted gamma rays from a radioactive source only 2 will be detected by the collimated camera. This lack of sensitivity is one reason for the "ignorance" of the observer; most photons are simply wasted in the detection process, and the resultant image appears fuzzy due to a lack of events. The camera is approximately two orders of magnitude less sensitive than a PET scanner.

The number and type of its collimators generally determine the practical limitations of a given camera. Four standard collimator varieties are designed for general use in the clinic: pinhole, converging, diverging, and parallel hole. Each term describes the collimator as seen from the camera's point of view: for example, converging means that the septal openings seem to converge as viewed from the camera looking toward the radioactive object. This design would magnify the object and would be useful, for example, in imaging a relatively small animal such as a mouse or a small clinical organ such as a thyroid. Pediatric studies may involve use of converging collimators. Note, however, that the amount of magnification depends on the distance from the end of the collimator to the object. A similar detriment occurs with the diverging case whereby the minification (magnification < 1) factor changes with separation from the collimator face. Pinhole designs suffer from both problems depending on precisely where the radioactive subject is located relative to the small (typically several mm) opening of the collimator. In addition, the pinhole, by its very design, has extremely low sensitivity since photons can get to the NaI(Tl) crystal only through this single small hole. To ease the constraint, some investigators have engineered a collimator design having multiple pinholes such that a number of projections are produced on the camera crystal face. These designs have had a continuing application in cardiac imaging.

Presently, a majority of all clinical camera applications employ parallel-hole collimators, which have a magnification of unity independent of the camera–object distance. In inventing parallel-hole systems, the designer has to make a trade-off between sensitivity and spatial resolution. By placing multiple septae relatively close together, the sensitivity is lost since more gammas will strike the Pb separators, and fewer will get to the crystal. Yet the spatial resolution would be quite good using the few photons that make the passage into the crystal. Collimators are often named following their intended usage. For example, the label LEHR is an acronym for low-energy, high (spatial) resolution. Its sensitivity would be inferior to the same manufacturer's LEGP (low-energy, general-purpose) device.

Determination of the location of radioactivity in the cancer or other patient is most commonly done using a matched pair of cameras supported on a single gantry. Some triple-head systems have also been developed. Two formats of image reporting are required by the radiologist. Of primary importance is the whole-body image pair where the two heads are translated along the bed from the patient's head to their thighs or beyond. One camera records the anterior and the other the posterior projection of the entire body. Here, we are making the assumption that there is overriding interest in finding all sites of tracer deposition. Such would be the case for the cancer patient as outlined already. If the organ or sites of interest are considered to be more limited spatially, the extent of the scanning procedure would be correspondingly reduced. If one or more positive uptake areas were seen in the whole-body survey, localized imaging obtaining more counts from the local areas would then be performed. Typically, such vignettes would also be taken with one camera in an anterior aspect with the other positioned posteriorly. More time and hence more counts would be recorded for the vignette than the corresponding segment of the whole-body image. Reasons for using a pair of opposed views will be described in Chapter 5. Most importantly, the geometric mean of the counts recorded by the pair may be used to quantitatively estimate activity in the tissue being imaged.

2.2.4 SPECT Imaging

Finding accumulation sites within a patient with the whole-body or vignette imaging is useful but can be enhanced with single-photon emission computer tomography (SPECT). In this case, the camera heads are simultaneously rotated around the patient at a given location along the scanning bed. A complete rotation is performed with each head being used for half the data acquisition. In a typical data situation, one may acquire 60 separate projections at 6° angular separation in a single SPECT study. The rotation may be done in one of two ways. The twin heads detect photons either continuously or in a set of fixed angular positions (step-and-shoot method). The outcome is a 3-D set of computer-generated images. Upon reconstruction, these data are usually viewed in the classical three orientations: coronal, sagittal, and axial. In standard SPECT, the numerical pixel values shown in these images are not proportional to the activity at that location in the animal or patient. Thus, these projections cannot be used to quantitate activity.

A number of reconstruction methods are in common SPECT usage. Probably the oldest and simplest of these techniques is the filtered back-projection method (FBP). Here, the algorithm takes each projection (each angular image set) and predicts the amount of radioactivity along that ray as it transits the unknown set of sources. Attenuation inside the object, due to soft tissue and bone, is corrected for in this algorithm. As more projections are taken into account, a sharper and sharper result generally takes shape inside the reconstructed volume. In most applications, these FBP calculations use a Fourier transformation of the data to expedite the process.

More recently, the ordered subset expectation maximization (OSEM) has become a favorite technique for SPECT image generation. In this case, certain projections are given greater emphasis in the processing due to their significant contributions to the clarification of the sources within the radioactive object (Hudson and Larkin 1994). Unlike FBP, the OSEM method may be iterated indefinitely to try to sharpen the sectional images. Usually a statistical criterion is invoked to end the computation—although the viewer may simply intervene when there appears to be little advantage in continuing. A simple rule would be to end the algorithm when counts seen in a given volume do not appear to be changing with increasing iteration number.

It should be noted that these two methods and their variations—as well as other techniques—may be applied generally to other types of projection images. Thus, besides application to the gamma camera, there is usage in PET. In the latter, one uses the simultaneous emission of the two 511 keV photons to provide the fundamental projection data.

2.2.5 PET Imaging

Emission of two back-to-back photons during e^+ decay leads directly to the possibility of PET imaging. If the two photons are recorded within a predefined narrow time interval, the observer knows that they were probably emitted by the same positron decay. No collimation is required due to the direction being defined by the detection of the pair of 511 keV photons separated by 180°. The associated detectors define a line (or mathematical ray) in space for that decay. Solid-state detectors are generally used in PET; bismuth germinate (BGO) has been the more common material, but recent designs have shifted to lutetium orthosilicate (LSO) due to its higher light output and shorter electrical pulse length. The latter feature allows greater count rates for the PET system. Each larger crystal is cut into segments, which are optically coupled into a set of PM tubes as in the Anger camera principle.

With sufficiently rapid electronics, it is possible to go one step further in the use of positron imaging systems. If the timing can be as quick as a fraction of a nanosecond or less, the difference in the arrival time of the two decay photons can be used to predict where, along the line of detection, the positron decayed. This strategy is termed time of flight (TOF) and is available on modern PET imagers. Recall that the speed of light is on the order of its rate *in vacuo* 3×10^{10} cm/sec so that a 511 keV photon travels 30 cm in one nanosecond. Thus, TOF implies time resolutions on the order of small fractions of a nanosecond to have spatial resolutions on the order of 1 cm.

PET scanners are usually circular arrays of detectors—multiple rings being common so that up to 15 cm of the patient may be imaged at a single time without moving the bed. Two geometric forms of detection are used in common practice. If each ring is separated by a collimator from all other rings in the system, the detection is defined as 2-D. This permits the best spatial resolution in the z-dimension (i.e., along the bed). Efficiency of detection of 511 keV photons is on the order of 4×10^{-3}. If, at the other extreme, all rings are allowed to have mutual coincidences with each other, the acquisition is defined as 3-D. The latter method will naturally be more sensitive but with somewhat poorer spatial resolution compared with a 2-D operation. In 3-D mode, the expected sensitivity is approximately 5×10^{-2}. By comparing with the gamma camera previously given results, one sees that the PET systems are one to two orders of magnitude more sensitive than cameras. Positron tomographic systems may operate in either mode. The radiologist will select a 2-D imaging set if greater spatial resolution is important. Efficiency will suffer by about a factor of 10 in this case because of the lack of coincidences between adjacent rings.

We have previously mentioned the logical concept of list data storage in connection with the gamma camera. While generally not available on present-day cameras, list mode is possible in all commercial PET systems. By using this facility, the observer can follow quick-targeting radiopharmaceuticals that move out of the blood and into tissues in times on the order of minutes or even seconds. Following these agents is particularly difficult if a new molecular tracer is being studied (i.e., an agent without a previous history). Associated with such rapidly accumulating tracers, the pharmacologist may elect to use positron emitters with commensurate half-lives on the order of minutes or shorter as described in Chapter 1. This type of label allows the absorbed radiation dose to be minimized. List mode acquisition will permit following novel situations and allows reformatting of the study in any sequence of time intervals that suits the kinetics of the tracer–target interaction. Reformatting may be done after the data are acquired and at the viewer's convenience.

The previous short discussion of PET efficiency should not lead the reader to assume that positron technology is the optimal answer for every nuclear medicine situation. As noted in Chapter 1, one difficulty with PET imaging is finding a radionuclide that suits the tracer agent in that its decay rate is comparable to or somewhat less than that of the physiological process of interest. If, for example, an intact antibody tracer requires several days to target a tumor, the half-life of the positron (or ordinary gamma emitter) must also be at least of comparable length. If the physical half-life is significantly less, the label will decay before the physiological process is completed, and the observer will, by definition, be blind to a majority of the entire targeting process. It is this lack of suitable emitters that limits the application of PET to RP evaluation.

2.2.6 SPECT–CT Hybrid Systems

Besides finding suitable labels to match the physiological clearance rates of various targeting agents, nuclear medicine has had another issue at its core ever since the imaging aspects of the modality were first developed in the 1950s. Conventional x-ray imaging could clearly demonstrate bone and, to some degree, differentiate soft tissues. CT imaging has carried this spatial resolution into sizes on the order of 1 mm. Thus, via CT scanning, the patient anatomy is always available to the radiologist and the referring physician. Yet nuclear images contain, in either planar or tomographic formats, only regions of high accumulation (hot areas) displayed against a relatively low accumulation zone (cold area or background). If the bright area looks like a liver or kidney, the tissue of origin of the activity is easily determined. In the general case, where no obvious anatomic structure is seen, a nuclear radiologist is often uncertain as to what organ was being imaged. Any resultant report could refer only to locations such as "the lower right quadrant of the abdomen." This careful statement, while correct and eminently diplomatic, often leads to frustration on the part of the internist who had to incorporate this calculated ambiguity into an eventual diagnosis.

Ambiguity of nuclear imaging can be largely resolved for those institutions having combined modality technology. A common commercial solution is to place the dual SPECT camera heads adjacent to a patient bed that is shared with a conventional CT device (Bocher, Balan, et al. 2000). The patient is then imaged, in succession, by the CT scanner and the gamma camera. Since the bed is common to both systems, there is no need to do complicated image fusion postscanning. There is an additional advantage in using the CT data to correct the SPECT images for patient attenuation and scatter. With such systems, it is conventional to show the CT image as black and white, whereas the nuclear is rendered in color. An example is given in Figure 2.1.

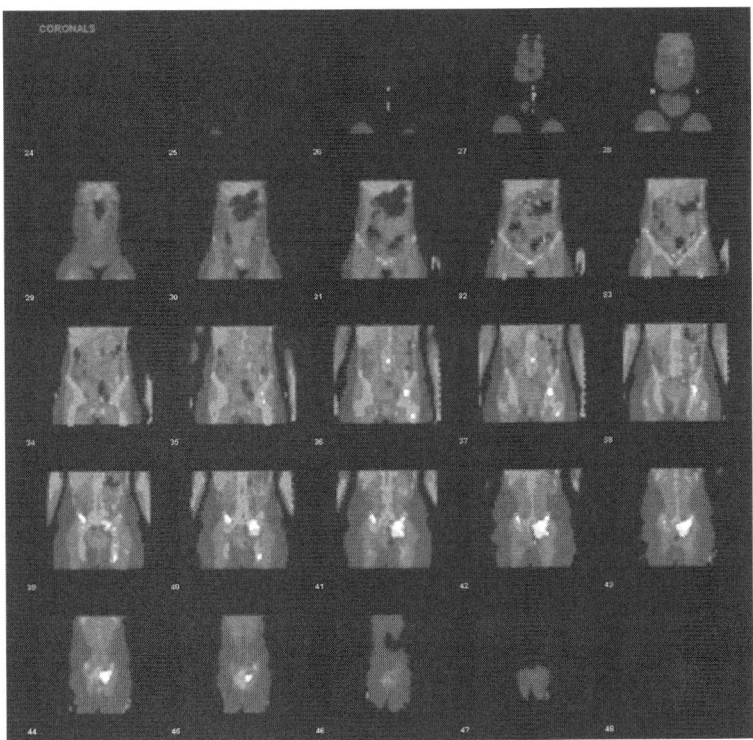

FIGURE 2.1 Hybrid (SPECT–CT) human coronal image set. The agent is metaiodobenzyl-guanadine (MIBG) labeled with ^{123}I. Multiple pheochromocytoma locations are seen in these sections.

Some limitations occur with hybrid nuclear medicine systems. Patients are imaged on each modality sequentially, so there may be slight changes in posture and anatomic positioning from one sequence to the other. Probably of greater concern is that the CT image is done using breath holding, whereas the nuclear, since it requires far greater time, must permit a patient to breathe normally. Difficulties in comparing lung volumes between modalities are expected. A similar limitation occurs in imaging the liver. With these restrictions in mind, SPECT–CT is an ideal way to obtain quantitative data on the amount of activity at depth in the patient. This topic is described more extensively in Chapter 5. Similar hybrid systems occur in positron imaging.

2.2.7 PET–CT

A PET–CT hybrid is analogous to that described for the gamma camera–CT case (Beyer, Townsend, et al. 2000). Again, the patient bed is shared by the multiring PET machine and a CT imager. Sequential scanning is performed with the CT data being taken first to allow a better estimation of the attenuation and scatter inside the patient for the subsequent PET reconstructions. Caveats that were described for the SPECT–CT also hold here. Breath holding versus free breathing may cause some differences in the two image sets.

While not available as commercial devices, hybrid SPECT–MRI and PET–MRI scanners are currently undergoing development. These compound machines present an unusual problem in that the electronic circuitry of the PM tube is sensitive to ambient magnetic fields. This follows since the tube uses electrons to record and amplify the visible light signal. Solid-state devices may be a more prudent choice for scintillation detection in such future systems.

2.2.8 Miniature Gamma, SPECT, and PET Cameras

The nuclear medicine research group has the option to use small-animal versions of the standard (clinical) gamma camera and PET scanner (Green, Seidel, et al. 2001). These devices have sizes on the order of one-third to one-tenth that of the clinical units. Both imagers are available for use with mice and rats to provide in vivo information on the uptake of the RP in preclinical studies. Animals much larger than these two species are generally imaged using clinical systems—but after-hours in the department. There are, in addition, miniature CT scanners that can provide corresponding anatomic information at the size scale of small rodents.

The advantages of small-animal imaging systems are several. First, because of the reduced size of the detectors, spatial resolution may be improved to values on the order of 1 mm for either device. This is quite an improvement (10x) over the clinical system. Second, there is now the possibility of following a given animal sequentially rather than having to sacrifice multiple animals at each time point. For example, let us assume that a rapid change in the biodistribution of an agent appears to be going on at a specific time. The observer can simply bring the group of test animals back to the imaging lab at this juncture and commence collecting data. There is no worry that one will "run out of animals" before the end of the experiment. Such flexibility is important in finding details of the RP targeting as well as fitting a model to the data. Until relatively recently, multiple animal sacrifice has been the standard method of determining where the RP has gone over a period of time. A certain number of animals (typically 5 to 10) had to be killed at each predetermined time. There may still need to be sacrifice of the imaged animals at the end of the biodistribution experiment to calibrate, or at least corroborate, the absolute activity in the organ systems. The next section provides a more complete discussion of the animal experiments.

2.3 ANIMAL BIODISTRIBUTIONS

After the inventive phase of RP development, the second step in testing a novel agent is the animal biodistribution. Investigators will typically use a mouse or rat as the test species. If tumor targeting is of interest, either a spontaneous animal tumor may be grown or a rodent obtained that has no thymus gland. In the latter case, the animal's immune system lacks T cells and thus cannot reject xenografts. Such animals are termed *nude* mice or rats since a cofeature of the athymic animal is an essential lack of body hair. Two types of data recording are done in animal work: a sequence of images and direct bioassay via sacrifice. Both of these data acquisitions use the set of counting and imaging tools listed previously. Ideally, imaging and bioassay are performed at the same time. It is statistically superior to use a few animals and to perform serial quantitative images. In that way, the experimenter can follow accumulation in tissues in sequential form and generate corresponding curves for each subject. Sacrifice destroys the animal and requires that many more be used in the experimental project. It may be necessary, however, to calibrate image data using the bioassay results. For example, the blood data as derived from a cardiac image may be normalized to that obtained by direct sampling of the tail vein.

Various parameters relevant to the biodistribution may be generated with the imaging and biosampling data. Several of these variables are shown in Table 2.1. Of primary importance is the traditional uptake value (u) in units of percent injected dose/g (%ID/g). Although the authors may not make it clear, almost all data in the literature are in this decay-corrected format. Data as taken [$a(t)$] are given in units of percent injected activity/g (%IA/g) and are generally not shown in a report. Both functions depend explicitly on time, on the RP and its target, and on the species being used as a test subject. It is important to understand that $u(t)$ is a constructed parameter whereas $a(t)$ is the directly measured quantity. One generally thinks of $u(t)$ as the uptake of the pharmaceutical. As noted in Chapter 1, however, the radioactivity may have come off the original agent to make this assertion less than factual. Remember that the analyst almost always follows the trail of radioactivity and does not investigate the integrity of the RP over time.

As we will see in Chapter 4, the $A(t)$ parameter is relevant for absorbed dose rate estimates. Here, the observer is measuring the total activity in an organ or in some part thereof (e.g., a particular voxel). For the physicist, $A(t)$ and its integral are needed for the absorbed dose rates and values, respectively. Total absorbed dose will be proportional to the integral of A out to sufficiently long times. A common rule of thumb for the integration is out to 10 physical half-lives. Most animal or patient data sets end long before this time point, so extrapolation of biodistribution results is generally necessary using some type of mathematical model as described in Chapter 6.

TABLE 2.1 Biodistribution Parameters

NAME OF PARAMETER	UNITS	CORRECTION	APPLICATION
Uptake (u)	%ID/g	For decay	Biodistribution
Specific activity (a)	%IA/g	none	Modeling of data
Activity	%IA	none	Modeling of data
Cumulated activity ($\bar{A}$)	%IA hours	none	Absorbed dose estimates

2.3.1 Specific Targeting in Vivo

To show specificity of targeting, $u(t)$ is initially compared with the numerical value that would arise if the agent were able to distribute completely uniformly in the animal. If we consider a 25 g mouse with all (100%) of the injected dose still on board, this ratio would be

$$100\%\text{ID}/25 \text{ g} = 4\%\text{ID/g} \qquad (2.1)$$

Biopsy or quantitative imaging uptake values significantly above the limit of Equation 2.1 would be indicative that some specific localization is seen in the mouse model. In a 70 kg patient, the corresponding uniform distribution value is 1.4 %ID/kg. This three-order-of-magnitude difference arises out of the corresponding thousand-fold variation in the total body mass of mouse versus man. Its small magnitude for humans should not be used to lament poor tracer accumulation in patients. In Chapter 11, comparisons of uptakes and kinetics are made between various mammalian species injected with the same tracers. In this allometric context, mice and humans are shown to have very different uptake (u) values, as one might expect, but are seen to be similar in the rate of tracer movement. In other words, the kinetics of a given RP are often much closer in the two species than their very disparate body masses or lifetimes would indicate.

The value shown in Equation 2.1 may not be achievable in some tumor-targeting molecular agents due to sequestration in normal tissues before the tracer can get to the lesion sites. Following, examples will be provided where the excretory routes of some constructs are so rapid that maximum tumor targeting is comparable to or even less than the value of 4 %ID/g of Equation 2.1. When this occurs, the pharmacologist may justify the low value by citing the ratio (r) of the target tissue's uptake to that found at the same time in blood. This ratio has become a common indicator of the usefulness of the agent. Some limitations to this simple criterion are described in the next chapter.

2.3.2 Biodistributions in Mice

To illustrate both the magnitude of uptake values in vivo and the variations seen with the various generic agents described in the previous chapter, we now report on several biodistributions and their associated images. Most of these data are from mice since they are the most economical animal to use in preclinical trials. As our first example, we describe liposomal targeting to murine tumors.

One of the more interesting aspects of targeting is that the process may involve more than a single step—or a single pharmaceutical. My colleagues and I (Proffitt, Williams, et al. 1983) used this strategy to increase the uptake of 40–50 nm diameter neutral ^{111}In-labled liposomes in EMT6 murine tumors in BALB/c mice. Table 2.2 contains the uptake results at 24 h after the mice had been previously given positively charged amino-mannose (AM) unlabeled liposomes as a liver-blocking agent. Notice that the tumor uptake increased from 18.5 to 28.7 %ID/g whereas most other tissues, except for the spleen, were not affected by the previous injection of unlabeled AM

TABLE 2.2 Biodistribution of Neutral ^{111}In-Liposomes in BALB/c Mice at 24 Hours after Blockade with a Preinjection (1 Hour Earlier) of Unlabeled Amino-Mannose Liposomes

ORGAN	AFTER RES BLOCKADE (%ID/g)	WITHOUT BLOCKADE (%ID/g)
Blood	6.8 ± 1.5	6.6 ± 0.8
Muscle	1.0 ± 0.3	0.49 ± 0.03
Lung	6.8 ± 1.5	6.0 ± 0.8
Liver	14.0 ± 2.1	14.6 ± 0.8
Spleen	13.7 ± 1.6	18.1 ± 1.6
Kidneys	6.3 ± 0.7	6.8 ± 0.3
EMT6 tumor	28.7 ± 2.6	18.5 ± 2.4

Source: Proffitt, R. T. et al., Science, 220, 1983. Reprinted with permission.

liposomes. The majority of the organs showed uptake values comparable to the result of Equation 2.1—that is, near 6% ID/g. Three tissues demonstrated values above this limit: spleen, liver, and tumor. The first two of these accumulations were not surprising since liposomes are well known to target the reticuloendothelial system. The overall result was that the tumor was higher than those RES uptakes and that the tumor's value could be manipulated to be even larger by using the blocking concept.

Gamma camera results, obtained with a single pinhole collimator on a clinical camera at 24 h, are given in Figure 2.2. These images show that the tumor is vividly evident on the flank of the blocked mouse but is clearly higher in uptake than the liver in the right-hand image (b). In animals given a single liposome injection, the liver is the more dominant tissue in the associated (left) image (a). As noted earlier, the gamma camera images are not always clear as to what tissues are being emphasized. In these two images, the outline of the animal permits a relatively good guess as to the organs being seen.

As a second study, consider the intact form (Williams, Wu, et al. 2001) of the radioiodinated cT84.66 chimeric antibody to CEA. Since ordinary mice do not produce CEA (it is probably a purely human protein), the relevant animal is the nude mouse into which human colorectal tumors (LS174T) had been xenografted. These tumors produce the desired CEA molecule. Data taken at 0, 48, and 168 hours are shown in Table 2.3. Here, tumor uptakes approached 42 %ID/g—a value that is characteristic of intact antibodies going to human tumor implants in this model. Note also that the blood curve was monotonically decreasing whereas the tumor uptake went through a maximum and then began to decline at long times. As we will see later, this decrease may, at least in part, be an artifact of the iodine label given to the antibody. If we had used an indium tag and an appropriate chelator, the tumor uptake would probably have remained much more constant after achieving a maximum value. In addition, unlike the liposomal case, the tumor is by far the dominant tissue by 48 h.

Use of a specific antibody binding to a tumor-associated antigen, as in the case of cT84.66 becoming attached to CEA, has become one of the standard strategies in targeting primary and metastatic lesions. Unlike the liposome example, the observer knows

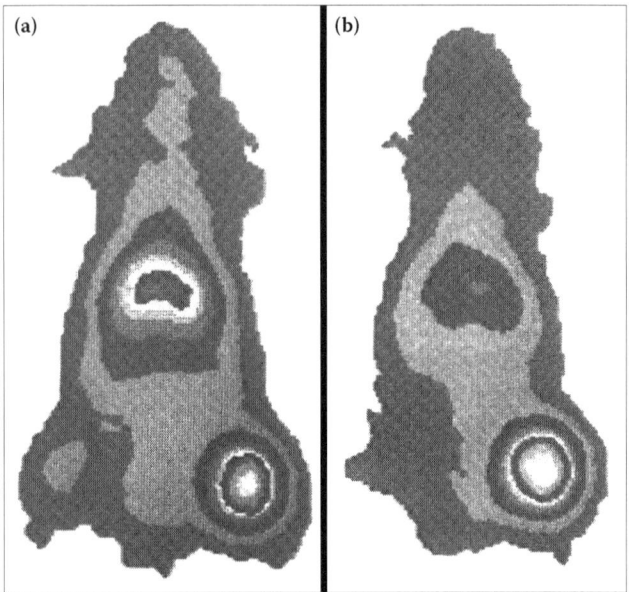

FIGURE 2.2   Set of two dorsal images from BALB/C mice having EMT6 tumors on the right hind limb. Image (a) shows the control result with the 50 nm DSPC:CH = 2:1 liposomes labeled with ^{111}In. Image (b) is of a mouse on that has been previously blocked with a prior injection of AM-modified liposomes (DSPC:CH:AM = 8:3:1) that were unlabeled. Notice the reduction in liver accumulation in this animal compared to the control case (image a). (From Proffitt, R. T., L. E. Williams et al. *Science*, 220, 4596, 1983. With permission.)

TABLE 2.3   Biodistribution (%ID/g) of ^{131}I-Intact cT84.66 Anti-CEA Antibody in Nude Mice Bearing LS174T Human Colorectal Tumors

ORGAN	t = 0	48 h	168 h
Blood	38.32 ± 4.3	13.2 ± 2.0	2.3 ± 1.1
Lung	15.24 ± 1.4	5.7 ± 1.3	1.1 ± 0.5
Liver	7.9 ± 1.0	2.7 ±0.2	1.1 ± 0.3
Spleen	3.5 ± 0.5	1.9 ± 0.2	0.48 ± 0.2
Kidneys	6.4 ± 0.9	2.6 ± 0.4	0.64 ± 0.29
LS174T tumor	0.9 ± 0.1	42.6 ± 9.1	15.2 ± 3.4

Source: Williams, L. E., Wu, A. M., Yazaki, P.J., et al., *Cancer Biother. Radiopharm.*, 16, 2001, published by Mary Ann Liebert, Inc. Reprinted with permission.

that the outcome depends on the association between antibody and antigen molecule. In the case of a nonspecific antibody, resultant tumor uptakes would be close to the value cited in Equation 2.1—approximately 6 %ID/g. This specificity (along with the possibility of a negative control) is probably the single most attractive feature of antibody usage. One aspect of the use of intact antibodies (Abs) is that the targeting may require several days—with associated needs of long-lived radiolabels and extensive time devoted to imaging. Use of large (150 kDa) proteins has the additional probability of causing an immune response in a patient if the Ab injection is repeated. It would, therefore, be useful if the targeting could be done more rapidly and with a lower molecular weight (MW) agent that does not elicit an immune response from the animal or patient.

Let us consider the use of the small aptamer TTA1 (Hicke, Stephens, et al. 2006) as our third case in point. This RNA construct was discovered by the systematic evolution of ligands by exponential enrichment (SELEX) process (Gold 1995) to bind to tenacsin-C with nanomolar affinities in lab testing. Tenacsin-C is a large protein associated with a number of human tumors including breast, lung, colon, and glioblastoma. The operational strategy was to use this 13 kDa agent instead of an intact protein to more rapidly exit the blood and more quickly target to tumor tissue. Small lengths of RNA are also thought to be nonimmunogenic. Because of the need of human tumors in the test animal, nude mice were implanted with clinical glioblastomas having masses between 200 and 350 mg. Two labels were investigated: ^{99m}Tc and ^{111}In. In the former case, extensive clearance equally by liver (50%) and kidneys (50%) was observed. By using indium, the activity was seen to rapidly transit the kidneys and vanish from the animal via the urine. Table 2.4 gives the biodistribution results for TTA1 with the ^{99m}Tc label.

In this aptamer example, clearance from the blood was very fast, going from 50 %ID/g at $t = 0$ h (not shown in Table 2.4) to only 0.1 %ID/g at 1 h. Tumor uptake was not that high, however, due to this rapid movement of TTA1 out of the circulation. Hicke et al. (2006) cited the very high tumor-to-blood uptake ratios (u_T/u_B) as indicative of the high specificity of the aptamer. A control RNA construct, not specific to tenacsin, showed no long-time residence at the tumor site and uptake ratios on the order of unity. Chapter 3 discusses the use of ratios to describe targeting and how other, more elaborate, indices may be used to better indicate the effectiveness of the tumor targeting. It will turn out that a u_T/u_B ratio greater than 1 is only a necessary—not a sufficient—indication of good tumor targeting. This concept of tumor/blood ratios > 1 may be extended to other possible "background" tissues such as muscle, kidneys, or liver.

One important strategy of tumor targeting with antibodies is the concept of multistep or so-called pretargeting. More than one injection is required, much as in the case of the previous liposome example. As in that example, only one targeting construct has a radioactive tag. In pretargeting of antibodies, the experimenter first gives an unlabeled hybrid protein that includes both an antibody to the original antigen as well as

TABLE 2.4   Biodistribution (%ID/g) of ^{99m}Tc-TTA1 Aptamer in Nude Mice Bearing U251 Human Glioblastoma

ORGAN	$t = 2\,m$	$10\,m$	$60\,m$	$180\,m$
Blood	18.3 ± 1.1	2.27 ± 0.25	0.11 ± 0	0.03 ± 0
Lung	9.0 ± 1.2	2.13 ± 0.08	0.16 ± 0.01	0.05 ± 0.01
Liver	9.1 ± 0.53	12.5 ± 1.3	1.23 ± 0.09	0.40 ± 0.08
Kidneys	44.4 ± 4.3	18.8 ± 0.9	1.51 ± 0.04	0.29 ± 0.03
U251 tumor	4.47 ± 0.41	5.94 ± 0.59	2.69 ± 0.31	1.88 ± 0.10

Source: Hicke, B. J. et al., *J. Nucl. Med.*, 47, 4, 2006. Reprinted with permission of the Society of Nuclear Medicine.

an additional antigen that will, sequentially, be targeted by a second, radioactive agent. In essence, one antigen is being exchanged for another as in the original description of the method by Goodwin, Meares, et al. (1988). Notice also that the first step may take as much time as is required biologically since no use of radioactivity is involved.

In the second step, activity may be placed at the objective sites with a relatively rapidly cleared material so that the blood exposure is minimized. A short-lived label is ideal, and ^{99m}Tc is suggested as the best radionuclide for associated imaging studies. Multistep antibody targeting advocates remark, as in the previously cited aptamer case, on the improved u_T/u_B ratio as the basic advantage of their method. As the next chapter shows, this improvement is only a necessary but not a sufficient condition for improving the tumor image relative to blood background. Generally, the radiolabeled material is a relatively small molecule to expedite clearance—for example, via the kidneys (K). Other normal organs may also be used to sequester the second injection. Uptake of the label in these normal tissues, however, may make imaging of any nearby lesions problematic. In this case, the u_T/u_B ratio should be augmented by u_T/u_K or other ratio to take into account the new background organs.

The report by Zhang, Zhang, et al. (2003) is an example of a two-step targeting using antibodies against lymphoma antigen CD25. This protein is found extensively on the surface of dividing B cells and is the receptor for IL-2. In this work at NIH, the tumor model is SUDHL1—a human anaplastic large cell lymphoma xenografted into nude mice. Two targeting methods were compared in this work. First was the traditional use of the intact humanized anti-Tac antibody (HAT) injected as a radioactive compound. The second method used a cold (unlabeled) fusion protein of single-chain Fv antibody fragment combined with streptavidin as the initial step followed by a radiolabeled biotin to place activity at the tumor site. In this strategy, a clearing cold dose of synthetic agent was employed to eliminate most of the still-circulating fusion protein. Both methods used ^{111}In as the radionuclide.

A comparison of the two targeting methods is given in Table 2.5. It is seen that the tumor/blood uptake ratio became very large in the case of the pretargeting method—approaching values in excess of 1000/1 by 48 h. At that time, the traditional method was able to demonstrate only a ratio of 2.2. Moreover, there were no tissues that have uptakes comparable, at any time, to that of the tumor in the pretargeting example.

As a final example of tumor-seeking agents, consider a novel form of hybrid nanoparticle developed at the University of California–Davis (UC–Davis; DeNardo, DeNardo, et al. 2007). This agent is interesting in that it consists of three separate parts integrated to achieve an eventual clinical objective. The overall idea of their strategy was to provide a means to induce hyperthermia at tumor sites in vivo. Such local heating has a long—probably prehistoric—application in oncology. This topic is described more extensively in Chapter 10. Heating can act as a therapeutic modality on its own or as an adjunct to ionizing radiation (Hildebrandt, Wust, et al. 2002). To provide heating at malignant sites, iron oxide particles were transported to the tumor locations, and an oscillating magnetic field was later applied to the whole body of the murine test animal. A variable magnetic field is equivalent to an electric field so that electrical currents were induced in the particles. These currents caused temperature levels in the tumor to achieve values in the range of 41 to 46°C (normal mammalian body temperature is around 37°C). Some real-time measure of the amount of heating is necessary in both

TABLE 2.5 Comparison of One-Step and Pretargeting Approaches in Lymphoma Targeting with Anti-CD25 Antibodies

TIME (h)	PRETARGETING			
	BLOOD (%ID/g)	KIDNEYS (%ID/g)	LIVER (%ID/g)	TUMOR (%ID/g)
0.5	1	2.4	0.8	15.2
2	0.5	1.6	0.4	11.2
24	—	2.8	0.4	6.8
48	—	2.8	0.4	6.4
	ONE-STEP (TRADITIONAL)			
6	28	8.0	8.8	12.0
24	25	6.4	9.6	28.2
48	16	5.6	9.6	28.2
96	8	5.6	8.0	17.6

Source: Zhang, M. et al., *Proc. Natl. Acad. Sci. USA*, 100, 4, 2003. Copyright 2003 National Academy of Sciences USA. With permission.

the lesions as well as in the adjacent normal tissues. These data were obtained using thermal probes.

In this construct, iron oxide spheres 20 nm in diameter were conjugated to the Lym-1 antibody. To allow the nanoparticles to circulate for an extended time and to permit relatively high levels of accumulation at the tumor site, polyethylene glycol (PEG) was placed on the exterior of each iron oxide sphere. This covering is another application of the "stealth" methodology. By coating the alien-looking sphere, the experimenter reduces the likelihood that they will be quickly sequestered from circulation by the RES. Notice that the overall agent consists of three parts: the iron oxide spheres, their PEG coating, and an attached targeting intact antibody. The reader should appreciate the complexity of this construct as well as the reasons for each segmental aspect. Ultimately, the specificity of the agent still depends on the intact antibody combining with the original tumor-associated lymphoma antigen.

The UC–Davis experimenters (DeNardo et al. 2007) were able to document that the delay in tumor growth correlated with the magnetic field intensity. Temperature probes inside the animal's skin and rectum recorded values approaching the 42°C value generally thought to be fatal to mammalian cells. Biodistribution values for the nanoparticle construct are given in Table 2.6. This application is interesting in that ionizing radiation is intentionally not used, although it might be considered as an eventual adjunct to the hyperthermia, as discussed in Chapter 10.

There are clearly multiple strategies to target molecules within the animal or patient. In summary of the tumor data shown in these five examples, lesion uptakes are found to approach values of 50 %ID/g or more for single-step intact antibody targeting in mice. By using several steps with a final administration of a labeled agent, the amount in the tumor would probably be less than this value, but the ratio of tumor/blood uptake would be much improved and may approach values of over 1,000. One can also realize that forms of therapy other than that of absorbed dose are possible. While earlier discussion

TABLE 2.6 Biodistribution (%ID/g) at 48 Hours of ^{111}In -cL6-PEG-Iron Spheres in Nude Mice Bearing Human HBT 3477 Breast Tumors

ORGAN	UPTAKE AT 48 h (%ID/g)
Blood	10 ± 2
Lung	3.8 ± 0.8
Liver	20 ± 2.5
Kidneys	6.3 ± 0.8
Spleen	14 ± 5.0
Marrow	6.7 ± 2.5
HBT 3477 breast tumor	13.3 ± 2.0

Source: DeNardo, S. J., G. L. DeNardo et al., *J. Nucl. Med.*, 48, 3, 2007.

emphasized imaging, the chapter has alluded to the switch of label to other radionuclides to effect radiation therapy by internal emitter. We have, so far, completely omitted the possibility of a chemotherapeutic agent being involved in the treatment process. These last two issues will be brought into sharper focus in Chapters 4 and 10, respectively.

Given promising results from in vitro testing and animal biodistributions, the clinician will be led to an interest in one or more of these agents for clinical evaluation. Such trials, however, will require increases in both the amount of data required as well as the number of regulatory steps needed for approval. Financial cost and ethical issues also now become very important.

2.4 LOGISTICS OF HUMAN TRIALS

Given sufficiently encouraging animal results, such as the biodistribution data in Tables 2.2 to 2.6, clinical evaluation of the tracer will follow after regulatory approval. While relatively few parameters are recorded in animal biodistributions, multiple different data formats are required in patient studies. Table 2.7 summarizes various data types with a short description of each. This table emphasizes the RP aspects of a clinical trial. Notice that blood draws, planar whole-body scans, SPECT images, and CT scans are routine parts of the patient study. Ideally, as mentioned already, SPECT–CT hybrid systems are preferred since the true size and location of the patient's major organs (and tumors) may be obtained directly. These data allow patient-specific absorbed dose estimates as outlined in Chapter 4.

If lesion localization is seen at some time in the whole-body images, vignette imaging will also be performed immediately on those anatomic regions. Generally, camera imaging will be based on a dual-headed device where one head is placed on either side of the patient. Typically, these are the anterior and posterior projections. Whole-body activity can be tracked using the whole-body images or with a separate probe that is sufficiently far away so that it can encompass the entire patient.

TABLE 2.7 Empirical and Derived Information Involved in a Radioimmunotherapy (RIT) Clinical Trial

DATA TYPE	DESCRIPTION
Personal information	Demographics of the patient, such as height, weight, organ sizes. The latter are determined from CT or MRI image information.
Blood uptake	Percent injected activity/ml (%IA/g) in the whole blood or in plasma
Urine accumulation	Percent of injected activity/liter in the urine
Immune data	Patient plasma samples tested for reactivity against the Mab. Looking for human antimouse antibodies (HAMA) or analogous entities against other frameworks such as HACA for human antichimeric antibody. If present, these antibodies may make repeat injections impossible.
Nuclear medicine image information	Pixel-by-pixel activity in NM planar images
CT and MRI images	Pixel-by-pixel Hounsfield or MRI numbers in tomographic images
Activity curves	Percent injected activity per organ over time
Area under the concentration–time curve (AUC)	Integrated area under the activity curve for organs and tumors
Absorbed dose values	cGy per MBq (rad/mCi) injected activity for the patient in question; vector set of doses for clinically relevant tissues including tumor
Clinical blood assay	Toxicity measurements such as determination of platelet and white cell concentrations in the patient's blood pre- and post-targeted radionuclide therapy (TRT)
Tumor size	Three external dimensions (x,y,z) of the tumors versus time using CT or MRI image results; spatial integration by summing slices
Gamma counts	Raw sample data from gamma cameras and counters
Clinical lab report	RN progress notes
Alerts	Notification of errors and critical conditions

If a positron emitter is being used, planar images are often generated using originally transaxial PET data. As in the case of the gamma camera, a PET–CT hybrid is preferred to attempting subsequent fusion between PET and separate CT images. Restrictions on PET application to pharmacokinetic studies center on the issue of available radiolabel. If the RP has long clearance times from the blood and other tissues, many cyclotron-produced labels such as ^{11}C, ^{13}N, and ^{15}O are not acceptable as they decay long before the agent has targeted. Even ^{18}F may not be suitable due to its 110 m half-life.

Besides the imaging and activity-measuring aspects of RP trials, other testing must be done both before initiating the trial as well as during the research study. Table 2.8 includes some of these other evaluations and a general timeline for the entire investigation. Taken together, Tables 2.7 and 2.8 indicate the extent and complexity of data acquired and generated for a given research trial involving a radioactive drug. There is a clear need for a data-handling system that would provide real-time monitoring of this process. At lowest level, such review is important to the various data-recording personnel ranging from the protocol nurse to the physicist. More importantly, the principle

TABLE 2.8 Possible Timeline for a Clinical Protocol Involving Antibody Therapy

TEST	BASELINE	6 h	24 h	48 h	72 h	144 h	168 h	6 WEEKS	3 MONTHS	1 YEAR
H and P	x							x		
Height	x									
Weight	x								x	
Tumor size	x							x	x	x
CBC (complete blood testing)	x							x	x	
CEA (carcinoembryonic antigen)	x								x	
NM imaging		x	x	x	x	x	x			
EKG (electrocardiogram)	x									
Creatinine clearance	x								x	x
Antibody assay	x							x	x	
KPS (Karnofsky performance scale)	x							x		
PFT (pulmonary function test)	x									
HIV	x									

investigator (PI) would very much like to follow how the latest patient is being handled in the protocol. For example, "Has the blood been drawn at the first imaging time?" may be a question that cannot be answered today unless the PI picks up a phone and queries the protocol nurse.

Availability of the patient study results to all required clinical workers in real time is an additional objective of any monitoring system. One of the important features of clinical work involving radioactivity is that absolute clock time is of great significance. All samples are counted and then referred back to the start of the protocol to provide $u(t)$ values. Times of imaging, blood draws, urine samples, and other variables will depend on the specific study. To prevent confusion and to accomplish all that is required, the course of a given patient should be observable, with appropriate security safeguards, from multiple computer stations throughout the research institution. Another feature of any data acquisition and computation system is the possibility of immediate data tabulations. Such summaries may be prepared for the PI as well as regulatory bodies ranging from the local institutional review board (IRB) or radiation safety committee up to the U.S. Federal Drug Administration (FDA). These tabulations would make unnecessary the yearly summaries that are presently required by the IRB and the FDA. Instead, regulatory bodies could simply tabulate data into the formats of interest to the oversight body at any time. Such immediate summaries would be very important if questions arose regarding toxicity of a RP during a clinical trial involving multiple sites.

Prompting personnel and keeping track of various other sampling events is often of high importance in clinical monitoring. Urine and feces may need to be counted to find the routes of excretion and their associated kinetics. Given such samples, well counters, high-performance liquid chromotography (HPLC) columns, and mass spectrometry could be used to evaluate the magnitude of the excretion routes as well as the MW of the excreted compounds. It is important to determine if these molecules are still radiolabeled; generally, there is some loss of a radiometal from the tracer when being excreted. Radioiodine losses are expected to be even higher due to enzyme action in the various tissues.

The outline shown in Table 2.8 involves multiple tasks that must be performed at the correct time by a variety of staff. Additionally, a given institution may have several clinical trials going on simultaneously and using the same clinical staff and equipment. These clinical trials may involve the same PI to add to the possible complexity. Thus, it is necessary to control the execution of each protocol and to make sure that all appropriate tests are performed on time and are logged into the system. Otherwise, the clinical data analyst (CDA) will discover a protocol violation. No commercial software system presently exists for these oversight, prompting, and review features.

2.5 COST OF HUMAN TRIALS

Clinical evaluations are complicated and involve numerous individuals and measurement events. It is no surprise that human trial costs are high when such extensive sequences of imaging and exotic pharmaceuticals are involved. By taking the protocol of Table 2.8

to the City of Hope accounting office, the estimated radiology cost per patient is found to be $40,000. This includes the imaging radiolabel ^{111}In and the therapy label ^{90}Y for eventual treatment. We have neglected completely the cost of the antibody since it is manufactured locally and covered by overhead funding. In the two current commercial lymphoma trials, where labeled anti-CD20 monoclonal antibodies (Mabs) are in use (cf. Chapters 8 and 10), the antibody cost to the institution is an additional $20,000 per patient. If we add these components, total costs exceed $60,000 per patient. Notice that no salary or institutional overhead has been included in the analysis.

No trial could reach valid conclusions using a single subject. To prove that the tracer is effective in finding lesions in cancer patients, for example, 20 to 50 individuals may be required by the statistical oversight group. Upon multiplying those numbers by the cost per patient, the research study total expense easily exceeds $1 million and may be as high as $3 million. For many institutions, the requisite number of patients to be entered may be a majority of the total patients with the relevant diagnosis being admitted over 1 to 2 years.

Another limitation to conducting clinical trials is that many patients—and their families—do not wish to be involved with experimental studies. Often they feel that the protocol has no direct benefit to them and that their final time on Earth is being wasted. Such an attitude is particularly understandable in the case of imaging trials since the outcome is not necessarily of any benefit and may reveal lesions that are greater in number and extent than those found in earlier (conventional) examinations. It has been stated that approximately 3% of adult cancer patients volunteer to participate in any level of clinical trial (Ramsey and Scoggins 2008). Clearly, medical researchers have a problem of focusing the trial on relevant issues and communicating the reasons for the medical research to the patient and their family.

Two facts emerge. Clinical studies are expensive and may use a large fraction of the total number of local—and willing—patients. If the experimental agent is not effective or is less effective than another treatment available locally, these costs and subjects have essentially been wasted. Extensive time may also have been lost since the institution will not have been able to execute other research studies on this patient population using alternative methods. Conflicts arise since different clinical investigators at the same institution may wish to use the identical patient population. Thus, as noted in Chapter 1, the selection of the best agent *a priori* is very important and should be done using animal and other data before beginning any patient study. Methods to select the optimal agent based on animal results are described in the next chapter.

2.6 SUMMARY

Preclinical evaluation of a possible novel RP involves determination of the amount of activity per gram (%ID/g) of the agent in a test animal. Both sacrifice and imaging methods are used in these studies. Nuclear medicine personnel use relatively primitive instruments to obtain biodistribution data. Two methods of imaging are possible: gamma camera and PET scanning. In these two methodologies, the PET devices are approximately one to two orders of magnitude more sensitive due to the need for a collimation system on the gamma camera. Hybrid devices for both systems are now conventional so that the clinical images may be SPECT–CT or PET–CT in which the anatomic data are provided by the CT scanner that shares the patient couch with the nuclear imaging 3-D device. Miniature gamma and PET imagers are available for small-animal studies involving rodents.

Several examples of murine biodistributions show that a variety of agents and molecular targeting strategies are possible. Complicated engineering tasks and multiple-step localization of the radiolabel are two of the methods used to maximize the uptake and the contrast between tumor and the background tissues. The multiple types and strategies of targeting agent imply that some selection process must be used before clinical trials are initiated.

Finally, we recognize that a clinical trial is complicated and expensive. Costs may exceed several million dollars for a 20- to 50-patient study of a single RP. Oversight of the trial in real time is not yet possible. Future developments must include a surveillance system that permits immediate status of the patient within the clinical context. One output of such a software package would be report generation for the PI as well as the oversight bodies both within and without the institution.

REFERENCES

Anger, H. O. 1952. Use of a gamma-ray pinhole camera for in vivo studies. *Nature* **170**(4318): 200–1.

Beyer, T., D. W. Townsend, et al. 2000. A combined PET/CT scanner for clinical oncology. *J Nucl Med* **41**(8): 1369–79.

Bocher, M., A. Balan, et al. 2000. Gamma camera-mounted anatomical X-ray tomography: technology, system characteristics and first images. *Eur J Nucl Med* **27**(6): 619–27.

Cassen, B. 1957. Radioisotope scanning instrumentation. *Int Rec Med Gen Pract Clin* **170**(3): 139–43.

DeNardo, S. J., G. L. DeNardo, et al. 2007. Thermal dosimetry predictive of efficacy of 111In-ChL6 nanoparticle AMF-induced thermoablative therapy for human breast cancer in mice. *J Nucl Med* **48**(3). 437–44.

Gold, L. 1995. The SELEX process: a surprising source of therapeutic and diagnostic compounds. *Harvey Lect* **91**: 47–57.

Goodwin, D. A., C. F. Meares, et al. 1988. Pre-targeted immunoscintigraphy of murine tumors with indium-111-labeled bifunctional haptens. *J Nucl Med* **29**(2): 226–34.

Green, M. V., J. Seidel, et al. 2001. High resolution PET, SPECT and projection imaging in small animals. *Comput Med Imaging Graph* **25**(2): 79–86.

Hicke, B. J., A. W. Stephens, et al. 2006. Tumor targeting by an aptamer. *J Nucl Med* **47**(4): 668–78.

Hildebrandt, B., P. Wust, et al. 2002. The cellular and molecular basis of hyperthermia. *Crit Rev Oncol Hematol* **43**(1): 33–56.

Hudson, H. M. and R. S. Larkin. 1994. Accelerated image reconstruction using ordered subsets of projection data. *IEEE Trans Med Imaging* **13**(4): 601–9.

Kim, E. M., H. J. Jeong, et al. 2005. Hepatocyte-targeted nuclear imaging using 99mTc-galactosylated chitosan: conjugation, targeting, and biodistribution. *J Nucl Med* **46**(1): 141–5.

Proffitt, R. T., L. E. Williams, et al. 1983. Liposomal blockade of the reticuloendothelial system: improved tumor imaging with small unilamellar vesicles. *Science* **220**(4596): 502–5.

Ramsey, S. and J. Scoggins. 2008. Commentary: practicing on the tip of an information iceberg? Evidence of underpublication of registered clinical trials in oncology. *Oncologist* **13**(9): 925–9.

Williams, L. E., A. M. Wu, et al. 2001. Numerical selection of optimal tumor imaging agents with application to engineered antibodies. *Cancer Biother Radiopharm* **16**(1): 25–35.

Zhang, M., Z. Zhang, et al. 2003. Pretarget radiotherapy with an anti-CD25 antibody-streptavidin fusion protein was effective in therapy of leukemia/lymphoma xenografts. *Proc Natl Acad Sci U S A* **100**(4): 1891–5.

Selection of Radiopharmaceuticals for Clinical Trials

3

3.1 INTRODUCTION

From the previous chapter, we see that there are limited numbers of available and willing patients for radiopharmaceutical (RP) trials. Yet, as is emphasized in Chapter 1, there are multiple kinds of RPs as well as an almost astronomical number of variations on each generic type. Only a very restricted subset of the total possible number of radiopharmaceuticals can be taken into the clinic. Additionally, the cost of a clinical imaging or therapy trial involving each novel agent is high and may be wasteful of both patients and funds. There is also the aggravating possibility that the truly optimal RP may be missed completely in the selection process, or, in any event, delayed significantly in arriving in the clinical context. Thus, there is a need for methods to choose, from in vitro and animal data, the best agent for a given tumor type. Beyond oncology, the optimal agent should also be selectable for any particular molecular target. In the case of the RP, this choice must be made in both imaging and therapy. For ordinary pharmaceuticals, therapy is the most probable clinical objective although there is the possibility that novel contrast agents may be developed for future use in computed tomography (CT), magnetic resonance imaging (MRI), or ultrasound.

In the following, imaging and therapy figures of merit (FOMs) are generated that may be used in the RP selection process. These mathematical constructs are compared with traditional figures of merit in both applications. For the therapy case, analogous consideration of nonradioactive agents is made. The FOMs are expressed in a numerical form that may prove useful to the researcher or the pharmaceutical house developing the agents in question. Importantly, the numerical format allows the developer to compare, in a quantitative way, one variant after another during in-house drug development. Sometimes it is hard to see if the molecular changes being made are "going in the right direction" as per the objective of the targeting process. A figure of merit can help to follow the effectiveness of the changes being made sequentially to the class of agent being studied.

One caveat must be emphasized at the outset. It may be—for a given RP—that animal data may prove not to be directly indicative of eventual human results. Such differences may arise for a number of reasons. Most obvious are the variations in size between species as well as the specifics of the major organ systems in man versus animals. Issues may also occur, as is the case of many of the antibodies, because the agent was developed in a laboratory species and later applied in the clinic. Reaction of the patients' immune system to murine-origin monoclonal antibodies (Mabs) has been an ongoing problem in tumor-targeting studies (Tjandra, Ramadi, et al. 1990).

Chapter 2 described the three-order-of-magnitude variation in the absolute magnitude of $u(t)$ [%ID/g] because of the larger mass of the patient compared with a mouse test species. This disparity is to be expected in any uptake measurements; such differences due to body mass variation are generally not of concern in the comparison of animal and human data sets. There remains the fundamental issue of respective kinetics in the two mammals. Are both species handling the RP agent in essentially the same manner? To answer this question requires a comparison of the two biodistributions and an extraction of kinetic parameters for each species using relevant mathematical models. This type of numerical comparison will be described to some degree in this chapter but more extensively in Chapter 6 (Modeling) and Chapter 11 (Allometry). At this stage of radiopharmacology, the question must be answered on an agent-by-agent basis since no general results are available. Presumably, the eventual answer will not be a generic one but rather a result that is determined for each class of novel RP that is developed. Thus, we may expect future allometric rules for segments of mRNA that may differ from those for selective high-affinity ligands (SHALs), for example.

3.2 TUMOR UPTAKE AS A FUNCTION OF TUMOR MASS

Before we consider the development of figures of merit for imaging and therapy, one topic needs to be introduced. In the early work on liposomes and intact antibodies, our group discovered that animal tumors generally showed a variation of tumor uptake u_T (%ID/g) with lesion size m (g). The most salient feature of this correlation was that the larger the tumor, the smaller the uptake. In our first publication (Williams, Duda, et al. 1988) on the phenomenon, we used three possible mathematical representations to try to fit these data sets: linear, exponential, and power-law. Explicitly, at a given time, the relationship was assumed to be one of the following forms:

Linear: $u_T(m) = bm + u_0$ (3.1a)

Exponential: $u_T(m) = u_0 \exp(bm)$ (3.1b)

Power-law: $u_T(m) = u_0 m^b$ (3.1c)

TABLE 3.1   Tumor Uptake (%ID/g) in Mice versus Tumor Mass (g)

AGENT	LINEAR	EXPONENTIAL	POWER LAW	NUMBER OF TUMORS
Liposomes	$u = bm + u_0$	$u = u_0 \exp(bm)$	$u = u_0 m^b$	20
u_0	18.8	17.9	10.6	
b	− 5.81	−0.390	−0.278	
ρ (correlation coefficient)	−0.617	−0.676	−0.823	
Specific Mab				
u_0	8.68	8.45	1.70	15
b	−0.878	−0.165	−0.305	
ρ	−0.723	−0.827	−0.882	
Nonspecific Mab				
u_0	4.46	3.82	0.93	15
b	−0.405	−0.134	−0.486	
ρ	−0.314	−0.538	−0.807	

Note: Three analytic formats are included.
Source: Williams, L. E. et al., J. Nucl. Med., 29, 1, 1988. Reprinted with permission from the Society of Nuclear Medicine.

Each of these possible formats is or can be converted into a linear format to provide a unique linear least-squares solution to the relationship between u_T and m. As Chapter 6 will show, other representations, such as multiexponentials, do not give determinate solutions in such fitting algorithms. In those examples, a hunting algorithm is applied to the fitting process, and a "best fit" solution is obtained; such results are not unique and may depend on starting conditions.

Table 3.1 gives the results for liposomes injected into mice with Lewis lung carcinoma as well as specific (T-101 Mab) and nonspecific (PSA399 Mab) against a lymphoma target growing in nude mice (Patel, Tin, et al. 1985). Only agents labeled with ^{111}In were considered in these analyses as an iodine label was assumed to be not as stable at the tumor site. In these examples, all three representations were evaluated using their correlation coefficient (ρ) values. As the table shows, the highest correlations in all cases were those derived from the power-law format. Exponents (b) in these three examples were essentially contained in the interval −0.3 to −0.5.

Because of the narrow range of the exponent in the power-law correlations, it is natural to try to find a physical picture to justify this parameter. Figure 3.1 shows two alternative geometric forms that could be used to predict numerical values consistent with those found by fitting the uptake variation with a power-law format. In the first part of the figure, a spherical lesion is imagined with blood flow only to the external side. Tracer then has access to the outside of the sphere since the agent is assumed to have a relatively large size or a sufficiently great molecular weight (MW) so that extensive diffusion in the time of the experiment is not possible. This model leads, through terms of lowest order, directly to an exponent (b) of −1/3. The result follows since the accessible zone is essentially the surface of the sphere while the total mass (m) is that of the

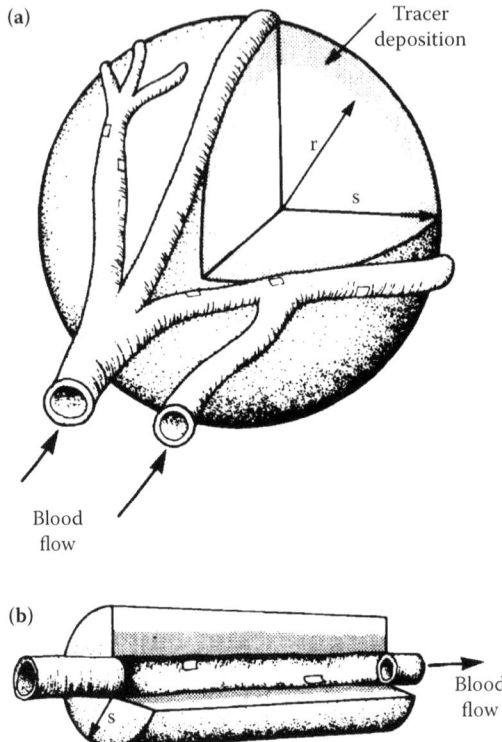

FIGURE 3.1   Schematic drawing of the perfusion of a spherical (a) and a cylindrical (b) tumor. Due to limited access of the blood supply, tracers may be limited to a relatively small zone near the openings in the capillary bed. This picture leads to power-law relationships between u_T(%ID/g) and tumor mass (g). (From Williams, L. E. et al., *J. Nucl. Med.*, 29, 1, 1988. Reprinted with permission from the Society of Nuclear Medicine.)

entire sphere. A cylindrical lesion is shown as a possible alternative in the second half of the figure. In this case, the lesion grows outward from the nurturing blood vessel. Thus, perfusion is from the inside only such that the zone of agent deposition is the inner surface of the cylindrical tumor. Doing the geometrical analysis to lowest order, one finds that the resultant exponent is –1/2. While no real tumor is likely to be as simple as either of these two idealized cases, it is reasonable that most lesions fall somewhere between the spherical and cylindrical examples—or are perhaps a combination of both types.

Chapter 6 will discuss the issue of what occurs in the analysis as the mass of the lesion approaches zero. Clearly, the power-law variation shown in Equation 3.1c cannot be used in that limit as the uptake would be unbounded (as $b < 0$). Instead, the very smallest lesions are perfused more uniformly due to the diffusion being sufficient to provide tracer at depth. Such small structures are not generally recovered at animal (or human) biopsy, so this result is predicted but difficult to prove.

Since these early results of our group and others, there have been several examples of the inverse correlation between tumor uptake and tumor mass. As Chapter 11 will show, some mass-law evidence also exists for clinical tumors. Not all lesions, however, are expected to have limited perfusion as shown in Figure 3.1. As noted in our original publication (Williams et al. 1988), there is at least one human tumor (Clouser) that is known to be uniformly perfused as it develops. In such examples, the inverse relationship between u_T and tumor mass does not hold and tumor uptake is independent of tumor size.

Mass dependence has important implications in at least two areas. First, every experimental group must include, in its biodistribution results, mass values for the lesions whose uptake is being measured. If they do not include these data, the investigators must show that their tumor model has no $u_T(m)$ variation. A second application is in the estimation of absorbed radiation dose. As the next chapter will show, the absorbed dose rate is directly proportional to uptake. A similar argument presumably is implied for chemical toxicity. Thus, if the mass-law holds, smaller lesions present more amenable targets for therapy. As the relationship indicates, this therapy will presumably go from the perfused side of the lesion toward the distal aspects that have lower probability of being targeted. A familiar metaphor from external beam therapy is the concept of "peeling the onion" whereby the radiotherapist indicates that the spherical lesion is destroyed from the perfused exterior inward to the more necrotic core region. As this process occurs, oxygenation of the inside increases with a correspondingly greater sensitivity to ionizing radiation (Stewart and Gibbs 1982).

Having found a likely relationship for tumor uptake versus tumor mass, we may now consider traditional as well as novel figures of merit to try to differentiate one tracer from another using preclinical data. Because we consider both imaging and therapy, these FOMs must be generated for both RP contexts.

3.3 DERIVATION OF THE IMAGING FIGURE OF MERIT

Determining an imaging figure of merit (IFOM) can be derived from a simple scenario that would arise in planar application of a gamma camera. One imagines that the observer is given an animal or clinical situation whereby a tumor is located within a background region. Because of the usual practice of injecting the RP into the venous supply, we initially use the blood as this background. Other tissues or organs are also possible, and muscle (or carcass) is a reasonable alternative. We can also expand the list to include organs that may compete with the tumor for agent accumulation. The most common of these tissues are kidneys and liver. Sometimes these competitions are surprising to the investigator.

Because of its historical importance and continued application to biodistribution results, we need to point out initially that the traditional figure of merit for imaging a lesion is the ratio of tumor to blood uptakes (Colcher, Bird, et al. 1990; Milenic, Yokota, et al. 1991):

$$r = u_T(t)/u_B(t) \tag{3.2}$$

where u_T and u_B are the tumor and blood uptakes, respectively, in units of %ID/g. Both of these parameters are typically measured using well counter techniques following animal sacrifice and are corrected for decay. As we will see in Chapter 5, there is a further possibility of determining such uptake values using either quantitative gamma camera or positron emission tomography (PET) imaging. In these cases, there is no need to sacrifice the animal during the course of the biodistribution experiment. Use of the tumor/blood ratio is based on the philosophy that the observer wishes to evaluate how much larger the tumor uptake is relative to the background level found in the blood of the animal. One may make a logical comparison to the signal-to-noise ratio used by electrical engineers. In the nuclear medicine case, blood acts as the "noise" in the system, whereas the tumor activity is the desired "signal." This analogy is not exact since blood acts as an input into the lesion; i.e., if blood has relatively low uptake, so would the tumor. Thus, blood uptake is always of interest (or should be) to the preclinical investigators.

It is important to realize that any imaging figure of merit, whether traditional or otherwise, must make a prediction as to the optimal time of imaging. The presence of time (t) in the right-hand side (RHS) of Equation 3.2 indicates this feature. Yet the traditional figure clearly has limitations since it cannot be used to find an optimal time. This follows since the absolute amount of activity in the tumor and the type of label are inherently lost in the analysis by taking a ratio of the uptakes.

Use of ratios has a long history in medicine, and this strategy leads directly into restrictions on application of Equation 3.2 in practical cases that occur in animal—and patient—biodistribution analysis. The most important of these limitations is that the absolute amount of RP in the tumor is not available to the reviewer since only a ratio of uptakes is being listed. Imagine a situation where the ratio of Equation 3.2 is very large due to a lack of activity in the blood; the tissue in the denominator of r. This could occur, for example, at a sufficiently long time after injection of the RP. Yet there may be so little simultaneous activity in the tumor target that the lesion cannot be imaged in any realistic time frame. In particular, the statistical certainty of the tumor being present becomes a fundamental issue in the imaging process. By realistic time frame, we mean a value in the range of 10 to 20 min, an interval during which patients can remain immobile under the camera or inside the PET scanner.

A second (and similar) limitation of Equation 3.2 is that the type of radiolabel is completely suppressed by using a simple ratio of uptakes. As seen in Chapter 2, the measured variables [$u_T(t)$ and $u_B(t)$] required in the definition of r are both corrected for radioactive decay of the label used in the original biodistribution experiments. Thus, if we assume other radiolabels, their ratio would still have the same numerical value; that is, the traditional figure of merit is, by definition, independent of radionuclide. Here, we assume that the biodistribution does not explicitly depend on the tag; this conjecture may be only approximately true. However, it is a good initial assumption for the analysis that follows.

Yet the usefulness of an agent for imaging depends directly on the radionuclide used as its tag. For example, we could consider a case where the half-life of the label was relatively short such that no significant number of counts is obtained in a limited counting interval using a gamma camera. Specifically, one could imagine a situation whereby a ^{99m}Tc label (6 h half-life) is attached, but the ratio of Equation 3.2 is maximum at 48 h postinjection. At this relatively long time (compared with the half-life of the label), the ^{99m}Tc would have gone through eight half-lives such that the amount of activity would be decreased by a factor of 256 due simply to radiodecay. Finding the tumor in the image might prove problematic in any reasonable time frame allotted to image acquisition.

Another disconcerting aspect of Equation 3.2 is that the traditional figure of merit predicts a paradoxical dependence on tumor mass. As seen earlier in this chapter, many animal tumors exhibit a power-law variation of u_T with size of the lesion. Clinical evidence for a similar behavior in humans is limited, but some examples are given in Chapter 11. Essentially, the uptake (u_T) decreases as the mass (m) of the lesion increases. In some relatively rare lesions, the perfusion is uniform, and there is no measurable variation of uptake with mass. For the sake of discussion, the variation can be approximately represented by

$$u_T = u_0 m^{-0.33} \tag{3.3}$$

as is the case for a spherical lesion. There is no dependence on tumor size for the blood uptake as long as lesion mass, or the number of molecular targets, is only a small fraction of the total body mass, or the total number of targeting molecules. This topic is discussed further in Chapter 12. Assuming the variation of lesion uptake with mass is given by Equation 3.3, substitution into Equation 3.2 reveals that r is directly proportional to the inverse of tumor size. In other words, using the r index, one would expect that the smaller the lesion the more readily it can be detected. Conversely, as the tumor increases in size, the less likely it is to be imaged in a blood pool background.

Prediction of such a mass-dependence result associated with the use of the traditional imaging index r is counterintuitive. The converse hypothesis is more likely to be the true state of reality; larger tumors should be more clearly defined with all other variables being equal. Lesions, as their size decreases, are expected to become progressively harder to find with any imaging device. This is the third example of why the r parameter is not a valid figure of merit for a radiopharmaceutical. We will show that $r > 1$ is only a necessary—but not a sufficient—condition for positive tumor imaging. Other factors clearly need to be added to the analysis to find a more comprehensive imaging figure of merit or IFOM.

It should be pointed out that, in both single photon emission computer tomography (SPECT) and PET imaging, there is a well-known problem of finding small lesions in vivo. As Chapter 5 will show, as the size of the tumor becomes close to or smaller than approximately two times the resolution of the imaging system, there is a reduced chance of seeing its structure against a background such as blood. A count-recovery coefficient (CRC) must be generated to correct the measured activity value upward to give the true radioactivity at the site (Boellaard, Krak, et al. 2004). Such results are, again, not consistent with the prediction of the traditional figure of merit, r. Thus, a novel indicator is called for.

In the following, we follow the argument used originally to explore the concept of a more logical IFOM at City of Hope but add several new considerations. Imagine a tumor lying adjacent to a blood volume in a planar imaging context. The investigator then obtains a camera image where n_T and n_B are the respective numbers of counts for tumor and blood recorded in time Δt.

Explicitly

$$n_T = \varepsilon \Delta t u_T \exp(-\lambda t) Vol A_0 \tag{3.4}$$

where ε is the camera–collimator efficiency factor, Vol is the tumor's volume (or mass as we assume a density of unity), and A_0 is the total injected activity. The counts recorded from the blood in this time interval are given by a similar equality but with the subscript B rather than T. Notice that the decay factor has been multiplied by the tumor uptake so that we are dealing with the activity as measured at time t. We must include the A_0 factor since u values are in units of %ID/g such that the total activity amount has been lost. For simplicity, we assume that these two volumes (tumor and blood) are equal. This means that the image reader looks at similarly sized regions on the image and associates one region with tumor and an adjacent identically sized region with the blood background. The net counts (z) detected are then the difference between tumor and blood:

$$z \equiv n_T - n_B = \varepsilon Vol[u_T - u_B] \exp(-\lambda t)\, \Delta t A_0 \tag{3.5}$$

We now introduce the statistical aspects of the analysis. Since these two n values result from radiodecay, they ideally would follow a Poisson distribution with a standard deviation equal to the square root of their magnitudes. The variance of z is the sum of the variances of n_T and n_B so that the coefficient of variation for z is determined using the ratio:

$$CV(z)^2 = Var(z)/Mean(z)^2 = [Var(n_T) + Var(n_B)]/[n_T - n_B]^2 \tag{3.6}$$

Yet these count variables are given by Poisson statistics so that $Var(n_T)$ is simply n_T and $Var(n_B)$ is n_B. Substituting these results into Equation 3.5, we realize

$$CV(z)^2 = [1 + 1/r]/\{\varepsilon \Delta t Vol u_T \exp(-\lambda t) A_0 [1 - 1/r]^2\} \tag{3.7}$$

To "find the lesion," the observer has to have confidence that the difference is significant—that is, that $CV(z)$ achieves a certain predefined (sufficiently small) value. Let us fix that number at some arbitrary amount, say CV_0. Then the superior imaging RP will be one that uses the shortest time to achieve the predefined value. To explicitly amplify this dependence, we set the IFOM equal to the inverse of the time required (Δt):

$$IFOM \equiv 1/\Delta t = CV_0(z)^2\, \varepsilon u_T \exp(-\lambda t) A_0 Vol[1 - 1/r]^2/[1 + 1/r] \tag{3.8}$$

The figure of merit of Equation 3.8 (Williams, Liu, et al. 1995) has several features that cause it to be mathematically distinct from the traditional r previously discussed. First, if r is unity, then IFOM is zero; that is, distinguishing a tumor from blood background is not possible in finite time if both tumor and blood have the same uptake value. Next, IFOM is proportional to the volume of the lesion and the efficiency of the camera system. Recall that both of these parameters drop out of the ratio r. Finally, if r approaches infinity, then IFOM remains finite. This last result will be of interest when we compare the two indices for a set of five cognate antibodies developed against carcinoembryonic antigen (CEA).

In the original derivation of IFOM, we did not consider agents where the background tissue (blood in the previous discussion) would continue to have greater uptake than the target tissue (tumor) throughout the period of the experiment. Since radiologists do not report "negative lesions" for agents that target to tumor sites, we suppressed showing IFOM values < 0 for these situations. If one now wishes to analyze such cases, it is important to allow IFOM to become negative. In this case, we take the sign of IFOM to be the sign of the term $u_T - u_B$. Mathematically, this is equivalent to $z/|z|$. In the previous considerations, we have used the variance of this difference so that the sign has been lost in the algebra. We will consider an example of this type of negative contrast for an agent with extensive kidney uptake. The augmented equation would now be written as

$$\text{IFOM} = CV_0(z)^2 \, \varepsilon u_T sign(u_T - u_B) \exp(-\lambda t) A_0 Vol[1 - 1/r]^2/[1 + 1/r] \qquad (3.9)$$

We should, at this point, discuss the result of Equation 3.9 if we allow r to approach zero. Physically, this corresponds to the earliest time points whereby there is material injected into the blood supply but little or no agent has yet gotten to the tumor site. If we let the traditional ratio r be represented by a small number, say δ, the appropriate variation in Equation 3.9 is then the term

$$\text{IFOM}(r \ll 1) \propto \frac{[1-1/\delta]^2}{[1+1/\delta]} sign(u_T - u_B) \qquad (3.10)$$

Since $1/\delta$ is much larger than 1, we may neglect the unity values in both numerator and denominator. The sign of $u_T - u_B$ is negative as the tumor has much less activity per gram than does incoming blood. Thus, the IFOM at low r values is approximately

$$\text{IFOM}(r \ll 1) \propto -1/\delta \qquad (3.10a)$$

If we take the limit as δ approaches zero, IFOM goes to negative infinity. Physically, this result implies that the tumor would appear as a deep void within the blood background. We should point out that the precise value of zero for r is not technically possible since we usually restrict our measurements of uptakes to periods of time after the blood is at least at a uniform value throughout the body (i.e., after allocating time of mixing). At this point, some activity would have gotten to the tumor site.

3.4 APPLICATION OF IFOM TO FIVE ANTI-CEA COGNATE ANTIBODIES

As an example of the use of Equation 3.9, we consider which of five cognate cT84.66 antibodies against CEA is optimal for imaging at a given time (Williams, Wu, et al.

2001). Here, since the murine biodistribution data were obtained with a radioiodine, we use both ^{131}I and ^{123}I as possible imaging labels. Note that we may conceptually change the label to see if using a different radioiodine can improve our imaging capability. However, we must be careful that we do not shift to a radiometal as a label since this change would possibly alter the biodistribution. Also, in these considerations, we do not allow the camera–collimator system (ε), volume of the lesion (Vol), or total activity (A_0) to change.

All of the five cognates, developed by Anna Wu and Paul Yazaki at the City of Hope, recognize the same antigen site on the CEA molecule. Figure 1.5 shows a cartoon version of these chimeric proteins where the dark areas are heavy chains and the light areas the light chains. Except for the single-chain variant (scFv), all have two reactive sites that recognize the antigen (i.e., are so-called bivalent). The MW range of the set is included in the figure and varies between 28 kDa and 150 kDa. A primary interest of the biodistribution study was to find which of the five was superior at a given time point and with a given radioiodine label. The tumor used in the experiments was the human LS174T colon cancer, which expresses CEA. Lesions were implanted in a nude mouse model and allowed to grow for several days until they reached sizes of approximately 100 mg. In these early biodistribution experiments, only radioactive forms of iodine (^{125}I and ^{131}I) were used as labels.

In the analyses, both ^{123}I and ^{131}I were considered as possible imaging labels. In addition, a speculative change was also assayed by shifting to ^{18}F as the radioactive marker. This was so that nuclear medicine personnel might be apprised of an optimal time for possible PET imaging. Note that we have to assume that the biodistribution will not depend on changing from iodine to a fluorine label. Due to its appearance in the same column of the periodic table as iodine, fluorine appears to be a realistic choice in this regard, but this result is yet to be verified.

To evaluate the cognates at every time point, each of the five is fitted as a set of biexponential functions over their measured biodistribution times (0 to 168 h). Table 3.2 contains the resultant curve parameters for each of the five cognates (Williams et al. 2001). With regard to the fitting procedure and the rationale for use of two exponentials, the reader should refer to Chapter 6. That section contains an extensive discussion of the application of multiexponential functions to represent biodistribution data. From Table 3.2, it is seen that the tumor uptake curve required that the second amplitude (A_2) be negative such that the curve essentially went to zero at the initial time point. Exponential fitting allowed IFOM values to be generated at all times within the experimental temporal range (interpolation) as well as values beyond it (extrapolation). Such broad-range results are important since it is likely that the best times of imaging may not be one of those chosen by the experimenters *a priori* for data collection.

Figures 3.2 and 3.3 show the IFOM results for ^{123}I and ^{131}I labels, respectively. In the graphic, we set CV_0, ε, Vol, and A_0 equal to constant values, and let their product be unity. This allows for a simpler plot. We relax some of these constraints in the later sections of this chapter. To be fair, each graph should be considered by itself without comparison with other IFOM tabulations since the detector efficiency may change with emission energies. Thus, we do not cross-compare the ^{123}I and ^{131}I-labeled results but compare only one ^{123}I-labeled entity with another and so forth.

TABLE 3.2   Double-Exponential Representation[a] of the Five cT84.66 Cognates Tumor and Blood Data (Labeling with Iodine)

COGNATE	A_1 (%ID/g)	A_2 (%ID/g)	k_1 (h^{-1})	k_2 (h^{-1})
Blood (u_B)				
scFv	14.4	3.77	4.7	0.21
Diabody	33.0	6.42	2.76	0.24
Minibody	18.8	30.0	1.05	0.13
F(ab')$_2$	12.7	26.4	1.41	0.11
Intact	15.4	22.9	0.19	0.010
Tumor (u_T)				
scFv	4.58	−2.18	0.18	22.4
Diabody	12.6	−11.0	0.034	10.41
Minibody	30.1	−28.5	0.021	0.57
F(ab')$_2$	50.0	−47.6	0.040	0.19
Intact	98.8	−96.8	0.0098	0.040

Source: Williams, L. E. et al., *Cancer Biother. Radiopharm.*, 16, 2001, published by Mary Ann Liebert, Inc. With permission.

[a] $u = A_1 \exp(-k_1 t) + A_2 \exp(-k_2 t)$. For tumors $A_2 < 0$, whereas for blood $A_2 > 0$.

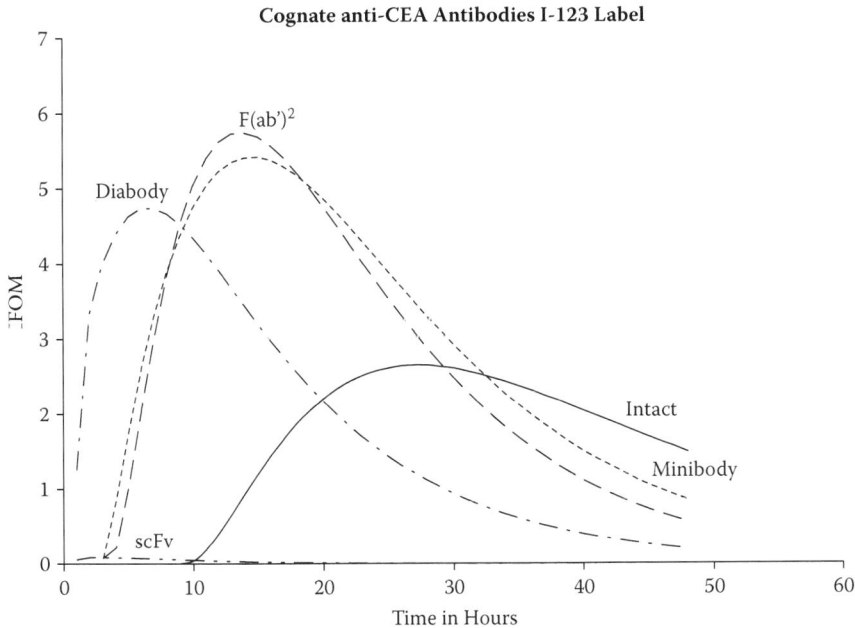

Cognate anti-CEA Antibodies I-123 Label

FIGURE 3.2   Imaging figure of merit (IFOM) for the five cognates using an ^{123}I label. Higher IFOM values imply reduced times to acquire an image with a given certainty of tumor being present in a blood background (Cognates are listed in Figure 1.5). (From Williams, L. E. et al., *Cancer Biother. Radiopharm.*, 16, 2001, published by Mary Ann Liebert, Inc. With permission.)

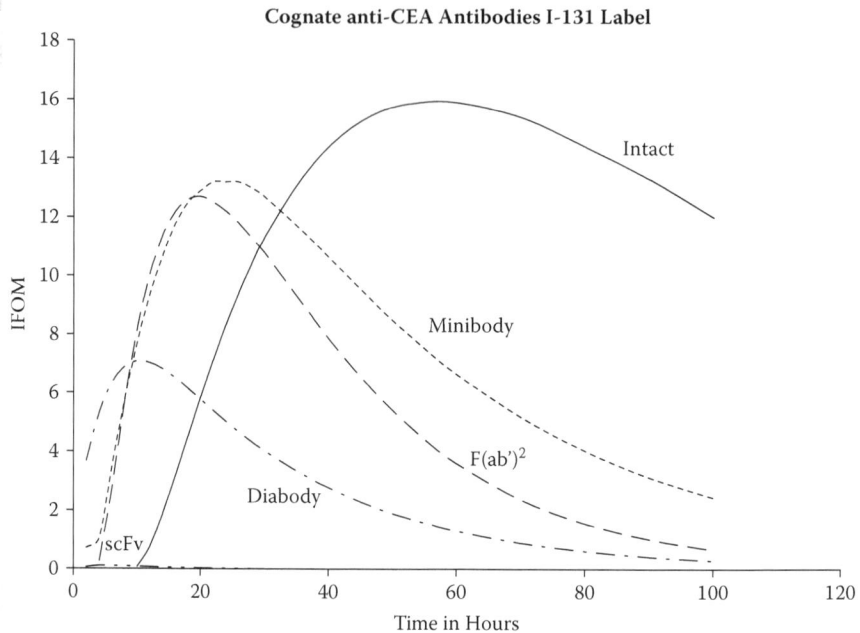

FIGURE 3.3 IFOM for the five cognates using a ^{131}I label. Otherwise, as in Figure 3.2. (From Williams, L. E. et al., *Cancer Biother. Radiopharm.*, 16, 2001, published by Mary Ann Liebert, Inc. With permission.)

In the example in Figure 3.2 of a ^{123}I-labeled cognate set, this shorter-lived iodine isotope (13 h = $T_{1/2}$) gives similar figure of merit results for diabody, minibody, and $F(ab')_2$ proteins. Maxima in IFOM occurred between 8 and 15 hours. The intact cognate was approximately a factor of two less in IFOM meaning that, to obtain a comparable image of the tumor, one would require approximately twice the imaging time with the intact antibody. Also, the maximum in the intact was at around 30 h postinjection. Finally, the single chain cognate was essentially of little value as an imaging label as the contrast between tumor and blood was very small.

Using ^{131}I as a label (Figure 3.3) demonstrates that the intact was now the optimal cognate protein, although it was only about 20% greater in IFOM magnitude than the minibody and $F(ab')_2$. The intact cT84.66 protein also showed persistently high imaging figure of merit values out to 100 h and beyond due to its long-lived label. Thus, the experimenter would have some latitude as to when gamma camera pictures may be taken. This is in contrast to the case of the diabody, in particular, whereby the IFOM decreases by a factor of two by 30 h into the study. For this 50 kDa protein, imaging must be accomplished by 2 days postinjection. A similar result may be seen in Figure 3.2 with the ^{123}I label.

Figure 3.4 gives the IFOM values for the five cognates with a hypothetical ^{18}F label. Here, the only possibly useful imaging agent is the diabody; the minibody is lower by a factor of almost one order of magnitude. Intact cT84.66 is not competitive due to the short half-life of radioactive fluorine (110 m). These results show the importance of

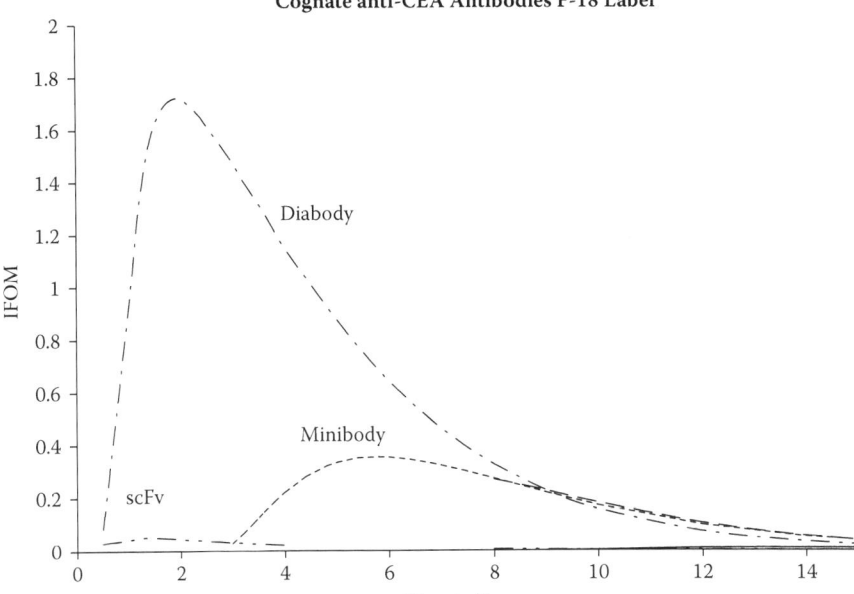

FIGURE 3.4 IFOM values for the five cognates using an ¹⁸F label. Otherwise, as in Figure 3.3. We assume that the noniodine label does not affect the biodistribution results. (From Williams, L. E. et al., *Cancer Biother. Radiopharm.*, 16, 2001, published by Mary Ann Liebert, Inc. With permission.)

matching the physical lifetime of the label with the kinetics of the protein. Again, the single-chain antibody is very limited in its usefulness in distinguishing LS174T tumor from blood background.

Using the two-exponential parameter sets of Table 3.2, one may compare the predictions of the traditional figure of merit (r) against those of the IFOM. A summary of the important features is given in Table 3.3 for general issues, as previously described. In addition, using the five cognate antibodies against CEA, there is another paradoxical result that occurs with r but not with IFOM.

With regard to the five cognates, it was found that r becomes unbounded as time goes to infinity. To put this surprising result in a different way, r is seen to be a monotonically increasing function of time for each of the five anti-CEA proteins. Physically, this growing ratio reflects the fact that the tumor holds onto the radioactive label relatively well while the agent is continuously being cleared from the blood. It is important to see why this happens mathematically as well.

For an explicit example, consider the case of the diabody member of the cT84.66 antibody set. The traditional figure of merit is given by u_T/u_B. By taking the tabulated biexponential representations for both these functions, it is seen that

$$\frac{u_T}{u_B} = \frac{12.6\exp(-0.034t) - 11.0\exp(-10.41t)}{33.0\exp(-2.7t) + 6.42\exp(-0.24t)} \tag{3.11}$$

TABLE 3.3 Comparison of Traditional(*r*) and IFOM Values for Imaging

CONDITION	TRADITIONAL IMAGING FIGURE OF MERIT (*r*)	IFOM
Radiolabel variation	No dependence	Proportional to $\exp(-\lambda t)$
Camera efficiency	No dependence	Proportional to efficiency (ε)
Amount of activity, A_0	No dependence	Proportional to A_0
Lesion size if u_T proportional to $m^{-0.33}$	Proportional to $m^{-0.33}$, hence smaller lesions are more easily detected	Proportional to $m^{0.67}$, hence larger lesions are more readily detected
Lesion size if u_T is independent of mass	No dependence	Proportional to m
Limit as $t \to \infty$	The *r* value is monotonically increasing with time for each of the five cT84.66 cognates; thus, $\lim_{t \to \infty} r = \infty$.	Goes to zero with $\exp(-\lambda t)$ for all agents, including the five cT84.66 anti-CEA cognates

As *t* increases, the larger exponential terms will become zero relatively rapidly so that we need to consider only the smaller rate constants in both numerator and denominator. In this approximation, *r*(*diabody*) becomes

$$Lim_{t \to \infty} r = \frac{u_T}{u_B} \cong \frac{12.6\exp(-0.034t)}{6.42\exp(-0.24t)} = 2.0\exp(0.21t) \tag{3.11a}$$

Because of the net positive exponent, the RHS of this equality goes to infinity approximately as exp (0.21*t*) as time becomes large. The interested reader can show similar results for each of the other four cognates using the Table 3.2 data set. By invoking *r* as a sole imaging index, the experimenter would expect to find the optimal imaging at infinite time for each cognate. As in the case of several other unusual predictions made using *r* as cited earlier, such expectations are hardly reasonable physically. In this cognate example, we thus find an additional particular reason for not using the *r* index to evaluate imaging agents. Use of the traditional figure of merit for imaging, although very common in the literature, appears to lead to a number of predictions that are essentially incorrect in both the general analytic case as well as in the specific five cognate examples studied here.

3.5 IODINE VERSUS INDIUM LABELING

It is important to point out that there are generally differences observed between iodine and radiometal biodistributions of the same agents. As described in Chapter 2, these disparities are generally accounted for by the endogenous enzymes whose function is to free iodine from proteins so that the thyroid will be well served. An example of this

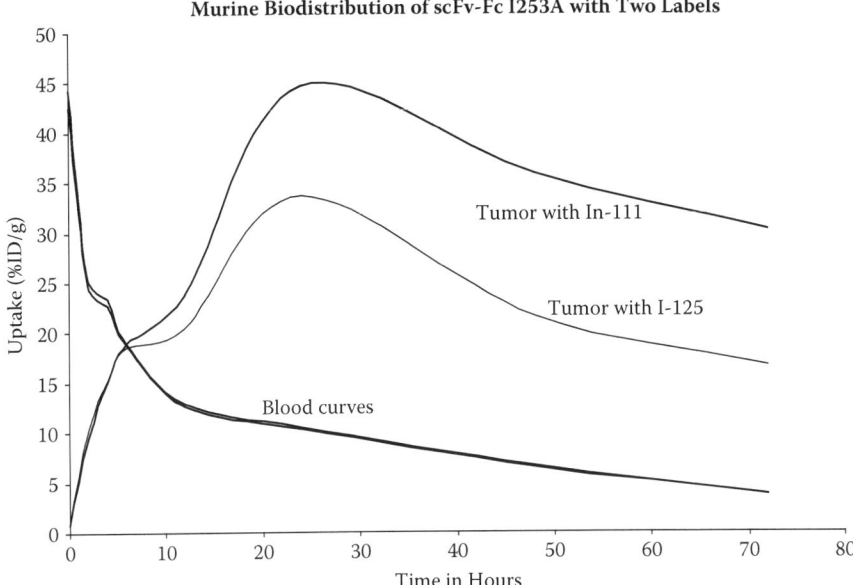

FIGURE 3.5 Biodistribution results for the engineered anti-CEA antibody fragment scFv-Fc I253A with both ^{111}In and radioiodine (^{125}I) labels (Kenanova et al. 2007). Notice that the blood curves are similar, but tumor uptakes begin to decrease as soon as 12 h postinjection for the iodinated form due to dehalogenation at the lesion site. In this experiment, the same animals were injected with two agents at the same time. The longer-lived ^{125}I label allowed a second bioassay at a time when the indium had decayed.

difference is shown for the engineered scFv-Fc anti-CEA antibody (Kenanova, Olafsen, et al. 2007) in Figure 3.5. The radiometal in this case is ^{111}In, whereas the iodine was [125]I. Generally, three features are found in any such comparison when the mass of the LS174T tumor is being controlled in both biodistributions. First, the two blood curves are essentially the same within error bars. This equality holds out to long times—in this case to 72 h postintravenous (post-IV) injection. Apparently, dehalogenation enzymes are not present in the blood of the animals. Tumor curves, on the other hand, begin to differ at times as short as 10 h and the uptake difference can be a factor of two at times such as 48 to 60 h. Such disparities are presumably due to the dehalogenation enzymes in this CEA-producing lesion, which is a malignant descendant of normal colon cells. One assumes that the original colon cells also have this attribute. Finally, the kidney curves differ at very early times as the metal-labeled minibody begins to lodge in the renal system whereas the iodinated form passes through that normal tissue readily. As times approach 24 h or longer, the difference between the indium and iodine renal uptake (%ID/g) can easily be a factor of two.

Dehalogenation essentially leads to two effects. Tumor uptake is higher for the radiometal-labeled protein, but renal uptake will generally be less for the iodinated form. This trade-off can also be expressed in terms of an IFOM.

3.6 PET APPLICATION OF THE IFOM

Originally, the IFOM concept was defined only for planar imaging using the gamma camera. However, it was later realized that 3-D imaging depends on a sequence of such 2-D acquisitions so that the concept was extended to PET imaging. Recently, working with the groups at Stanford and University of California–Los Angeles (UCLA; Cai, Olafsen, et al. 2007), we have shown the advantages of switching labels on the diabody developed by Wu and Yazaki. In this work, we allowed the diabody protein to have either an ^{18}F or an ^{124}I label. Here, due to a lack of data, we assumed that the biodistributions were independent of the label. Equivalently, we assume that fluorine and iodine are equally labile. In this case, we had to take into account the different probabilities of positron emission (cf. Chapter 1) for the two radionuclides. The diabody IFOM results, shown in Figure 3.6, demonstrate that fluorine is the superior label, at least at times out to 4 hours. These predictions were substantiated by the imaging done using the diabody and fluorine label (ibid.). For the minibody, the converse is the case with ^{124}I being a superior label for the heavier minibody agent (Figure 3.7). Here, ^{18}F is a very poor choice since the targeting is so slow such that the radioactive fluorine has essentially decayed even before the time of maximal tumor uptake at approximately 20 h.

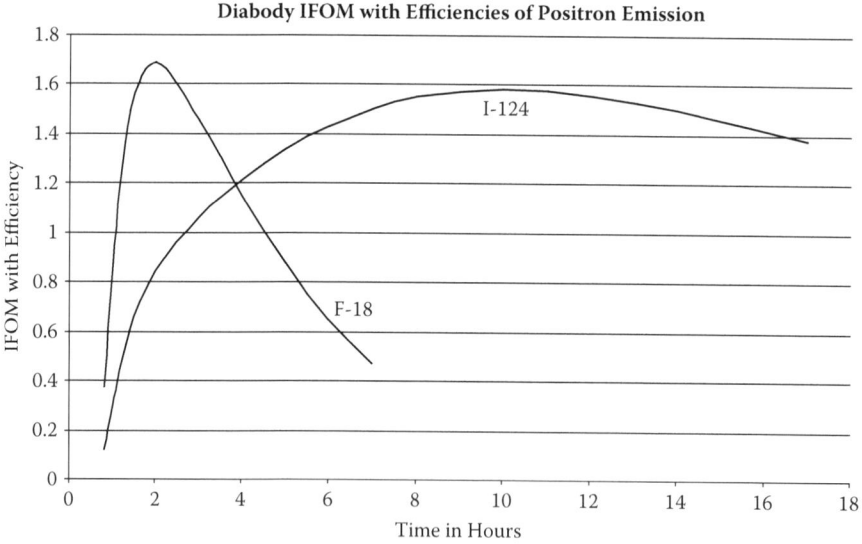

FIGURE 3.6   IFOM values for the diabody using an ^{18}F or a ^{124}I label. Fluorine is the superior label out to times of 4 hours.

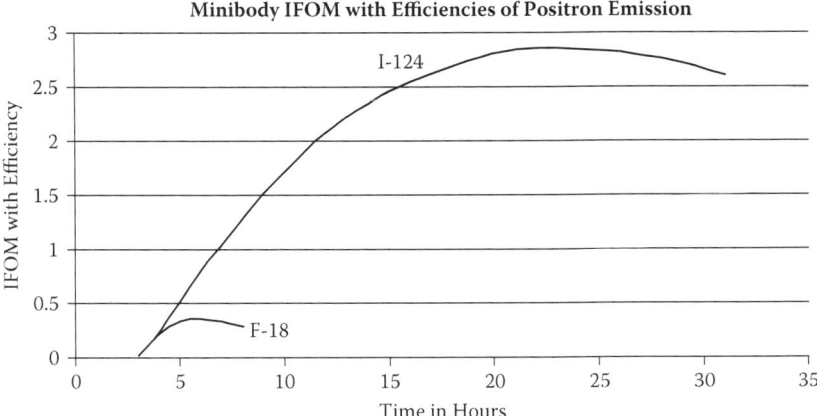

FIGURE 3.7   IFOM values for the minibody using an ^{18}F or a [124]I label. In this case radioiodine is a superior label due to the relatively slow targeting of this more massive cognate.

3.7 VERIFICATION OF THE IFOM

While IFOM seems superior to the traditional r index, it is still important to show that better images are associated with increasing values of this new figure of merit.

In a direct attempt to demonstrate the use of the IFOM, a Monte Carlo calculation can be performed on a simple geometry. In Figure 3.8, the situation is represented with a 1.0 cm diameter tumor, given as a sphere, lying within a blood pool. A nominal gamma camera–collimator system was defined for the Monte Carlo algorithm, which was implemented with the MCNP4 software (Williams, Lopatin, et al. 2008). Approximately 10^6 histories were run separately for the diabody and minibody cognates. In either protein, the tumor and blood uptakes are represented by the biexponential functions of Table 3.1. Results are given in Figures 3.8 and 3.9, respectively, for the two agents. Values of the r parameter (given as R in the figures) are included to show how the traditional imaging figure of merit varied with time after injection. It is seen that the magnitude of the IFOM corresponded, in at least an approximate fashion, with the clarity of seeing the tumor in the blood pool background. A 20-minute data acquisition was used in the simulations with a 2 μCi source in the MC phantom. This activity corresponded to the total amount of the ^{123}I radionuclide injected into a mouse during the determination of uptake of a member of the cT84.66 cognate set.

We conclude that an improved imaging figure of merit can be derived from consideration of the statistics of perceiving a tumor within a blood background. This IFOM shows reasonable variation with its primary parameters and approaches zero as time goes to infinity due to the $\exp(-\lambda t)$ factor. None of this behavior is seen in the traditional r index, which, at least for these five cognates, unfortunately goes to infinity as time

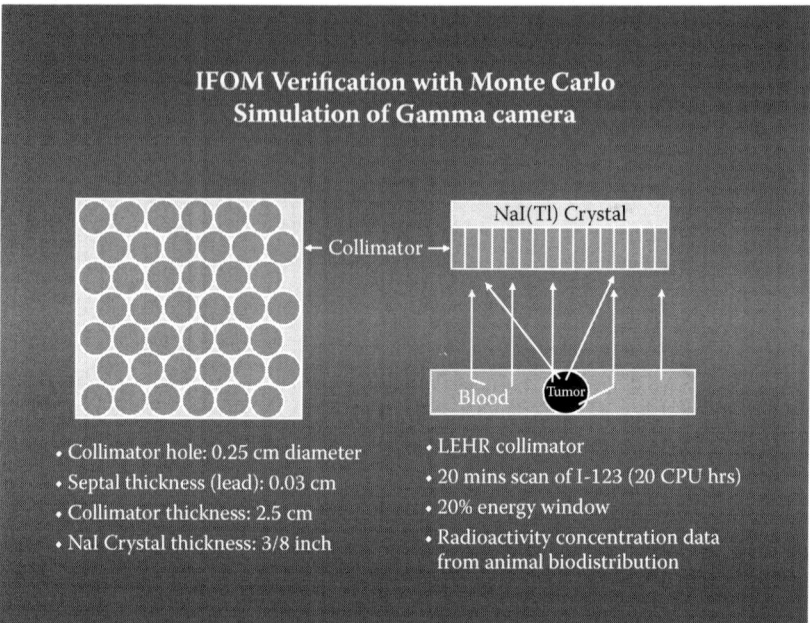

FIGURE 3.8   Simulation of the gamma camera for Monte Carlo calculations of *r* and IFOM. (From *Proceedings of the World Congress on Medical Physics and Biomedical Engineering,* Paper 737, Sydney, Australia, August 24–29, 2003. With permission.)

increases arbitrarily. This last result implies that the best time of imaging is beyond the reach of even the most dedicated experimenter.

One application of the use of the IFOM is making predictions regarding imaging for combinations of agent and labels that have not yet been made physically. As we show for ^{18}F, the investigator can conceptually add a novel radiolabel to an antibody whose biodistribution is known to make predictions about future murine imaging experiments. We must emphasize that when these predictions look promising, the developer must establish validity by a biological experiment on suitable animals. This last testing phase is nothing more than good engineering practice.

Engineers have long known that no single FOM can be applied to a complicated problem of optimization. We see that the IFOM presents certain advantages that are missing in the traditional ratio of tumor-to-blood uptakes. That is not to say, however, that IFOM is the ultimately optimal indicator for imaging. Other derivations are possible and obviously will be generated in the future. Such evolution is expected in the analytic toolkit available to the RP developer.

An associated aspect is that the IFOM (and *r*) both depend on only two biodistribution variables: tumor and blood uptakes. Yet we could have selected other backgrounds or even a group of normal tissues for inclusion in the analysis. Of particular interest would be liver, kidneys, and muscle. Each of these tissues, if their uptake were sufficient, could interfere with lesion imaging. The interested pharmacologist can think of other cases as well. If we wish to detect tumors in the prostate, the normal gland

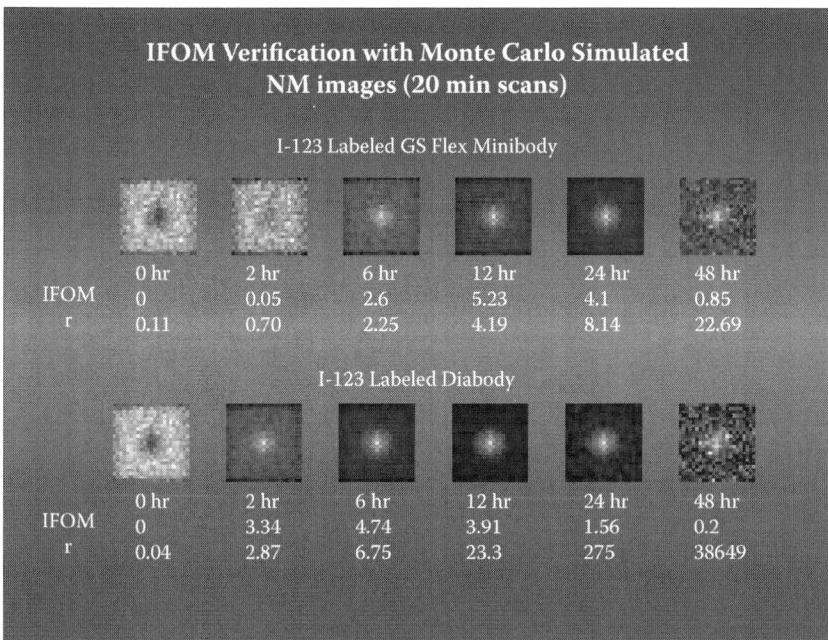

FIGURE 3.9   Monte Carlo simulations of imaging with two cognate anti-CEA antibodies (minibody and diabody) with an[123]I label. Included in the figure are both the IFOM and *r* values at the time points listed. Note the monotone increase of *r* with time while IFOM goes through a maximum at finite times. A 20-minute imaging interval for the gamma camera was assumed. Notice also that the *t* = 0 image shows the "hole" at the tumor location. This negative contrast is quite deep and has been omitted in the IFOM images. Radiologists will generally not call such negative sites as an indication of tumor. (From *Proceedings of the World Congress on Medical Physics and Biomedical Engineering,* Paper 737, Sydney, Australia, August 24–29, 2003. With permission.)

would be a suitable background tissue. It may be that a number of IFOM comparisons would need to be made with each of these results pertaining only to a single pair: tumor and normal tissue. Only after several of these analyses were done would the observer be more certain as to which of several possible clinical agents was the superior RP.

There is an additional caveat associated with a single IFOM. We have heretofore assumed that the data set containing u_T and u_B was all that could be determined for a given agent. Yet it may be that one or more of a set of putative RPs has difficulty getting to the tumor due to sequestration—and possible excretion—by normal tissues elsewhere in the body. It could be that one of the five cognates may be optimal at interacting with the molecular target (CEA) within the tumor volume. This situation can be resolved only by isolating the tumor from the body in which it survives so that normal tissue interference is eliminated. Clearly, this is a difficult project physiologically. Such a separation can be done mathematically, however, by deconvolution of the tumor curve given the blood curve. We now implement this type of analysis to see if, indeed, one of the five cognates has a superior accumulation and persistence inside the CEA-positive lesion.

3.8 FINDING POTENTIALLY USEFUL IMAGING AGENTS BY DECONVOLUTION

It is important to realize that the blood curve is itself a net result of multiple RP interactions throughout the body of the animal or patient. It may be that the blood concentration at the lesion is relatively quickly reduced because of sequestration of the RP by one or another normal tissue. Thus, while directly injecting the agent into the lesion volume at the arterial side might be quite effective in accumulating within the tumor, the RP of interest simply cannot arrive at that site due to being taken out of the bloodstream elsewhere. Such an interception by the normal organs may lead to decreased IFOM (and r) values that cause the investigator to overlook a potentially useful agent. Chapter 8 discusses such direct interventions in the therapy of liver disease in clinical oncology practice.

Let us imagine that the item of major interest to the bioengineer is how the tumor actually responds to an arbitrary input of the radiopharmaceutical agent arriving via the bloodstream. Conceptually, the simplest such function is a spike injection on the arterial side of the lesion. This perfect bolus is, in mathematical terms, the Dirac Delta function (Dirac, 1981). As noted in Chapter 6, such rapid procedures on the arterial side of a normal tissue or tumor are difficult to perform and may be relatively dangerous to the subject. Instead, the experimenter attempts to eliminate the effect of the blood curve's shape on the tumor curve by solving the following convolution equation for $h(t)$, the impulse response function.

$$u_T(t) = \int_0^t h(t-\tau)u_B(\tau)d\tau \tag{3.12}$$

A resultant $h(t)$ function represents what the tumor response would be if a Dirac Delta injection could be made at the arterial side of the lesion. Notice that, in the notational system shown here, h has units of inverse time. In this segment, emphasis also is placed on tumor impulse response functions; normal tissues have their own $h(t)$ representations. At present, this is an area of extensive research interest. Other convolutions, in spatial coordinates, are given in Chapter 4. A general consideration of the concept of convolution is found in Chapter 6 on modeling.

The physical meaning of convolution is that any tissue's time–activity curve is given by time-delayed product of some unknown function $[h(t)]$ and the driving input. In the case of tumors and normal tissues, the input is the blood curve coming into the arterial side of the organ or lesion. Since direct measurement of arterial blood is itself difficult and hazardous, it is typical that one uses the venous blood as a surrogate for the arterial. After a few mixing times, this assumption appears to be reasonable.

Five anti-CEA cognates were analyzed via Equation 3.12 to solve for their respective impulse response functions. The $h(t)$ values are given as graphical results in Figure 3.10. One of the striking outcomes of the analysis was that the diabody appeared to have the optimal impulse–response function. By this we mean the resultant $h(t)$ was higher for

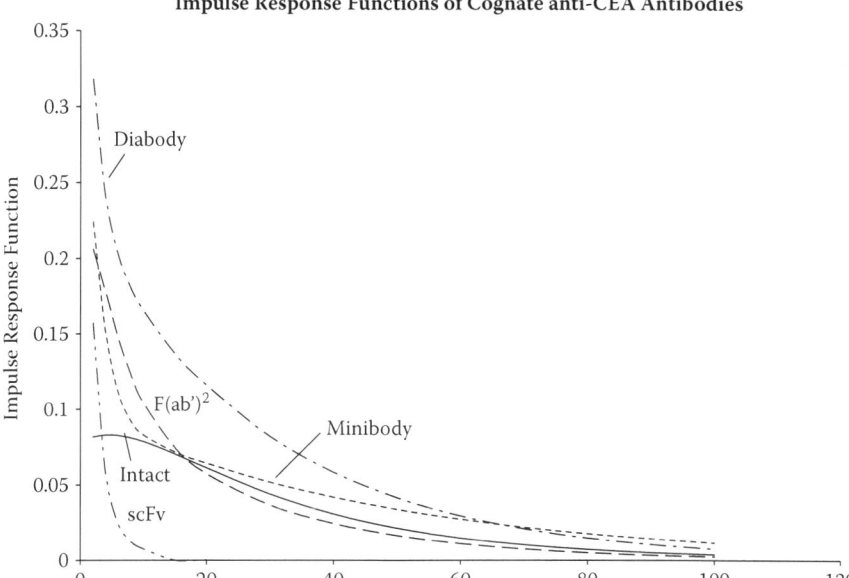

FIGURE 3.10 Impulse response functions determined for the five cognate set of cT84.66 anti-CEA engineered antibodies. The diabody has both greater amplitude and greater width than any of the other cognates. This result enables the prediction that this molecular species would be the ultimate superior imaging agent of this cognate set if its sequestration by the renal system could be reduced. (From Williams, L. E. et al., *Cancer Biother. Radiopharm.*, 16, 2001, published by Mary Ann Liebert, Inc. With permission.)

diabody, and its resultant temporal width also was greater. Thus, for identical bolus injections of all five cognates, the diabody would demonstrate the best tumor response in terms of both magnitude and duration after the RP delivery. As in the IFOM analysis, the scFv cognate had the poorest $h(t)$ function. Presumably, this result again reflects the single CEA-binding site on the single-chain molecule.

However, IFOM values of the diabody were competitive but not that outstanding in our prior iodine analyses as shown in Figures 3.3 and 3.4. It appears that the diabody has a manifest limitation in that its blood curve goes down too rapidly and thus results in relatively inferior accumulation of activity in the tumor. If one investigates the entire normal organ biodistribution, it is seen that the diabody blood curve is indeed relatively quick to decrease with time and has commensurate accumulation in the kidneys. This targeting is due to passage of this relatively low MW compound into the renal system. In other words, low tumor accumulation is a consequence of rapid sequestration of the diabody by the kidneys.

Issues of low MW compounds appearing to target the renal system are well known. A number of strategies have been advanced, including the use of lysine (Kobayashi, Yoo, et al. 1996) to attempt blocking of the renal cells. This concept is analogous to that used to get more activity into the tumors by blocking labeled liposomes from the reticuloendothelial system (RES) using unlabeled liposomes of a similar but not identical type as described in the previous chapter (Proffitt, Williams, et al. 1983). In either case, the investigator is using the partial saturation of the sequestering organ to keep the blood curve elevated for a longer period of time. Such interference with the "natural" deposition of the RP will probably continue to be of great interest in the future. This strategy invokes the use of more than one pharmaceutical in the test animal or patient at a given time. In the current example, one of these agents is labeled, and one is not. This topic is given a more extensive review in Chapter 10.

Radiopharmaceuticals have possible therapeutic as well as diagnostic applications. While optimal imaging times are important, it is also significant if one can select the best agent for possible cancer treatment. A therapeutic figure of merit (TFOM) is therefore useful in preclinical analyses.

3.9 THERAPY FIGURE OF MERIT

As in the case of imaging, there is a preexisting TFOM, which is the ratio (R) of tumor area under the curve (AUC) to blood area under the curve. Such integrals must be done with the data uncorrected for decay since we are computing the radiation dose per each tissue. It is significant that this indicator can be used for RP as well as ordinary pharmaceuticals. In the latter case, we assume that the total therapeutic or, for normal tissues, toxic effect is proportional to the total area under the curve. In either case, the ratio represents the benefit–risk quotient for the agent with the numerator being proportional to the therapeutic effect or benefit and the denominator to the risk to the blood tissue. For radioactive drugs, the risk is well known and involves a number of organs—particularly the red bone marrow (RM). Here, as seen in Chapter 4, the investigator is using the blood curve as a surrogate for the agent's movement through the RM.

Following the original City of Hope derivation (Williams et al. 1995), we set an absorbed dose difference equal to the tumor dose minus the blood dose. Each of these, in turn, is the product of the area under the curve of uptake values modulated by the $\exp(-\lambda t)$ factor. The coefficient of variation of the dose difference (z) is then found as

$$CV(z)^2 = \frac{1+1/R}{k[1-1/R]^2 \, \text{AUC}[u_T \exp(-\lambda t)]} \tag{3.13}$$

With R being the traditional therapy index

$$R = \frac{\text{AUC}[u_T \exp(-\lambda t)]}{\text{AUC}[u_B \exp(-\lambda t)]} \tag{3.14a}$$

where the tumor area under the curve implies the integral

$$\text{AUC}[u_T \exp(-\lambda t)] = \int_0^{\infty} u_T \exp(-\lambda t) dt \tag{3.14b}$$

with a similar definition holding for the blood. Notice that we have to explicitly enter the decay factor since uptake (u) values, as published, are corrected for decay. In effect, we are using the $a(t)$ parameters of Chapter 1.

To maximize the certainty in the dose difference between tumor and blood, the TFOM was then set equal to the inverse of the $CV(z)^2$ with the result

$$\text{TFOM} = \frac{k[1 - 1/R]^2 \text{AUC}[u_T \exp(-\lambda t)]}{1 + 1/R} \tag{3.15}$$

The constant k in Equation 3.13 and Equation 3.15 contains the product of the injected activity A_0 times the emitted mean energy per decay ($<E>$) times the probability (ε) that the energy is absorbed in the tumor (or blood) volume. An additional factor is required to convert from energy density to absorbed dose. The unit of absorbed dose is the gray, which is 1 joule per kilogram. Notice that R, the traditional therapy figure of merit, is contained within TFOM as a factor of $1/R$. This dependence is discussed more completely following.

As in the case of the IFOM, comparisons can be made between the traditional R therapy parameter and the newer concept of TFOM. Use of a simple ratio as in Equation 3.14a is limited for the same sort of reasons as the traditional r index is for imaging. Namely, the ratio format has caused the user to lose track of the magnitude of the AUC for the tumor as well as the energy of the emitter. For example, a large R value may be due simply to a very small blood area under the curve. In such cases, the dose to the tumor may also be so minimal that it requires very large amounts of activity (A_0) to achieve a given radiation dose to the lesion. This amount of radionuclide may not be acceptable clinically or economically. A second, and somewhat similar, limitation of R is that the energy of the emitter has been lost due to the use of a ratio. This is analogous to the loss of the decay constant in the r parameter. What if the radiolabel, for example, emitted a very low energy particle such that the absorbed dose was quite small? Yet R could be very large in this circumstance and not be indicative of the restrictions of the possible therapy due to a relatively poor choice of radionuclide. Both of these limitations are not found in TFOM. The newer index explicitly depends on the AUC of the tumor and the energy of the (alpha or beta) emitter chosen.

Other difficulties occur with the use of R. One could have comparison situations where two agents have very similar or even identical R values. No choice could be made in such instances with the traditional indicator, yet TFOM would emphasize the one agent having higher AUC(tumor) or higher energy of the emitter. Finally, as in the imaging example, one could imagine situations where R becomes very large due to the small magnitude of the denominator, that is, the AUC(blood). If we consider the theoretical limit of R approaching infinity, the traditional indicator becomes useless. Yet, in such cases, TFOM remains finite and becomes equal to the AUC(tumor) times the constant k.

TABLE 3.4   Comparison of Traditional (R) and TFOM Values for Therapy

CONDITION	TRADITIONAL THERAPY FIGURE OF MERIT (R)	TFOM
Large numeric value for the ratio (R)	Agent having larger R value would be chosen	Agent may not be useful because the AUC(tumor) is too small. Too much activity is required to treat to a given absorbed dose level.
Radiolabel variation	No dependence	Proportional to mean energy of the emitter
Situation when R = 1	Not excluding from consideration for therapy	Becomes zero—that is, no benefit in this situation
Limit of large ratio of AUC(tumor)/ AUC(blood)	As R becomes unbounded, it is difficult to choose one agent versus another if both have very large R values.	Is finite in the limit of large R. Favors the agent with the higher AUC(tumor) and higher emission energy.

TABLE 3.5   Traditional (R) and TFOM Values for the Five Anti-CEA Cognates with a ^{131}I Label

COGNATE	AUC (BLOOD) (%ID h/g)	AUC (TUMOR) (%ID h/g)	R (PURE NUMBER)	TFOM (ARBITRARY UNITS)
scFv	20.7	24.9	1.20	0.38
Diabody	38.3	334.3	8.73	235.1
Minibody	242.4	1175.3	4.85	613.8
F(ab')$_2$	241.4	901.6	3.73	381.3
Intact	1766.8	5166.9	2.92	1667.3

Note: Areas under the curve are also included.

We would therefore recommend use of TFOM in lieu of R. A summary of the reasons for this selection is given in Table 3.4.

Representative R and TFOM values for the five cognate set are given in Table 3.5. Here, we assume a ^{131}I label and set the multiplicative constant in the definition of IFOM equal to unity since all members of this protein set have the same radionuclide as their radioactive marker. The results of this analysis were interesting. From the prospective of the R (traditional) parameter, the diabody is the superior agent with the largest value of R of the five members of the set. This result reflected the large value of the ratio of areas under the curve for the diabody. For a given AUC(blood), the diabody generated the largest corresponding AUC(tumor) of any of the five cognates. Yet, because of the sequestration of diabody by the renal system, its effectiveness as a therapeutic agent was compromised. Indeed, the TFOM indicator was maximum for the intact version of the cT84.66 antibody due to this agent's extended time—and correspondingly enhanced absorbed dose—inside the human tumor xenografts. Such a large AUC(tumor) for the intact form was noteworthy; this area is almost an order of magnitude larger than that of the diabody and is the primary reason that the TFOM was maximal for the intact form. Nature appears to favor the native form in this therapeutic consideration.

One example of such differences can be seen with the single-chain antibody. TFOM values for the scFv are less than unity due to its very low tumor AUC. The tumor area, in fact, is only slightly larger than that of the blood for this 28 kDa agent. As in the imaging analyses, there seems little reason to use a single-chain cognate for therapy of a colorectal tumor. Intact protein TFOM value is more than three orders of magnitude larger than that of the scFv.

3.10 SUMMARY

Because of the multiple types of agents that may be designed, preclinical selection is an important feature of RP development. Tumor-targeting agents, in particular, require that the developer compare targeting to malignant lesions with that found in normal organs. Blood is primary among these latter tissues for several reasons. One is that, for imaging, the tumor must be distinguishable from a blood background. Some of the material must always be in the blood due to the general use of IV injections for the imaging agents. Second, in the case of radioactive agents, blood acts as a surrogate for the red marrow—a tissue having probably the highest sensitivity to ionizing radiation. Thus, figures of merit must be designed, just as the drugs themselves must be designed, to account for the differences between the tumor and blood background.

Before developing the indictors, it is important to discuss a salient feature of drug accumulation, measured in %ID/g, in tumors. In a number of animal studies, it has been found that the uptake goes as an inverse power of the lesion mass. Such power laws have been shown in this chapter to have exponents in a relatively narrow range of –0.3 to –0.5. This interval in fact can be predicted by invoking a simple spherical or cylindrical model of the tumor. A more complete discussion of power-law models occurs in Chapter 6. As is pointed out in that segment, the model cannot hold as the mass of the lesion approaches zero. In that limit, the uptake becomes more uniform due to diffusion becoming the dominant process of accumulation throughout the lesion.

Two figures of merit, IFOM and TFOM, are described from earliest beginnings to their latest elaborations. In the former case, the imaging figure of merit leads to assertions that are in reasonable concordance with reality. In particular, the IFOM depended directly on the amount in the tumor at a given time as well as the total activity injected. Neither of these dependencies was present in r, the ratio of u_T/u_B. We discuss the variation in IFOM when tumor mass changes; here, the novel indicator predicted variation with mass raised to an exponent greater than zero. Yet the mass dependence for r was paradoxical: mass raised to a negative power such that smaller lesions should be more visible.

Similar difficulties occurred for the TFOM indicator and for largely the same reason: use of a simple quotient. While somehow beloved by a number of medical analysts, use of a ratio neglects completely the amount of RP in the tumor and the energy of the attached emitter radionuclide. Both of these must be included in a realistic indicator of therapeutic potential.

Both indicators are applied to the analysis of five iodinated engineered cognate antibodies to the CEA molecule. It is seen, for imaging, that certain time intervals are best. These ranges of values depend on the kinetics of the cognate as well as the half-life of the radiolabel. In the analyses, two isotopes of iodine are studied to see when each has its optimal time for imaging, In addition, in speculating regarding use of ^{18}F as a possible tag, it is seen that only faster targeting agents such as the diabody and minibody are appropriate for usage. These last analytic results show the application of IFOM to a situation that has not yet been studied experimentally. Thus, a possible label may be tested without having to go to the lab and generating a new biodistribution. While fluorine may not be identical in its labeling to iodine, this speculative swap of labels is at least reasonable.

Finally, the use of deconvolution was described and applied to the five cognates. Diabody targeting was not found to be optimal by IFOM analysis using tumor and blood uptake data as directly recorded by observers. The diabody was superior, however, if one could inject all five agents with the same sharp bolus at the arterial side of the CEA-positive tumor. One could speculate that if the blood curve could be maintained in the animal (or patient?), the tumor uptake curve would be optimal in the case of a diabody agent. The practical limitation of using this cognate turns on the issue of its accumulation in the kidneys of the test animals. These uptake results, having magnitudes on the order of 10x greater uptake than the tumor, must be reduced so that the diabody can achieve its possible clinical potential. Such work is now ongoing using a variety of methods including stealth technology (Li, Yazaki, et al. 2006).

REFERENCES

Boellaard, R., N. C. Krak, et al. 2004. Effects of noise, image resolution, and ROI definition on the accuracy of standard uptake values: a simulation study. *J Nucl Med* **45**(9): 1519–27.

Cai, W., T. Olafsen, et al. 2007. PET imaging of colorectal cancer in xenograft-bearing mice by use of an 18F-labeled T84.66 anti-carcinoembryonic antigen diabody. *J Nucl Med* **48**(2): 304–10.

Colcher, D., R. Bird, et al. 1990. In vivo tumor targeting of a recombinant single-chain antigen-binding protein. *J Natl Cancer Inst* **82**(14): 1191–7.

Dirac, P. A. M. 1981. *The principles of quantum mechanics*. Oxford: Clarendon Press.

Kenanova, V., T. Olafsen, et al. 2007. Radioiodinated versus radiometal-labeled anti-carcinoembryonic antigen single-chain Fv-Fc antibody fragments: optimal pharmacokinetics for therapy. *Cancer Res* **67**(2): 718–26.

Kobayashi, H., T. M. Yoo, et al. 1996. L-lysine effectively blocks renal uptake of 125I- or 99mTc-labeled anti-Tac disulfide-stabilized Fv fragment. *Cancer Res* **56**(16): 3788–95.

Li, L., P. J. Yazaki, et al. 2006. Improved biodistribution and radioimmunoimaging with poly(ethylene glycol)-DOTA-conjugated anti-CEA diabody. *Bioconjug Chem* **17**(1): 68–76.

Milenic, D. E., T. Yokota, et al. 1991. Construction, binding properties, metabolism, and tumor targeting of a single-chain Fv derived from the pancarcinoma monoclonal antibody CC49. *Cancer Res* **51**(23 Pt 1): 6363–71.

Patel, K. R., G. W. Tin, et al. 1985. Biodistribution of phospholipid vesicles in mice bearing Lewis lung carcinoma and granuloma. *J Nucl Med* **26**(9): 1048–55.

Proffitt, R. T., L. E. Williams, et al. 1983. Liposomal blockade of the reticuloendothelial system: improved tumor imaging with small unilamellar vesicles. *Science* **220**(4596): 502–5.

Stewart, J. R. and F. A. Gibbs, Jr. 1982. Prevention of radiation injury: predictability and preventability of complications of radiation therapy. *Annu Rev Med* **33**: 385–95.

Tjandra, J. J., L. Ramadi, et al. 1990. Development of human anti-murine antibody (HAMA) response in patients. *Immunol Cell Biol* **68** (Pt 6): 367–76.

Williams, L. E., R. B. Duda, et al. 1988. Tumor uptake as a function of tumor mass: a mathematic model. *J Nucl Med* **29**(1): 103–9.

Williams, L. E., A. Liu, et al. 1995. Figures of merit (FOMs) for imaging and therapy using monoclonal antibodies. *Med Phys* **22**(12): 2025–7.

Williams, L. E., G. Lopatin, et al. 2008. Update on selection of optimal radiopharmaceuticals for clinical trials. *Cancer Biother Radiopharm.* 23: 1–10.

Williams, L. E., A. M. Wu, et al. 2001. Numerical selection of optimal tumor imaging agents with application to engineered antibodies. *Cancer Biother Radiopharm* **16**(1): 25–35.

Absorbed Dose Estimation and Measurement

4

4.1 INTRODUCTION

Radiopharmaceuticals (RPs) have a unique attribute of emitting ionizing radiation that enables both diagnosis and therapy in nuclear medicine. Local ionization is in addition to any other physical, chemical, or immunological effects of the RP. The emitted particles deposit energy in various organs and may cause various forms of destruction. These results can occur relatively soon after irradiation (direct damage) or sometime later (delayed effects). Historically, the lowest-order estimate of the likelihood of such damage has been the concept of absorbed radiation dose. Rather than repeat this long phrase, we will sometimes use the term *dose* as an alternative in the following discussions. Dose is essentially an energy density in the target tissue. It is important, therefore, that methods for estimation and even measurement be considered for RP absorbed dose in vivo. Likewise, we must realize that direct dose measurement, because of its invasive nature, is generally not applicable in either animals or patients. Mathematical methods of estimating absorbed dose are seen as the fundamental practical technique for use in determining dangers associated with ionizing radiation. Several such methods are described in this chapter. In addition, we will show that some measurements are possible using humanoid phantoms or with skin detectors on patients.

4.2 ABSORBED DOSE

In the following, we describe the concept of and the corrections to absorbed dose in biological contexts. We will see that dose will generally be an estimated parameter; even with the calculations, there will need to be both spatial and temporal enhancements to the simple numerical value to correlate with experimental data sets. Such correlations, in the context of living animals or patients, are often difficult to make. In the case of patients, only a very limited number of such correlations have ever been reported (Chapter 9).

4.2.1 Absorbed Dose as a Concept

Recall from Chapter 3 that absorbed radiation dose (D) is defined by the ratio of energy deposited in a target (ΔE) divided by the mass of that target (Δm).

$$D = \Delta E/\Delta m \tag{4.1}$$

If 100 ergs are deposited within 1 gram, the resultant dose is defined as 1 rad. The International System of Units (SI) of dose is the gray (1 joule/kg) and is equal to 100 rad. Dose, by its definition, does not depend on the ionizing particle or the material of the absorber. On the other hand, the biological effect of the dose may depend on multiple factors including precise location of the energy deposition, type of radiation, time course of the irradiation process, as well as the sensitivity of the material (or tissue) of the absorber. Because of these nuances, multiple factors have been conceptually engineered to correct (or "edit") absorbed dose of Equation 4.1 into a more useful concept in the life sciences. The general strategy is that, by itself, the numerical magnitude of the dose is necessary—but almost never sufficient—for the calculation of therapeutic and toxic effects of the radiation. These various corrections to the dose estimates are described in more detail herein.

There is an associated tradition in radiobiology to establish a dichotomy of effects induced by the radiation damage to the tissues. Direct effects are those produced relatively soon after the exposure to the ionizing radiation. Generally, these are related to dose in a direct way in that the result is often proportional to the amount of dose. Such results are then contrasted with "stochastic" or random results that may occur later—perhaps years after the exposure. The latter must be studied in a context of probabilities; in other words, the result of a given absorbed dose might not cause the same result in two different animals or patients. A stochastic situation might involve induction of a second cancer 20 years after a radiation treatment of some other (original) malignancy. A specific example is the probability of an eventual lung cancer in a patient originally treated by external beam radiation of the chest for a childhood lymphoma.

Absorbed dose is not a simple concept in an additional sense. One cannot anticipate that any biological effect will be a strictly linear function of the dose value. On the basis of cellular measurements, it is usually assumed that the dose–response effect will be in the shape of a sigmoid curve (Hall and Giaccia 2006). Such curves have both a "foot" and "shoulder" and are described more completely in Chapter 6. Thus, there may be little observable cellular or tissue response until the dose exceeds a certain minimum value. This result may be effectively seen as a threshold since there will generally be an ensuing linear portion of the curve until one observes saturation of the response. At this level of dose, there can be no greater effect—for example, all of the cells have been killed and no increase in ionizing radiation can produce any additional damage. An example (the probit function) of a sigmoid curve is given in Chapter 9 for lymphoma patient therapy outcomes.

Although awareness of its effects is a recent phenomenon, ionizing radiation is not a human by-product of the twentieth century. All life forms have had to contend with natural background levels over hundreds of millions—or probably billions—of years. Presently, in the United States, the whole-body value is on the order of 0.3 cGy per year. To prevent accrual of greater and greater damage from these natural sources, there have arisen a number of repair mechanisms for DNA within living cells. These enzymes allow biological systems to survive, genetically intact, at a sizable fraction of one cGy per annum. Several investigators have asserted, generally on the basis of epidemiological studies, that very low-level radiation doses are beneficial to animals and humans. This conjecture is termed *hormesis* (Calabrese and Baldwin 2000) and has a long history in medicine. Generally, and with regard to ionizing radiation in particular, the concept is very controversial. No definitive proof of the conjecture has appeared in the radiobiology literature. Advocates appeal to the idea that cellular DNA repair enzymes are always available but may be up-regulated by external ionizing sources. By giving a small incremental dose of ionizing radiation, these proteins are thereby primed into action and become available for other simultaneous insults caused by a variety of environmental factors such as chemical toxins and infectious viral particles. This strategy is equivalent to taking a minuscule amount of a known toxin each day to protect against the hardships of life. The method is well known in folklore: "What won't kill you will cure you."

Nuclear medicine imaging protocols produce maximum organ absorbed doses on the order of 1.0–5.0 cGy. Uncertainties in these values for a given individual may be several-fold due to patient geometry. Even with these uncertainties, such values are still comparable to yearly background measurements. Unlike putative hormesis effects, these results—or actually the images resulting from nuclear imaging procedures—can be directly demonstrated to benefit the patient. By undergoing a few such nuclear medical studies over a lifetime, the patient is not increasing his or her total dose significantly or, therefore, increasing susceptibility to untoward effects of ionizing radiation. Thus, the impact of nuclear medical diagnostic absorbed dose on a given patient is generally accepted as a benefit with little or no additional direct or stochastic risk.

In the case of targeted radionuclide therapy (TRT), however, the intended maximum absorbed dose value increases by several orders of magnitude. Using external beam magnitudes as a guide, the radiation oncologist wishes to deliver doses on the order of 10 to 100 Gy to each tumor site. Normal tissue irradiation, at the same time, must be strictly controlled to prevent toxicity (Emami, Lyman, et al. 1991). Hence, accurate therapeutic dose values are much more important than those found in diagnosis. These predictions should be required before the therapy is carried out. Such a process may be called treatment planning in analogy with external beam radiation therapy (EBRT).

It comes as a surprise to many clinicians and other medical professionals that the numerical TRT dose values for various tissues within a given patient (or animal) must be estimated using rather involved mathematical rules. Although the estimation process is colloquially called dosimetry, the term is an ironic misnomer. A correct descriptor is dose estimation. While dosimetry is logically possible in a water bath, inside an anthropomorphic phantom, or on the skin of a patient, it is generally going to be quite difficult—and probably unethical—to perform within living tissues. Practical issues of implementing true dosimetry are also characteristic of external beam work for the same reasons. Generally, neither internal emitter nor EBRT involves the application of dosimetry. Both instead require dose estimation for their respective treatment planning operations. As we will see, the absolute accuracy of external beam estimates is superior to those of internal emitters by approximately an order of magnitude due to more exact knowledge of the radiation source positions and strengths.

4.2.2 Geometry of Absorbed Dose Estimation

The difficulty of obtaining accurate absorbed dose estimation in the context of internal emitters follows from two issues: (1) determining source strength (i.e., the amount of radioactivity in MBq in the multiple sources); and (2) defining the target–source geometry. The latter is a particularly problematic since the estimator has to calculate the probability that an emitted particle can deliver an amount of energy (ΔE) in the target tissue. In external beam work, exposure rates from the accelerator focal spots are known, and their positions are well defined by the treatment plan. For internal emitters, radionuclide amounts at source points may be difficult to quantitate using any type of present-day technology. This topic is discussed further in the next chapter. In addition, the number of these geometric points will generally be very large. This is particularly the case if the dose estimator is considering a photon (gamma or x-ray) emitter so that multiple source organs contribute to radiation dose at a given target. It may also be that the estimator would consider a voxel-based algorithm such that multiple volumes within a source tissue could contribute to radiation doses in multiple voxels of the target tissues. Finally, any or all of the aforementioned sources could be a function of time since we are dealing with radiopharmaceuticals moving throughout the patient anatomy.

Two human geometries are of general interest in the computation of internal emitter absorbed dose. The older of these is defined by the use of a humanoid phantom that is appropriate for the age and sex of the patients to be enrolled in the RP study. Such phantoms have both defined organ masses and organ separations. This rigor permits a relatively simple analysis to be performed once the activity distribution is known. The more modern (and realistic) geometry is that of the individual patient being treated via the ionizing radiation. Using this geometry allows a patient-specific absorbed dose estimate. Until the advent of anatomic 3-D images such as computed tomography (CT) or magnetic resonance imaging (MRI) scanners, the information for such person-specific dose estimates was not generally available. This lacuna of knowledge has led to an overuse of phantoms and an overdependence on phantom-based estimates.

4.2.3 Biological Applications of the Dose Estimation Process

Three levels of dose estimation may be seen in life sciences. Biophysicists are generally involved at the cellular context in which cultures of cells are grown and exposed to ionizing radiation. The irradiated cells are then plated and observed for growth in units of surviving fraction (SF). Typically, the researcher looks at the cell survival as a function of a dose calculated from Equation 4.1. For a given cell type, these survival curves will appear to be different depending on the ionizing radiation.

The next level of dose calculation involves animal irradiations. Generally, these are performed by physicists prior to beginning a clinical trial of some type of RP-based therapy. Animal survival, tumor size variation, and organ toxicities are the primary outcomes in these experiments—often done with rodents as the test animal. Dose estimation is more difficult in these circumstances but is possible largely via the use of animal biodistribution results following sacrifice at various time points. An alternative format is to use imaging technology at the animal level to follow specific animals over the time course to determine the location and strength of the radiation sources. In such imaging studies, it is common to sacrifice all the test animals at the last time point to corroborate (or normalize) the imaging data obtained earlier. Description of micro single-photon emission computer tomography (microSPECT) and micro positron emission tomography (PET) imagers is included in Chapter 1.

Clinical trials are the ultimate application of dose estimation. This computation is the most difficult since activity localization is made more problematic due to the size of the patient and the usual lack of biological sampling information. Data are obtained indirectly using probes, gamma cameras, and PET imagers. As noted herein, it is preferred that anatomic information is available at the same time as the nuclear imaging. Without these spatial data, dose estimation uncertainties may be several hundred percent and thus may be relatively useless in treatment planning.

4.3 REASONS FOR CLINICAL ABSORBED DOSE ESTIMATION

There are three applications for dose estimation in nuclear medicine clinical practice. Probably the most common is prediction of absorbed dose values in the preclinical (regulatory) phase of RP development. If an investigator wishes to obtain regulatory approval of a possible labeled agent, for either diagnosis or therapy, a dose table will need to be made on an organ-by-organ basis. Most preapproval tables are constructed using animal data. As an alternative, an analysis may be made by analogy with another clinical agent that has a similar chemical structure and (presumably) similar biodistribution. For example, the applicant could use one intact antibody's absorbed dose results taken from a previous clinical trial to predict absorbed doses for a similar, yet clinically untested, novel intact antibody in a different patient trial.

In the United States, two possible regulatory agencies are involved at this stage. If the RP is intended as a physiological agent testing some metabolic pathway, a local radioactive drug research committee (RDRC) would probably be the appropriate regulator. An example of this might be the inhalation of a CO_2 molecule labeled with ^{11}C for PET imaging of gas exchange. When clinical trials are being considered—either locally or on a national scale—the U.S. Food and Drug Administration (FDA) would be the recipient of the application. In this case it is an investigational new drug (IND) application that is being submitted. Besides absorbed dose estimates, this document would have to address other pharmaceutical issues of safety and preparation as well as the proposed statistical analysis of the trial results.

A second reason for absorbed dose estimation is the possibility of direct comparison of one RP with another. This type of scientific analysis is presently very rare but should become more common due to the multiple new drug formulations being designed and discovered as noted in Chapter 1. Here, one imagines that there are two or more RPs developed for the same objective such as internal emitter radiation therapy of colorectal cancer. By comparing their absorbed dose profiles, a radiation oncologist or regulatory committee member could see which of the multiple potential agents is optimal. Given absorbed dose values, this comparison could be made using the traditional or the therapy figure of merit (TFOM) described in Chapter 3.

In the two applications of absorbed dose cited so far, the calculation will have to be based on an appropriate humanoid phantom. Only in this way can the regulatory agency—or the clinician—make a logical comparison with historical antecedents or present-day competitors. If these absorbed dose values had instead been computed using patient-specific information from a limited set of subjects involved in a clinical trial, any comparison becomes difficult since the sample sizes are relatively small and are not necessarily near the average of the general human population in terms of organ mass sizes or anatomy. Similar problems arise if an animal population were to be extrapolated directly to the human case without use of an anthropomorphic phantom. We call all of these estimates based on a specific phantom a Type I dose estimation. Such a calculation is appropriate for either a regulatory body or a scientific comparison of absorbed dose values.

The third application of absorbed dose calculation involves treatment planning for individuals in a clinical radiation therapy trial using an internal emitter. Here, the estimator must use the geometry as well as the activity distribution in a specific patient to make the organ absorbed dose calculation for that person. We call this estimation a Type II calculation. It may be that no phantom is a good representation of the geometry of the particular individual. Therefore, specific correction terms must be generated. These corrections are not small; at City of Hope their magnitude has been as large as a factor of twofold or threefold. We should also mention that humanoid phantoms do not generally contain tumor sites. Thus, the tumor absorbed dose estimates must be made on an individual patient basis. At the present time, the tumor is often assumed to be a sphere, although this simplification is clearly of limited use.

4.4 DOSE MEASUREMENTS

Previously we emphasized the necessity of estimating absorbed doses in a variety of contexts including clinical protocols. This is a mathematical exercise. We should not, however, entirely exclude the possibility of dose measurements in relevant animals or even patients. For example, one can conceive of a small thermoluminescent dosimeter (TLD) placed on the patient's skin or within a body cavity. More imagination—and probably a surgical protocol—is needed to implant a detector within a normal organ or tumor. Implantation can lead to a disruption of the tissues and possible untoward effects on the subject. Organ trauma, bleeding, and the effects of the required surgery are three issues that immediately come to mind. One can also postulate viable tumor cells being released from a lesion that has been penetrated by the dosimetric device. Such possible metastatic cells, moving through the blood or lymphatic systems, may have eventual negative impact on the patient's survival. These various reasons make it unlikely that the local institutional review board (IRB) would approve clinical implantation procedures. A similar outcome might be expected if the local institutional animal care and use committee (IACUC) were to be asked for approval of implantation of dosimeters in a test species such as a BALB/c or nude mouse.

4.5 CORRECTIONS TO THE DOSE ESTIMATES

Physicists are quick to point out that a simple energy density ratio such as Equation 4.1 is equivalent to a finite temperature rise in the absorbing medium. This association allows use of thermal dosimetry. In fact, one of the oldest methods to assay absorbed dose is via a water bath and a sufficiently sensitive thermometer. By converting the ionization energy to the mechanical equivalent of heat, 1 cGy is seen to be equal to a temperature increment of only 2.4×10^{-6}°C. Even a very large absorbed dose is a very small temperature rise; for example, 100 Gy is equivalent to 2.4×10^{-2}°C. Yet this calculation avoids the issue of energy heterogeneity within the target mass. In other words, the ionizing radiation presents a far greater danger than the gross temperature increment (simple dosimetry) would indicate. Otherwise, radiodecay would be no more harmful than a tissue "exposure" causing the same temperature rise made within a microwave oven, for example.

If we were to narrow our conceptual focus to a single ionization event within the target tissue, the resultant atomic temperature rise would be much larger than the macroscopic value given above by the absorbed dose. If a hydrogen atom (having 1 atomic mass unit [amu] of mass) were ionized within a protein molecule, the 13 eV energy deposition in this target is equivalent to a local temperature rise of 3.1×10^{5}°C. The result of the ionization is the production of a proton and an electron. This very large increment, raising the atom to a temperature higher than the surface of the sun, demonstrates the basic danger of ionizing radiation and the true significance of absorbed dose. Because of the heterogeneity of the energy deposition, several corrections to the dose value of Equation 4.1 may have to be employed in estimating risk to the tissues involved. These corrections depend on spatial variations (and therefore radiation type and energy), time duration of exposure, and target nonuniformity. These are discussed in the order given.

4.5.1 Ionization Energy Density and Absorbed Dose

Three types of emission are typically seen in dealing with radioactive sources. The triad includes alpha rays, electrons (beta+, beta−, and Auger electrons), and high-energy photons (x-rays and gamma rays). Alphas and electrons are charged particles that have a rest mass so that their path through matter is described by the spatial rate of energy deposition ($\Delta E/\Delta x$). The linear energy transfer to local electrons implies a finite range since eventually all the original energy must be lost by ionizing electrons in the target molecules. Photons, however, are without charge and can be described only via a half-value layer—that is, the distance over which half of the original number of photons are lost from the original beam. Their range is undefined so that a single photon might pass through the target molecule—or even the patient—and not cause ionization. After all, this "stochastic" passage of photons is what makes animal and patient imaging possible. Likewise, their finite range makes imaging using charged particles extremely difficult unless the sources are essentially at the surface of the subject.

Alpha rays are the most intensely ionizing in terms of eV deposited per μm of particle path length. Values on the order of 100 keV/μm are obtained for these helium nuclei that are seen in the decay of radionuclides near the upper end of the periodic table. Correspondingly, the range of the typical alpha is extremely short—often less than 50 μm, a value comparable to several cell diameters. This distance is less than the thickness of the epidermis, so alphas present a danger only from the inside of the patient. Alphas do have one interesting aspect: they travel in straight lines through the target. This is due to the much larger mass of the alpha ray compared with that of the electrons in the molecular targets (approximately $10^4/1$). As shown following, electrons moving through a medium will have very different spatial behavior as their mass is identical to that of the electronic targets.

Electrons originating from radiodecay have much lower $\Delta E/\Delta x$ values than alphas and can have ranges approaching 10 mm or more in soft tissues. Such relatively long distances imply that electrons may be useful in irradiating a tumor that has spatially variable molecular targets or regions of poor perfusion. Thus, by targeting the accessible sites, other regions of the tumor may be struck incidentally by ionizing radiation. I should also note that decay electrons may undergo large-angle scattering by the electrons in the target since the two interacting particles have the same mass (think of a billiard ball hitting another billiard ball). This scatter causes, among other effects, the fuzziness of a PET source in an imaging situation. It also produces a more uniform dose at the molecular level.

Photons present an altogether different entity than a particle emission. When gammas or x-rays are produced via a radiodecay, those photons pass through the tissue with three types of ionization usually possible. They may cause photoelectric effects, Compton scatter, and pair (e+ and e−) production if the energy is sufficient (1.02 MeV or greater). Relative likelihoods of these three processes depend on the energy of the photon and the electric charge of the target. Movement of photons through the medium is described by Equation 1.4 and requires an attenuation coefficient (μ) that depends on energy and medium.

To take into account the variation in energy deposition of the several particles and photons, a biophysical term has been introduced called the ionizing radiation's quality factor (QF). An alternative term is the radiation weighting factor (w_R). Originally, experiments were done on specific cell types to see what the QF value was for a novel form of ionizing radiation (Hall and Giaccia 2006). The researcher would measure, for example, survival curves for a given cell type using both a standard form (e.g., 250 kV$_p$ x-rays) and an experimental form of ionizing radiation. If the two curves could be made to overlie each other by multiplying the experimental dose by a certain constant, QF or w_R, that weighting factor would be assigned to the experimental radiation. Note that this analysis depends on the target cell and will probably change if one were to go from, say, cultured hepatocytes to colon cancer cells. The resultant modulated dose (in a new unit called the sievert) is called the equivalent dose (ED) since it indicates that the doses are now equivalent in causing a given biological effect such as cell death. While cultured cells may be measured in such experiments, the weighting factor for a specific patient's tumor may be difficult to determine.

$$ED(\text{sieverts}) = D(\text{gray}) \cdot w_R \tag{4.2}$$

By citing a result in sieverts, a researcher is stating that the dose of Equation 4.1 has been corrected (always upward) by a weighting factor to yield a biologically equivalent dose in some context. Of course, if the context were to change, the w_R must vary accordingly. Thus, it is generally a good idea to include both the result of Equation 4.1 as well as the result of Equation 4.2 in any dose estimation report. By only citing an equivalent dose, the researcher is not making clear what weighting factor has been applied in the analysis. Similarly, by giving only a dose in grays, the true effect of the ionizing radiation may be underestimated. Table 4.1 gives standard QF values that are generally assumed for the three types of emissions observed in radionuclide decays (Valentin and International Commission on Radiological Protection 2003). Note that these values have only one significant figure and may well be considered crude estimates. Quality factors determined in biophysical experiments on cells or animals generally have two

TABLE 4.1 Typical Quality Factors (QF or w_R) Used in Obtaining Equivalent Dose

EMISSION	QF
Alpha ray of any energy	20
Neutron (function of energy)	$f(E)$[a]
Proton	2
Electron (e, or β$^+$ or β$^-$) of any energy	1
Photon (x-ray or gamma ray) of any energy	1

Source: ICRP, *The 2007 Recommendations of the International Commission on Radiological Protection*, ICRP Publication 103, Ann. ICRP 37 (2–4), 2007. With permission.

[a] Values range from 2 to 20 over 1 kV ≤ E ≤ 1 MeV.

TABLE 4.2   Alpha/Beta Ratio Parameters for Late Effects

TISSUE	ALPHA/BETA
Normal tissues except CNS	3.0 Gy
CNS	2.0
Tumor except prostate	10
Prostate	1.5

or more significant figures but are applicable only in the precise context in which they were derived.

4.5.2 Temporal Variation in Dose Rate

In the case of external beam, ionizing radiation is delivered in short, intense bursts—typically on a 5-day-per-week basis. With radionuclide decay inside the tissues, there is continuous deposition of energy with an approximately multiexponential reduction of dose rate with time (cf. Chapter 6). Clearly, these are very different procedures, and some accounting should be taken of the disparity. From radiobiology, it has been conventional to define a pair of parameters (termed α and β) that can be used to express a given cell type's sensitivity to ionizing radiation (Fowler 1990). These two coefficients are known for certain cells in normal tissues and, to some degree, for tumor cells of various types. Table 4.2 contains a rough summary of what these values are estimated to be for several normal organs. It is unlikely that a specific patient's tumor alpha and beta values are available, however.

The SF of cells in culture is given by

$$SF = \exp(-\alpha D - \beta D^2) \tag{4.3}$$

where α and β have units of inverse dose and inverse dose^2, respectively. If we now let the resultant effect of the radiation be the logarithm of cell kill, the effect becomes the negative logarithm of the left-hand side (LHS) of Equation 4.3 with the result

$$\textit{Effect} = \alpha D + \beta D^2 \tag{4.4}$$

Notice that *Effect* is defined to be dimensionless. Yet conventional external beam irradiation uses a sequence of small (typically daily) doses (d) to spare the normal tissue (typically skin) so that the clinical outcome becomes

$$\textit{Effect} = n(\alpha d + \beta d^2) \tag{4.5}$$

By algebraic manipulation, one can establish

$$\textit{Effect} = nd(\alpha + \beta d) = nd\alpha[1 + (\beta/\alpha)d] \tag{4.6}$$

The *nd* term is the total dose (*D*) delivered over the time allocated to the entire treatment. If we finally divide through by the α factor, the result is

$$Effect/\alpha = D[1 + (\beta/\alpha)\, d] \qquad (4.7)$$

In traditional external beam work, the LHS of Equation 4.7 is termed the "biologically effective dose." Since *Effect*, by definition, has no units, this effective dose has units that are the inverse of alpha (i.e., Gy). The bracket on the right-hand side (RHS) of Equation 4.7 is called the relative effectiveness (RE) of the radiation on that cell type.

With that purely physical prelude, one may now introduce dynamics into the formalism by including two additional concepts into the analysis. First, following Dale (1996), one may consider simultaneous radioactive decay and biological clearance due to the agent being a radioactive drug. In that case, the *RE* factor becomes

$$RE = 1 + \frac{R_0}{(\mu + \lambda)(\alpha/\beta)} \qquad (4.7a)$$

where R_0 is the initial absorbed dose rate, μ is the cell replenishment rate, and λ is the effective clearance rate of the RP from the cell type of interest. For simplicity, clearance is assumed to be a monoexponential, as is the cell growth rate.

Any potential user of dose rate corrections will want to know what these various parameters in Equation 4.7a are and how well they may be determined in a given patient's disease. These data are generally not well known—particularly for entire organs that are ensembles of different cell types. An application of the method to kidney dose estimates and the correlation of these with toxicity data is given in Chapter 9. We should emphasize that such corrections were necessary before there was evidence of a positive correlation between dose estimates and renal toxicity.

4.5.3 Organ Heterogeneity

Target tissues are not uniform in their cell types. Among the more important types in the liver are hepatocytes and Kupffer cells. Kidneys have glomerular cells, proximal tubule brush border, and distal tubular cells. The brain has glial cells acting as supporting structures to local neurons. Thus, the dose estimator should ideally provide values for each of these types within an organ. Because of imaging resolution issues, this sophistication is not yet possible for most of the dose estimates used in nuclear medicine. However, as SPECT–CT and PET–CT scanners become more common, there will be the possibility of computing a dose to a specific subset of the tissue of interest. Here, the anatomic information can be used to provide more detail on the target tissue and hence more localized dose values. Alternatively, it may be possible that only a known subset of the organ's cells are targeted by the radiopharmaceutical; thus, only that cell type should be considered in the dose estimates that are obtained.

Absorbed dose is clearly a function of spatial position. While most contemporary estimates done in nuclear medicine involve the average dose to an entire organ or tumor, the voxel-by-voxel distribution of dose will eventually become the quantity of interest. One will want to map that variation against the different tissue types in normal organs and against the various levels of oxygenation in lesions. Only in this way will the actual effectiveness of a given radiation exposure become known. Following the lead of external beam physicists, the nuclear medicine dose estimator will probably construct histograms of these dose distributions; such displays are termed dose–volume histograms, and their application in the future will become more widespread.

4.5.4 Effective Dose

A third form of dose estimation is also used in medical practice—in particular in radiation protection. Here, the epidemiologist or radiation safety officer is interested in the stochastic danger of cancer induction, or other toxicity, over the lifetime of a cadre of exposed subjects. Effective dose is computed using weighting factors (w_{τ_i}) for the previously determined equivalent radiation doses for each tissue in the body

$$D(effective) = \sum_{i=1}^{i=N} w_{\tau_i} ED_i \qquad (4.8)$$

where ED_i represent equivalent doses (sieverts) obtained using Equation 4.2. Statistical weights for the various organs are estimated using the likelihood of cancer induction for that tissue. Table 4.3 contains a summary of these weighting terms for various major organs (Valentin and International Commission on Radiological Protection 2003). Application of this result to a specific person is not appropriate; the effective dose is intended to predict outcomes for large populations subjected to a given radiation protocol. Equivalent doses substituted into Equation 4.8, however, are those of representative

TABLE 4.3   Representative Organ Weighting Factors (w_T) for Effective Doses Due to Ionizing Radiation

ORGAN	WEIGHTING FACTOR w_τ
Stomach	0.12
Lung	0.12
Breast	0.12
Red marrow	0.12
Colon	0.12
Gonads	0.08
Thyroid, liver, bladder	0.04
Brain, skin, bone surfaces	0.01

Source: ICRP, *The 2007 Recommendations of the International Commission on Radiological Protection,* ICRP Publication 103, Ann. ICRP 37 (2–4), 2007. With permission.

individuals. Effective doses may then be used to compare the epidemiological result of one RP with another if both agents are used for the same clinical objective. This strategy can be carried one step further and be used to compare one radiographic or therapeutic procedure with another with respect to their ultimate hazard to society. In this last case, we are actually in the process of "comparing apples to oranges" since the various procedures may have nothing more in common than that they are both done in a radiology or a radiation oncology department. Mathematically minded readers will notice that the LHS of Equation 4.8 has no free index; thus, it is a single parameter, which is used to predict toxicity via an effective dose to the entire individual.

What if we wish to consider a specific patient, such as a young woman receiving a CT exam on a periodic basis because of lymphoma diagnosis? Then the risk for that individual must be computed separately and on an organ-by-organ basis. Probably the tissue of greatest radiological protection interest would be thyroid tissue and breasts, which are exposed repeatedly during the neck and chest segment of a neck, chest, abdomen, and pelvis (NCAP) CT exam. Dose to these organs would be a more relevant estimate of danger than referring to the effective dose since the latter distributes the radiation throughout the body. Such organ-specific absorbed doses could, in principle, be measured with dosimeters placed near the thyroid or breast—if these devices did not interfere with the images. In most cases, including nuclear medicine, the actual absorbed doses will have to be estimated using a mathematical algorithm. We now consider how that calculation is done for internal emitters.

4.6 METHODS FOR ESTIMATING ABSORBED DOSE FOR INTERNAL EMITTERS

This section describes several techniques for nuclear medicine absorbed dose estimation. The simplest of these is due to the Medical Internal Radiation Dose (MIRD) Committee of the Society of Nuclear Medicine (Snyder 1975). It is by far the most common method and has been employed in generating many reported results in the literature. Other techniques can be applied in certain circumstances wherein the source–target situation is amenable to a convolution argument. The concept of convolution, at least for temporal variation, was defined in Chapter 3 and is considered more formally in Chapter 6.

4.6.1 The Canonical MIRD Estimation Method for Internal Emitter Doses

There are essentially two factors involved in internal emitter dose estimates. First, the investigator has to determine how much radioactivity is present in all observable source organs. By observable, we mean detectable via nuclear probes and various imaging devices. Generally, this aspect is accomplished by using a gamma camera or PET imager to show organs that have substantial accumulation of the tracer—or, in any event, of its label. As noted in Chapters 2 and 3, the label is the item of interest to the physicist since its emissions drive the ionization processes. A physicist is technically not doing a calculation of absorbed dose for the RP—only for its radioactive label as it was deposited by the RP. The subtlety of this difference is sometimes lost on radiopharmacists. Finally, the amount of activity in any given source organ (A_i) will be a function of time so that the total number of decays is given by its integral ($\tilde{A}_i$) over time out to perhaps 10× the physical half-life or longer. The difficult art of measurement of activity in vivo is described in more detail in Chapter 5. Modeling techniques to expedite interpolation and integration of the activity over time are considered in Chapter 6.

The second aspect of dose estimation is a calculation of the probability per nuclear decay of subsequent energy deposition into a specific set of target tissues. In most calculations, the list of such tissues is extensive and may exceed 30 or more organs or parts thereof. For the sake of exposition, let us consider two organs with the source labeled with the index i and the target with the index j. Keep in mind that i and j may refer to the same organ as a tissue may give rise to an absorbed dose to itself as well as other organs in the body. The quantity of interest in the MIRD formulation (Snyder, Fisher, et al. 1969) is then

$$S_{i,j} = \sum_{k=1}^{k=N} n_k E_k \phi_{i,j,k} \Big/ m_j \tag{4.9}$$

where n_k is the number of emissions of energy (E_k) given off per each decay of the radiolabel. These possible decay channels are summed to the total number of emissions (N). The $\phi_{i,j,k}$ parameter gives the probability that the emission energy is deposited in the target organ of mass (m_j). This last entity is a sensitive function of the source–target geometry and is often estimated using a Monte Carlo (MC) method. Two representative adult phantoms used in such calculations are shown in Figure 4.1. For the sake of simplicity, we initially neglect any variation in target mass during the irradiation. Notice that the resultant summation is termed $S_{i,j}$. This quantity is an element of a matrix (**S**) that is, in practice, rectangular since the number of targets (or tissues of interest) is usually much larger than the number of detectable sources. Tabulations of **S** (cf. Chapter 7), however, are generally square matrices whereby the number of sources and targets are essentially identical (Snyder 1975). When the two indices are the same, one is referring to the diagonal elements in the matrix. In most examples, such elements dominate the estimation process. In other words, the largest component of absorbed dose in a given target organ is generally due to radiation originating in that organ itself. This domination of the self-dose result is particularly true for beta and alpha emitters because of their short ranges in tissue. In particle emissions, it is often assumed that the **S** matrix is indeed diagonal and that radiation doses due to nonself-tissues are literally nil.

Absorbed internal emitter dose is then estimated as the product

$$D_i = \sum_{j=1}^{j=total} S_{i,j} \tilde{A}_j \tag{4.10}$$

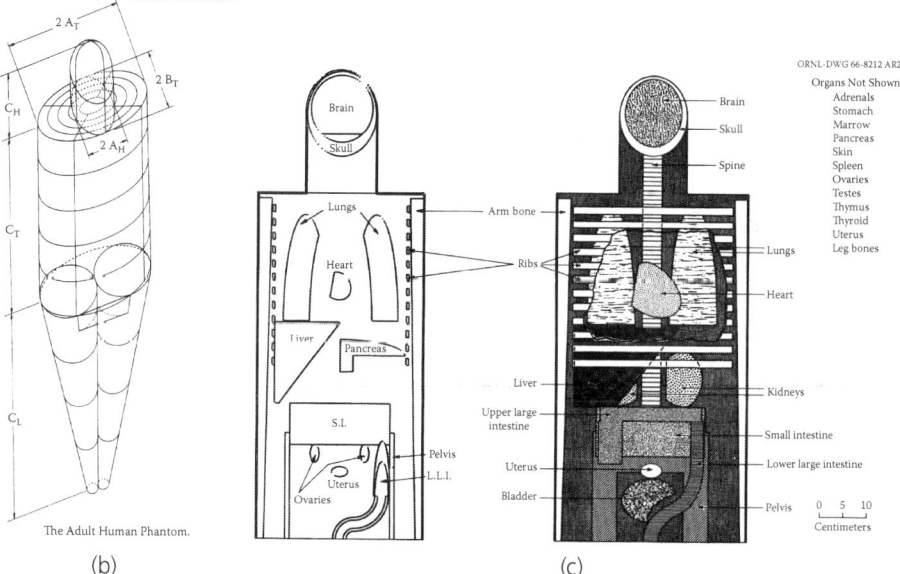

FIGURE 4.1 (a) The set of six representative MIRD phantoms for a family of individuals (Cristy 1987). (b) Expands the adult male phantom. Notice that both the size and separation of the various major organs varies in each geometric model. Organ sizes and location of their respective origins are given by mathematical functions to expedite Monte Carlo analysis. (c) The older adult male developed by Snyder et al. (1969). (Courtesy of the Oak Ridge National Laboratory/U.S. Department of Energy. With permission.)

with the sum going over the complete number (total) of source organs. Recall that this set of sources is discovered by the researcher upon looking at the probe or imaging results for the animal or patient. One fundamental result of Equation 4.10 is that there is a strict dichotomy between time effects ($\tilde{A}$) and spatial effects (**S**) of the ionizing radiation. This strict separation is the most common usage of the MIRD strategy.

If we need to consider the more general case of target mass variation during the radiation episode, then the equality of Equation 4.10 must be restructured into its more fundamental differential form:

$$dD_i/dt = \sum_{j=1}^{j=\text{total}} S_{i,j}(t)A_j(t)$$ (4.11)

Now, **S** is explicitly a function of time as well as emission energy and type due to mass and geometry variations. In such cases, we must matrix multiply by the activity at that time to find the corresponding dose rate in any target organ. To determine total target organ dose, Equation 4.11 is integrated over time. Implicitly, such a requirement implies an anatomic sequence of images, for example obtained with CT, that document the mass variation of m_j. Originally, this type of target variation was studied in the case of fetal exposure in which the mass of a target organ increased with gestational age. Mass variation is also of interest in the treatment of tumors whose size is presumably decreasing due to their response to the earlier radiation in the exposure episode. Two clinical examples from lymphoma therapy are reported in the literature (Hartmann Siantar, DeNardo, et al. 2003; Hindorf, Linden, et al. 2003) for variation of tumor masses over the course (1 to 2 weeks) of the radionuclide treatment times. Logically, a lesion mass could also possibly be increasing because of contravening tumor growth during therapy.

4.6.2 Types of MIRD Human Dose Estimates

As noted, two geometric idealizations are possible for the **S** matrix and the absorbed dose estimation process. Phantoms based on a standard geometry are the more traditional and can be justified by extensive prior appearances in the literature. As described already, we call such estimates Type I. The resulting dose estimates refer to the phantom and may be done for the benefit of either a regulatory agency (e.g., the FDA) or a scientific comparison as described. We must notice initially that, in this case, the $\tilde{A}$ value as obtained from an animal study or patient will not be correct since the distribution of biological tissue masses is not necessarily equal to that of any MC-generated **S** matrix. For example, if we have animal results from a biodistribution study and wish to predict human doses with a humanoid phantom:

$$\tilde{A}(phantom) = \frac{\tilde{A}(animal) \bullet [m(phantom)/m_{WB}(phantom)]}{m(animal)/m_{WB}(animal)}$$

(4.12)

where m is an organ mass, and m_{WB} is the symbol for the whole-body mass of the species. Here, we are correcting for the relative sizes (relative to the whole-body mass) of the various source tissues. If the animal has a greater proportion of its mass in a given organ such as the liver, for example, the animal's $\tilde{A}$ for that tissue must be correspondingly reduced before beginning the human dose estimate. Such corrections are made on an organ-by-organ basis and rely on the measurement of the experimental (not the theoretical) animal's mass distribution. In this case, the **S** matrix is taken directly from that given by the MC calculation for the phantom. Usually, this phantom will be either an adult male or adult female.

On the other hand, if we wish to compute an absorbed dose estimate for a specific patient, then the $\tilde{A}$ as measured for that person is inherently correct. Yet no previously generated **S** matrix may be appropriate since the patient will, in general, have a different geometry than any humanoid phantom available in the literature. We refer to these examples as Type II calculations. In such cases, the **S** matrix must be corrected to better represent the patient. This process is more complicated than that cited for $\tilde{A}$ correction. If we have an alpha or beta emission case wherein the ionizing radiation cannot escape the source (= target) organ, then the lowest-order result is

$$S_{ii}(patient) = S_{ii}(phantom) \bullet m_i(phantom)/m_i(patient) \tag{4.13}$$

where m_i refers to a given organ's mass value. Notice that we are explicitly neglecting cross-organ terms in this correction. As noted already, the magnitude of the correction factor in Equation 4.13 may be twofold to threefold for several reasons. One cause is simply the human population variation of organ mass—a quantity that has been largely overlooked in the dose estimation literature. A second reason for this disparity is disease or its therapy. A patient may have an increased (or decreased) organ size due to medical condition or subsequent intervention. Enhanced splenic size in lymphoma is a well-known example.

If the radionuclide has photon as well as particulate radiation, then the correction of Equation 4.13 must be augmented. No best method is known for performing this task, although several techniques are in the literature. These modifications are described in Chapter 7. Generally, such changes in **S** due to the photon-absorbed dose are not significant if the study involves therapy. Most analysts, therefore, do not invoke an **S** correction on the photon component of the radiation source.

The reader should also realize that, of necessity, some patient-absorbed dose estimates are neither type I or Type II. These estimates are mixed computations containing aspects of both calculations. Typically, a patient's absorbed dose array is estimated using that individual's $A(t)$ for each organ but with a phantom-derived **S** matrix. It turns out that such results are the most common in the dose estimation literature due to a lack of patient geometric information. In other words, the extensive organ-sized data set is not at hand for each patient in the study. In the early days of the MIRD Committee, CT and MRI scanners had not yet been invented, so geometric data for a patient were not available. In more recent clinical studies, the local IRB may not allow exposing the patient to CT imaging just to expedite dose estimates in nuclear medicine. Examples of such mixed estimates are given in Chapter 8.

Other techniques for dose estimation do not use the **S** formalism but still continue the concept of a set of sources and a set of targets. To determine the absorbed dose at the target sites, a convolution approach (cf. Chapter 6) is used. This strategy can be implemented in either a point-to-point or a voxel-to-voxel calculation depending on the type of information available to the user.

4.7 POINT-SOURCE FUNCTIONS FOR DOSE ESTIMATION

One of the earliest techniques for calculating internal emitter absorbed dose is a two-step process involving the exposure rate from a point-source being combined with the geometries of the source and target. In this case, one first determines, in a laboratory setting, the decrease of exposure rate as a function of distance from a point (i.e., small) source. If no attenuator were present, the result would be an inverse-square law. When attenuation is included, phenomenological models are built that represent the exposure rate as an explicit function of both distance and attenuating material. Because of the desire for simplicity, the material is usually assumed to be uniform and of a tissue-equivalent type. Several experimenters have generated such point-source functions (PSFs) for beta and even gamma radiation (Berger 1971; Leichner 1994). In the former case, the PSF is made to go to zero at distances beyond the range of the beta radiation. Examples of PSFs are given in Chapter 7.

Given the experimental point-source function, one then computes the absorbed dose at a point r_2 inside a finite-sized target organ from a source tissue via an integral

$$D(\vec{r}_2) = \int\limits_{\text{Volume of source}} PSF(\vec{r}_2 - \vec{r}_1)\tilde{a}(\vec{r}_1)A_0 dV_1 \qquad (4.14)$$

where the activity concentration integrated over time is given by $\tilde{a}A_0$, and the vector $\vec{r}_1$ moves over the source distribution. Dimensions of a are %IA/g. Recall from Chapter 2 that a is the fractional activity per unit mass: it is not corrected for decay. As Chapter 6 will show, the mathematical form of the equality is that of a spatial convolution. The estimator is convolving the known activity distribution in the source with the PSF to find the absorbed dose at a specific point in the target. Clearly, this could be a tedious computation as the size of the organs becomes large and complicated in shape. If one had to do only simple geometries, the process given in Equation 4.14 would be much more tractable. Before using Equation 4.14, it is important to consider a simpler method that permits organ-sized dose estimates in reasonable time durations. The following section describes one of these simpler cases in the example of a TLD lodged inside a tissue culture sample or organ of interest.

4.8 ABSORBED DOSE ESTIMATES USING VOXEL SOURCE KERNELS

Use of a PSF may entail a very long computational process. This follows from the small size of the calculation step; one may imagine that the step size used in a computer program may be 5% of the range of the beta radiation. If the latter is approximately 10 mm, the step size would be 0.5 mm. Yet the patient or animal will have organ dimensions on the order of 1 to 40 cm. Thus, many steps are needed to cover an entire organ even if there is no associated photon flux from the radiation source.

An important, and practical, alternative to the PSF is the voxel source kernel (VSK). In this style of analysis, the estimator uses a finite sized volume element as the source and uses Monte Carlo methods to find the absorbed dose in adjacent voxels. As in the PSF example, the medium is generally assumed to be uniform and of tissue quality. Several VSK functions are in the literature (Bolch, Bouchet, et al. 1999; Liu, Williams et al. 1998), with ^{90}Y being the radionuclide of greatest interest due to its involvement with targeted radionuclide therapy. Table 7.5 contains some of the elements of a voxel source kernel for ^{90}Y. Other VSK results are given in Chapter 7 for both mouse and human dose estimations.

4.9 MEASUREMENT OF RADIATION DOSE BY MINIATURE DOSIMETERS IN A LIQUID MEDIUM

The emphasis heretofore placed on absorbed dose estimation via calculations should not be taken as a sign that no effort is being made to directly measure energy deposition in culture media or even in tissue. From the mid 1980s onward, a technique has been undergoing development (Wessels and Griffith 1986) for beta dose measurement in vivo using miniature $CaSO_4$ (Dy) crystal detectors set within a Teflon matrix (Demidecki, Williams, et al. 1993). These TLD devices are used in the form of rectangular solids on the order of $0.4 \times 0.2 \times 5$ mm. To test the practicability of such small TLDs in internal emitter radiotherapy, dose measurements have been performed inside tissue-like media in which sources of beta radiation such as ^{90}Y have been dissolved. Here, we assume that ^{90}Y would be the generic radiolabel for tumor therapy due to its relatively high mean beta radiation energy.

To use any such miniature TLDs, four correction factors must be included in their calibration: dose linearity, beta energy dependence, time *in situ,* and finite size of the detector. Linearity of the response is evaluated by measuring light output as a function of ^{60}Co absorbed dose. Energy dependence is tested with beta beams of various energies including the radionuclide ^{60}Co. Use of ^{60}Co is justified since its mean Compton electron energy (0.6 MeV) is a good approximation to that of the mean energy of ^{90}Y beta radiation (0.93 MeV).

It has been seen that the first two of these four factors are not significant in the analyses but that time in medium- and finite-size corrections are of greater importance. In the former case, it was found that the dosimeters lose their light output (i.e., their apparent dosimeter reading) with time in a monotonically decreasing fashion. A measured result is shown in Figure 4.2. By using scanning electron microscopy (SEM), the surface of the dosimeters was seen to be porous. Crystals of the $CaSO_4$ (Dy) can be leached from inside the Teflon support matrix. Figure 4.3 shows the comparison of the Teflon before and after immersion in the medium. It was also found that the removal of the thermoluminescent material is accomplished not only by the fluid or gel in which the dosimeter is placed but also may occur due to mechanical stresses being placed on the Teflon matrix. For example, the stress could be due to flexing the TLD during implantation into tissue or by the motion of the tissues into which the device was inserted. Muscle implantation would be particularly susceptible to this form of loss. It is seen that the leaching effect is exacerbated by both elevated temperatures and low values of solution pH. By correcting for these factors and time in the appropriate medium, the user could correctly estimate the actual absorbed dose in that material. Because of loss of light output, that correction could be an order of magnitude or more as seen in the figure (Demidecki et al. 1993).

A second significant calibration factor involves the finite size of any TLD. Because the dosimeter is inserted into the radioactive medium, it follows that radioactivity is being excluded away from the center of the detection device. Activity that would have been at the site of the detector is displaced out of the space that is occupied by the TLD so that a reduced signal (hence apparent dose) is thereby detected. This effect may directly be estimated by using the previously described PSF and integrating over the volume of the dosimeter. If we consider a point within the TLD, the dose recorded there would be reduced by the integral of PSF over the volume

$$\Delta Dose = \int_{-z_1}^{z_1} \int_{-y_1}^{y_1} \int_{-x_1}^{x_1} PSF(x,y,z) \tilde{a} A_0 dx dy dz \tag{4.15}$$

where we have implicitly set the point of interest to be the center of the TLD at $(x,y,z) = (0,0,0)$. Limits of integration are given by the size of the rectangular dosimeter as already described. Similar computations can be made at other points inside the "void volume." If we limit ourselves to results at the coordinate origin, the finite volume correction factor is given by 1.08 for ^{90}Y and 1.72 for ^{131}I (Demidecki et al. 1993). The relative increase for radioiodine is due to its lower energy beta emissions, which led to a more significant effect at a given size of detector.

We should emphasize that both crystal loss and finite size effects can be corrected via a direct calibration in appropriate media. By showing the magnitude of these factors, one is given a feeling for the necessity as well as the required size of the corrections. The larger factor, due to leaching of the sensitive dosimetric material, can possibly be eliminated by using an impermeable barrier around the TLD. Some investigators are currently attempting to produce such devices with coatings that are

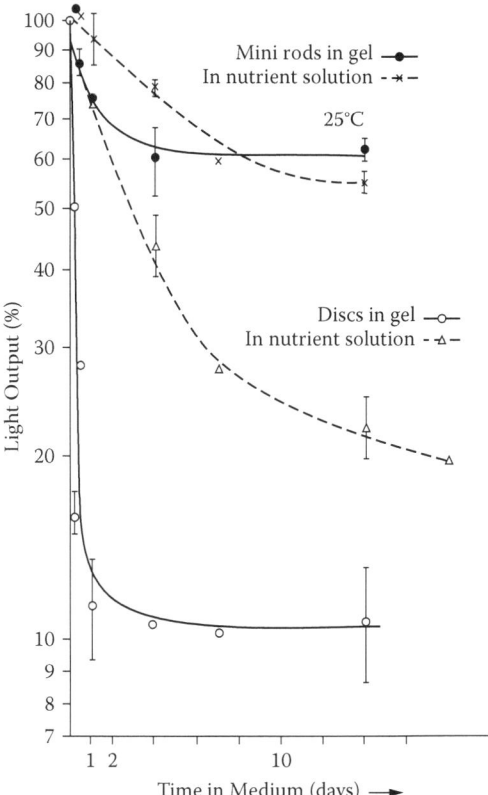

FIGURE 4.2 Loss of light output (signal) with time in solution for TLDs. A two-exponential model (cf. Chapter 6) is used to represent this loss, which can approach one order of magnitude over a several day-long period of immersion. (From Demidecki, A. J. et al., *Med. Phys.*, 20, 4, 1993. With permission.)

not toxic to the animal or human tissue. Finite-size corrections, however, are not as amenable to engineering and will continue to be made via exposure in a gelatin or other medium that has been uniformly impregnated with the radionuclide of interest. Because of possible coatings on the TLD in the future, such finite-size corrections will become more important.

4.10 MEASUREMENT OF BRAKE RADIATION ABSORBED DOSE IN A PHANTOM USING TLDS

A less stressful alternative to the use of TLD devices in a liquid milieu is their application to a dry humanoid phantom containing localized radiation sources (Williams, Wong, et al. 1989). An example of this type of work is a measurement of brake radiation

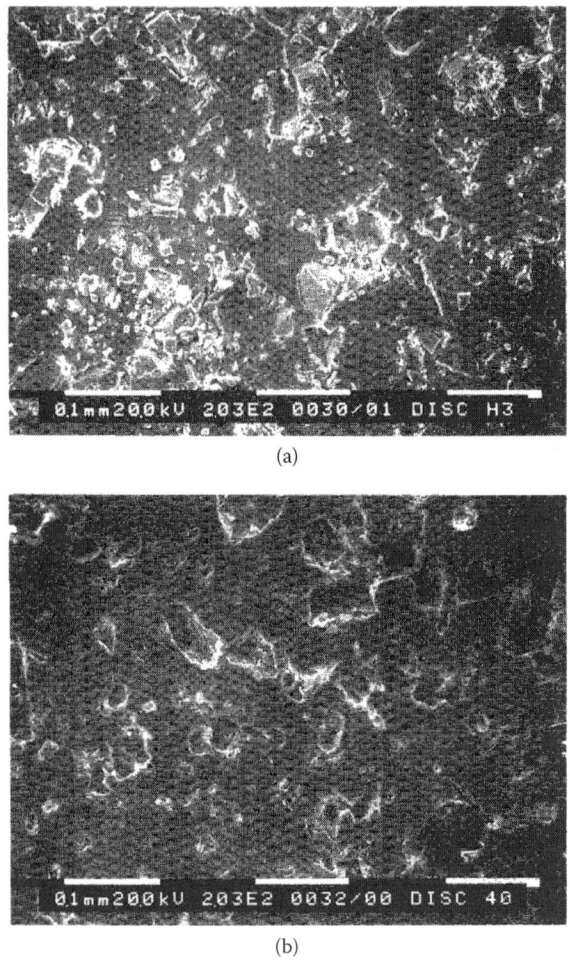

FIGURE 4.3   Electron microscopic scan of the TLD after immersion in a medium. The upper figure gives the original image; the lower figure shows the result after 11 days in a nutrient solution. Note the loss of sensitive [CaSO$_4$(Dy)] material out of the matrix and the resultant large holes leading into the dosimeter. This loss could be further exacerbated by physical bending of the dosimeter. (From Demidecki, A. J. et al., *Med. Phys.*, 20, 4, 1993. With permission.)

absorbed dose (Gr. Bremsstrahlung) in the RANDO phantom. Brake radiation is produced when charged particles, such as beta rays, are decelerated by their passage near atomic nuclei located in tissue or other materials. The effect is similar to the production of radio waves by accelerating electrons in a broadcasting antenna. A general statement is that accelerated electric charges (of either sign) must radiate. The importance of brake radiation arises from the fact that, in the case of pure beta emitters, it is the only cause of dose at distances beyond the range of the emitted electrons. Some investigators have, in fact, attempted to image distributions of pure beta emitters such as ^{90}Y or ^{32}P using Bremsstrahlung and a gamma camera (Fabbri, Sarti, et al. 2009).

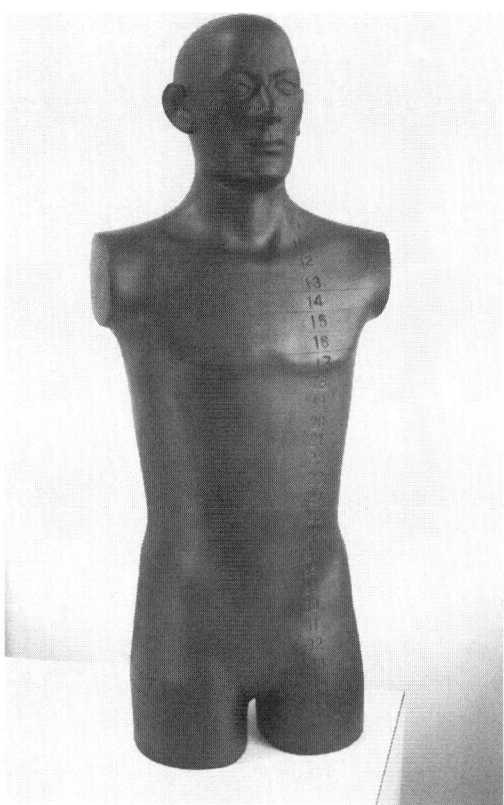

FIGURE 4.4 The male RANDO anthropomorphic phantom used in the brake radiation measurements. Each section of the phantom is 2.5 cm thick with positions for cylindrical TLD insertion being located in each slice. (Courtesy of Phantom Labs Inc. With permission.)

In measuring brake radiation, LiF TLDs were placed at various pseudo-organ sites within the RANDO phantom, which has both bones and lung-equivalent material. The phantom (Figure 4.4) is cut into 2.5 cm slices with holes in each slice to admit the dosimeters. In the original experiments (Williams et al. 1989), a single 10 mCi ^{90}Y source was sited in one of these holes in the bladder area with detectors at specific slices equivalent to location of the bladder (BL), uterus (U), lower large intestine (LLI), upper large intestine (ULI), small intestine (SI), kidney (K), and liver (L), respectively. The source was allowed to decay over several weeks to obtain as many events as possible.

In this context, it was also possible to generate the **S** matrix for brake radiation using Equation 4.9. Here, one blends the emission spectrum of the betas from ^{90}Y with the triangular spectrum of brake radiation to find **S** values given the source and target positions. MIRD φ values for these various geometries and photon energies were taken from Snyder et al. (1969). Table 4.4 gives a direct comparison of the estimates and the TLD-measured dose values. It is seen that the two values were comparable, within estimated uncertainties, for most cases. In some instances, such as the uterus and lower large intestine, locations were not unique as the organs had a finite size and were not

TABLE 4.4 Comparison of Estimated and Measured Brake Radiation Dose

TARGET ORGAN	ESTIMATED DOSE USING TABULATED Φ VALUES (mrad/mCi)	MEASURED DOSE WITH RANDO PHANTOM (mrad/mCi)
Uterus	130	71–260
Lower large intestine	65	47–71
Upper large intestine	14	16 ± 1
Small intestine	20	16 ± 1
Kidney	2.6	2.9 ± 0.3
Liver	1.6	1.4 ± 0.7

Source: Williams, L. E., Wong, J. Y. C., Findley, D. O., and Forell, B. W., *J. Nucl. Med.*, 30, 8, 1989. With permission.

in a single section of the phantom. In such situations, a range of TLD measurements is given in the table.

Agreement between the estimated and measured absorbed doses was generally within the errors of the measurement process. In particular, relatively large values—on the order of 1 rad/mCi were found for tissues containing a source of ^{90}Y. Until these values were measured, the magnitude of this effect had not been determined. There were also no adjustable parameters in the brake radiation calculation. It would be helpful to see more such measurements in which the MIRD format is tested in a realistic way.

4.11 SUMMARY

A unique aspect of radiopharmaceuticals is their ability to deposit ionizing energy at organ and tumor sites within the body. While intentionally small (1–10 cGy) in a diagnostic context, the designated absorbed dose for lesion therapy implies achieving values on the order of 10 to 100 Gy. Estimation of these dose levels requires the use of relatively complicated mathematical methods following the determination of the amount of activity [$A(t)$] at various body sites as a function of time. This activity may be in the whole body, individual organs, segments of organs (voxels), or even specific cell types. Thus, the dose estimation may be at one or more of several spatial levels. The MIRD canonical form for the requisite dose is given by $S*\tilde{A}$, where S is a matrix that contains energy and geometric information, and $\tilde{A}$ is the vector set of integrals of activity integrated out to long times. The concept of dose estimation was emphasized in the discussion. Unfortunately, this topic is most often referred to as *dosimetry* in the literature; use of such terminology is both illogical and ironic.

Absorbed dose, as estimated in Gy, is generally not sufficient to describe radiation effects at either the cellular or organ level. Because of the spatial distribution of the ionizing energy as well as its time course relative to cell cycle kinetics, correction factors must be brought into consideration. Spatial effects due to changes of dE/dx give rise to the quality factor (QF or w_R), whereas time effects are handled with the requisite α and β parameters. In clinical cases, the particular values for these corrections are often not known and thus become adjustable in fitting the biological outcome. Presently, we do not have sufficient information to see if there is consistency of such parameters across patients or even organ and tumor types. It is important to realize that such correction factors are usually quoted to only one significant figure. In addition, the factor may be quite sizable; values of w_R for alphas and neutrons can be as large as an order of magnitude or more. Thus, the conversion of dose to effective dose or other computed surrogate is both a large and uncertain change.

Because of the multiple correction factors to be applied to the concept of absorbed dose, a typical analyst may be hard-pressed while attempting to correlate tumor regression and an estimated dose in Gy. A similar situation arises in normal organ toxicity and absorbed dose estimations. At the present relatively primitive state of affairs, there is a manifest difficulty in finding internally consistent factors that may be applied to certain tissues and tumor types. This situation hopefully will improve with continued application of the various corrections. Some correlations between absorbed dose and clinical outcomes are included in Chapter 9.

Three types of clinical dose estimates are included in the discussion. The most common, by far, is a calculation of dose for an appropriate MIRD phantom given an animal biodistribution. In this context, the animal source organ activity data [$A(t)$] are standardized by means of correcting for the relative blood flow to each organ. Use of this computation is appropriate for both regulatory applications as well as for cross-drug comparisons. These two uses are called Type I applications of dose estimation. The third estimation is that done for a specific patient. In this case, the $A(t)$ data are taken directly to the computation, but the tabulated phantom **S** matrix values must be corrected to account for organ mass differences between the individual and the closest MIRD (or other) phantom. This application requires anatomic data for the individual patient and is a motivating reason for using hybrid SPECT–CT or PET–CT scanning equipment. Historically, the number of Type II estimates is limited due to the lack of anatomic information.

While dose estimation is emphasized, dosimetry is also possible. To find the magnitude of the brake radiation dose from ^{90}Y in a clinical context, City of Hope researchers measured the absorbed dose at various organ locations in an anthropomorphic RANDO phantom. Results were then compared with predictions based on assumptions regarding the geometry of the exposure. Agreement within the estimated errors was generally good, showing that the phantom resident inside the computer was a good approximation to the physical phantom.

REFERENCES

Berger, M. J. 1971. Distribution of absorbed dose around point sources of electrons and beta particles in water and other media. *J Nucl Med* Suppl 5: 5–23.

Bolch, W. E., L. G. Bouchet, et al. 1999. MIRD pamphlet No. 17: the dosimetry of nonuniform activity distributions—radionuclide S values at the voxel level. Medical Internal Radiation Dose Committee. *J Nucl Med* **40**(1): 11S–36S.

Calabrese, E. J. and L. A. Baldwin. 2000. The marginalization of hormesis. *Hum Exp Toxicol* **19**(1): 32–40.

Cristy, M. and. K. Eckerman. 1987. Specific absorbed fractions of energy at various ages from internal photon sources. *ORNL* ORNL/TM-8381 V1–V7.

Dale, R. G. 1996. Dose-rate effects in targeted radiotherapy. *Phys Med Biol* **41**(10): 1871–84.

Demidecki, A. J., L. E. Williams, et al. 1993. Considerations on the calibration of small thermoluminescent dosimeters used for measurement of beta particle absorbed doses in liquid environments. *Med Phys* **20**(4): 1079–87.

Emami, B., J. Lyman, et al. 1991. Tolerance of normal tissue to therapeutic irradiation. *Int J Radiat Oncol Biol Phys* **21**(1): 109–22.

Fabbri, C., G. Sarti, et al. 2009. Quantitative analysis of 90Y Bremsstrahlung SPECT-CT images for application to 3D patient-specific dosimetry. *Cancer Biother Radiopharm* **24**(1): 145–54.

Fowler, J. F. 1990. Radiobiological aspects of low dose rates in radioimmunotherapy. *Int J Radiat Oncol Biol Phys* **18**(5): 1261–9.

Hall, E. J. and A. J. Giaccia. 2006. *Radiobiology for the radiologist*. Philadelphia: Lippincott Williams & Wilkins.

Hartmann Siantar, C. L., G. L. DeNardo, et al. 2003. Impact of nodal regression on radiation dose for lymphoma patients after radioimmunotherapy. *J Nucl Med* **44**(8): 1322–9.

Hindorf, C., O. Linden, et al. 2003. Change in tumor-absorbed dose due to decrease in mass during fractionated radioimmunotherapy in lymphoma patients. *Clin Cancer Res* **9**(10 Pt 2): 4003S–6S.

Leichner, P. K. 1994. A unified approach to photon and beta particle dosimetry. *J Nucl Med* **35**(10): 1721–9.

Liu, A., L. E. Williams, et al. 1998. Monte Carlo-assisted voxel source kernel method (MAVSK) for internal beta dosimetry. *Nucl Med Biol* **25**(4): 423–33.

Snyder, W. S. 1975. *S, absorbed dose per unit cumulated activity for selected radionuclides and organs*. New York: Society of Nuclear Medicine.

Snyder, W. S., H. L. Fisher, Jr., et al. 1969. Estimates of absorbed fractions for monoenergetic photon sources uniformly distributed in various organs of a heterogeneous phantom. *J Nucl Med* Suppl 3: 7–52.

Valentin, J. and International Commission on Radiological Protection. 2003. *Relative biological effectiveness (RBE), quality factor (Q), and radiation weighting factor (Wr)*. Oxford: Pergamon.

Wessels, B. W. and M. H. Griffith. 1986. Miniature thermoluminescent dosimeter absorbed dose measurements in tumor phantom models. *J Nucl Med* **27**(8): 1308–14.

Williams, L. E., J. Y. Wong, et al. 1989. Measurement and estimation of organ Bremsstrahlung radiation dose. *J Nucl Med* **30**(8): 1373–7.

Determination of Activity In Vivo

5

5.1 INTRODUCTION

Using the canonical Medical Internal Radiation Dose (MIRD) format, radiation dose depends on the matrix product of $\mathbf{S}$ and $\tilde{A}$. In the computation, $\tilde{A}$ is the array of integrals of the time-activity curves $A(t)$ for each source organ. Prior to performing the integration, one must first find the activity values for these sources. After more than a half century of development, it is one of the most surprising aspects of nuclear medicine practice that no standard method exists to determine activity within a living animal or patient. Multiple techniques have been implemented to assist in particular circumstances. There is not yet an optimal method to solve the general problem. This quandary may be termed *the problem of nuclear medicine*. Because of this lack of information, not only are dose estimates difficult to make, but diagnostic reporting by the clinician is also similarly limited.

Due to this lack of quantitative results, radiologists have long had the standard practice of reading out activity levels in nuclear images on a purely subjective scale. Generally this is done with respect to a nearby strong emitter such as the liver, spleen, or kidneys. In liposomal and antibody work, the hepatic uptake is often quite vivid. Finding a lesion in the abdomen, for example, will lead to a report whereby the tumor uptake is given a value of +1 if detectable at all, +2 if tumor uptake is slightly less than liver, +3 if equal to liver, and +4 if tumor is apparently greater than liver. Yet if the patient has, for whatever reason, reduced (or enhanced) hepatic accumulation, this sort of subjective scaling can distort the entire reporting process. The patient should not be used as his or her own internal standard. It would be better if absolute knowledge of the amount of activity in various tissues as well as in the suspect region were known.

Several methods are available for activity measurement. These involve both non-imaging and imaging strategies. The clinical investigator will need to know what these various techniques are and when each can be used. Of equal importance are the magnitudes of the uncertainties in the various methods. In present-day practice, it is unlikely that any single method would prove sufficient for all data inputs in a given clinical study. These methods are not mutually exclusive, so several of them may be required in a given clinical protocol (Siegel, Thomas, et al. 1999).

5.2 ACTIVITY DATA ACQUISITION VIA NONIMAGING METHODS

5.2.1 Blood Curve and Other Direct Organ Samplings

Direct sampling of the activity in the blood supply is almost always possible. Evaluation of the blood curve may be the single most important measurement taken of the patient because of two factors. One is the necessity of obtaining knowledge of the kinetics of the radiopharmaceutical (RP) activity curve in the circulation. This curve is fundamental to understanding the overall interaction between the RP and the physiology of the animal or patient. Sometimes called the clearance curve, it is discussed more completely in Chapter 6. This curve has a long history in chemotherapeutic agent development as well. Second, the blood curve, as shown later in this chapter, acts as a surrogate for the red marrow time–activity curve in many clinical protocols. Thus, it is necessary to refer to it for estimating one of the most important organ doses. In many clinical protocols, RM absorbed dose is the limiting parameter on the amount of activity that may be given to the patient.

Typically blood samples (several ml) are taken by a nurse at multiple time points during the RP study. For the sake of simplified logistics, clinical investigators generally take such samples at each imaging time point. As Chapter 6 demonstrates, it is also a useful simplification if the blood samples are taken at the same time as the images of the tissues so that all data are synchronized for input into the software. If such sampling cannot be done, however, it is also possible to input blood data at their unique time points. Unlike most other activity evaluations, blood counts must be recorded as a ratio of activity per volume of the sample. In the language of Chapter 2, $a(t)$ in units of %IA/g is being evaluated for the blood and not $A(t)$. The $a(t)$ result may be for either plasma or whole blood. In some cases, both values are measured to see what fraction of the RP is associated with red cells.

Other direct tissue samples are also possible in principle. These organ segments may be obtained in the operating room from the surgeons involved in following the patient shortly postinjection. Again, tissue values, like the blood, will be in an $a(t)$ format (i.e., in concentration units). Unlike most blood values, there will probably be a lack of synchrony with imaging results. Operating room samples are generally rare; in any case, they will be of a very small subset of the total tissue, even for tumor specimens. This follows from the needs of the pathologist, who will want to assay almost all of any organ segment or tumor taken in surgery. Thus, such surgeon-derived values may be criticized as being only part of any tissue of interest.

Excreta are a third type of possible direct assay. Typically, urine or fecal $a(t)$ values are recorded by having the patient void into a container that is sampled and counted at specific time intervals. Typically, this is done every 24 hours, although more rapid accumulation sampling may be done at early time points. Notice that total sample volume is also needed to estimate the total activity $A(t)$ in the excreta. One additional item of interest in such assays is the molecular form of activity being excreted. As noted in

Chapter 3, the radioactive label may have moved away from the pharmaceutical, or the RP may have undergone some metabolic process in the patient. In such cases, a column assay may be employed to determine the molecular weight (MW) of the excreted molecules and which of these are carrying the label.

Counting of blood and other samples is generally done using a NaI(Tl) well counter to record gamma or x-ray decay events. Calibrations are done with standard (reference) sources to provide absolute activity values. Internal absorption of the radiation due to sample volume or tube thickness can be accounted for by phantom measurements in similar conditions. It may be that the sampled radionuclide is a pure beta emitter such as ^{32}P or ^{90}Y. For these emitters, brake radiation may be used instead of the gamma emission. The origins of brake radiation are described in Chapter 4. Cerenkov radiation from the high-energy betas is another possibility for assay of such cases.

5.2.2 Probe Counting

Sodium iodide [NaI(Tl)] or solid-state probes may also be useful in measuring activity inside the patient or, in some cases, on the skin surface. The former case is very common in RP measurements when the investigator wishes to find the total-body activity versus time curve. Here, the patient is asked to stand or be seated in a fixed, repeatable geometry so the probe can view the entire anatomy. By normalizing to the initial time value just after injection of the total activity, the observer is able to plot a curve of whole-body activity versus time. This method is simple and does not require use of gamma cameras.

Several quality assurance tests are in order for whole-body measurements by probe. First, the observer must be sure that the counter can detect activity literally from head to heel of the patient. This can be tested with a point-source placed sequentially at these points of the body in the standard geometry. A second test, which is important in therapy studies, is proof that the probe will not begin to saturate at the relatively high count rates expected at the maximum activity level given to the patient. This is sometimes referred to as a dead-time test since the probe and its associated electronics have a finite response time, on the order of microseconds, to register an ionizing event. If a second particle or photon strikes during that time interval, its impact will not be recorded.

Because of such effects, if GBq of activity are given to the patient, the probe may not be equally responsive to the total activity level as it is at diagnostic levels (MBq). Again, a test prior to the therapy study is called for. If dead-time effects occur at the therapeutic level, corrections to the count rate and total number of counts can be made with a calibration curve. This evaluation should also be applied to gamma camera and positron emission tomography (PET) imaging devices. Testing at high activity levels can also be used to evaluate environmental safety. The radiation safety officer and staff may be positioned around the injection space to monitor resultant exposure rates. Such measurements should be carried out before any patient therapy is attempted.

Surface lesions, such as sarcomas and melanomas, can be counted directly with the probe detector. Here, the situation is more complicated than whole-body work since there are background counts coming from deeper in the patient as well as lateral to the sources. One may also have scatter radiation from the patient, the floor, and walls of the room. These photons can confound the measurements if they are comparable in number to those coming from the organs and lesions of interest. An energy window set to exclude scatter proves useful in these cases.

Counting is not done with the probe in contact with the lesion since this makes the geometry very sensitive to positioning. In other words, sequential counts may be inconsistent due to the difficult of replacing the probe at precisely the same location. Instead, the collimated detector is placed at 10 to 50 cm away such that only one site is in view at a given time. Calibration of the probe is performed with the same geometry and radionuclide. The method is accurate to better than 10% if the scattered radiation and other sources are corrected for.

One other effect, well known to the external beam therapy physicist, is the presence of a so-called buildup factor (BF) in sources. Instead of a simple inverse-square law, the observer sees the count rate at a distance r from the small source of activity A as

$$rate(r) = const \bullet BF(d)A/r^2 \tag{5.1}$$

where $BF(d)$ is dependent on the emitter total thickness (d). This buildup factor must be measured for various underlying tissue thicknesses at the energy of the radiolabel. BF is > 1, and monotone increases with d. Let us consider a melanoma patient with various skin lesions. In such cases, the buildup factor for the hand would be less than that for the thigh, which, in turn, would be less than that for the abdomen. The buildup effect is due to gamma (or x-) rays coming from the radioactive source, striking deeper tissue in the body of the patient and then scattering back into the detector. In effect, the probe is recording photons from both sides of the patient—although it is seeing only a small fraction of those going in the opposite direction.

It would be an improvement if there were detectors on the opposite side of the patient to record as many emissions as possible. A strategy of attempting to observe the source from more than one side has been employed in many activity measurement techniques. These include both gamma camera and positron emission tomography. Historically, the most important of these techniques is the geometric mean method with a gamma camera.

5.3 ACTIVITY DATA ACQUISITION VIA IMAGING

5.3.1 Camera Imaging to Determine Activity

In general, the gamma camera is the standard instrument to determine activity in a tissue. Probably the most common mathematical technique with the camera is the geometrical mean method developed in the early 1970s by Thomas, Maxon, et al. (1976)

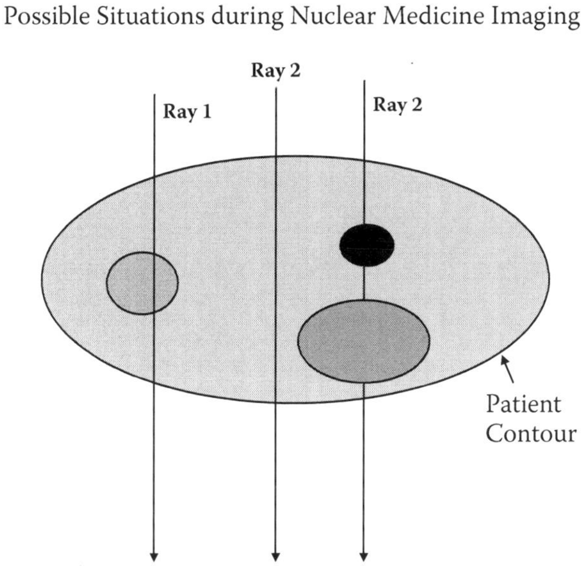

Possible Situations during Nuclear Medicine Imaging

FIGURE 5.1 Mathematical representation of several mathematical rays passing through the patient's anatomy. By viewing the patient from a single nuclear projection, it may not be clear which organs are along a particular direction. Anterior and posterior strong sources may be different tissues (cf. Ray 2) to confound the analysis. If anatomic imaging information is also present, these ambiguities may be largely removed.

and Hine and Sorenson (1967). We imagine that the same tissue is imaged from opposite sides of the patient. Typically, this is done using anterior (a) and posterior (p) views since most people offer a minimum thickness (and hence least attenuation correction) in this projection set. Likewise, significant source organs such as liver, kidneys, spleen, and heart are generally as well separated as possible in this a–p projection set. A pair of lateral views, on the other hand, would often find liver, kidneys, and spleen apparently coalesced into a single hot object. A schematic drawing of several mathematical rays is shown in Figure 5.1. These geometric lines are drawn by the observer to pass through various points on the patient's anatomy.

By integrating along a geometric ray passing through the organ of interest, the number of counts recorded in the anterior projection is

$$N_a = \Delta t \varepsilon A \exp(-\mu d_a)[1 - \exp(-\mu d_s)]/(\mu d_s) \tag{5.2}$$

where ε is the camera detection efficiency, μ is the linear attenuation coefficient in the soft tissue, d_a is the thickness of tissue over the source, and d_s is the source thickness. Notice that the overlying thickness is assumed to have no radioactivity. Counting time is given by Δt. For simplicity, we have neglected the buildup factor. The primary unknown is the activity (A) in the single tissue being imaged.

Unlike Equation 5.1, there is no need to include an inverse-square law in the camera count formula since the detector crystal views the patient via a parallel-hole collimator.

In essence, the camera is focused at infinity. In principle, one could stop with Equation 5.2 and attempt to determine A from knowledge of the distances and other parameters. In some clinical situations, this must in fact be done since no other imaging projection data are available to the observer. In other words, the object of interest can be seen in only a single projection. Thus, Equation 5.2 is the formalism for determining activity A at time t in a single view of the patient or animal.

5.3.2 Geometric Mean Imaging to Determine Activity

A simplification in the analysis occurs if we assume an equality similar to Equation 5.2 but for the posterior projection

$$N_p = \Delta t \varepsilon A \exp(-\mu d_p)[1 - \exp(-\mu d_s)/(\mu d_s) \tag{5.3}$$

By taking the geometric mean (GM) of $N_a N_p$, one may show

$$[N_a N_p]^{1/2} = \varepsilon \Delta t A \exp[-\mu(d_a + d_p + d_s)/2]\sinh(\mu d_s/2)/(\mu d_s)/2 \tag{5.4}$$

The sum of $d_a + d_p + d_s$ represents the total thickness of the patient along the mathematical ray. We designate this distance as L. The hyperbolic sine term $\sinh(\mu d_s/2)/(\mu d_s)/2$ can be represented by $HS(d_s)$.

By solving Equation 5.4 for A, the unknown, one finds

$$A = [N_a N_p]^{1/2} \exp[+\mu L/2][1/HS(d_s)][1/(\varepsilon \Delta t)] \tag{5.5}$$

The HS term is often close to unity, so Equation 5.5 essentially indicates that a single attenuation factor, $\exp[+\mu L/2]$, can be used to correct the geometric mean of the two counts and thereby determine activity in the source organ. Notice that the distance used in this correction is half of the total patient thickness (L) at the anatomic level of the source organ. At other sections of the patient, the magnitude of L will be a different value and must be remeasured. This distance can be determined by calipers, by passing another source of radiation through the patient at this level, or by using anatomic imaging such as a computed tomography (CT) or magnetic resonance imaging (MRI) scan. In the case of a transmission measurement, a relevant source is imaged with and without the patient in the beam. In that case, the result is

$$N(pt) = N(0)\exp(-\mu L) \tag{5.6}$$

where $N(0)$ is the counts obtained without the patient. By rearrangement, one can show that

$$\exp(\mu L/2) = \sqrt{N(0)/N(pt)} \tag{5.7}$$

Equation 5.7 yields the attenuation factor needed in the geometric mean Equation 5.5. Hence, a transmission exposure of the patient provides the correction for internal absorption of the radiation source. Because of the added patient absorbed dose and the

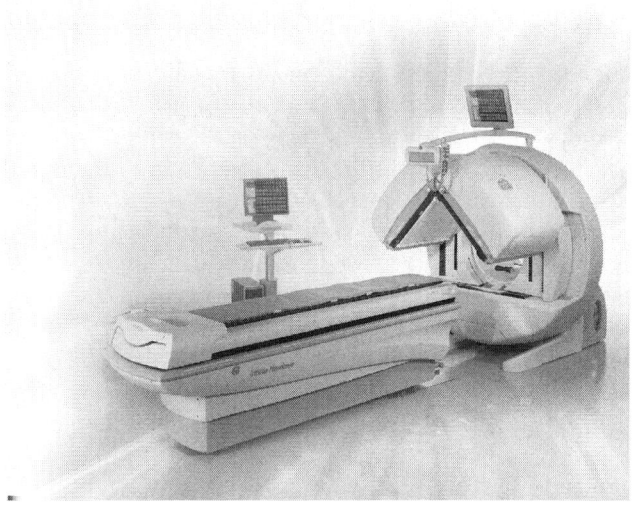

FIGURE 5.2 Dual-headed gamma camera and CT systems (Hawkeye) is shown. Typically, a standard flat (2-D) source of the radionuclide is also in the field of view to check camera sensitivity during the study. (Courtesy of General Electric. With permission.)

cost of said source, some experimenters are reluctant to use this strategy for attenuation correction. Figure 5.2 shows the arrangement of the two gamma camera heads in a 90° relative orientation.

Two (or more) overlapping organs in the field of view are limitations to an application of the geometric mean method. For example, the right kidney and liver are almost never sufficiently far apart to permit strict use of the GM method. The left kidney and spleen can be another difficult pair. In cases such as these, the investigator may elect to evaluate just part of an organ—namely, the part that is clearly separated from any other tissue sources—and to find its GM activity value. By estimating what fraction of the total organ the separated segment might be, the observer can increase the calculated activity by the inverse of that factor to correct to total organ mass. This sort of computation is most effective if a CT scan is available to determine the two volumes. Hybrid camera–CT units are ideal for this type of approach.

Most users of the GM technique allow for some background count contributions due to adjacent tissues. This is an attempt to alleviate effects due to overlying and underlying activities along geometric rays passing through the organ of interest. In this correction, the counts (N) are reduced by setting a region of interest (ROI) of a size comparable to the organ size over an adjacent segment of the patient or animal. Counts in this background region are then directly subtracted from N to give net organ counts. In single-projection work, such as the multiple gated acquisitions (MUGA) for cardiac studies, this type of correction is standard procedure.

A second limitation to the GM method is the presence of scatter radiation into the camera. These photons, as in the probe example, may come from tissue in front of and beyond the organ in either projection or may come from outside the camera field of view. An energy window set below the photoelectric peak can be used to exclude some but never all of these rays. Various methods are available including the setting of up to three energy windows (Macey, Grant, et al. 1995). Because of the finite energy resolution (±10% for ^{99m}Tc) of the NaI(Tl) crystal in the camera, some true count events are discarded through this mechanism. The method may be difficult to implement if two or more imaged photons are produced in a given radionuclide's decay such as for ^{111}In. In addition, there are manufacturer's limits to the number of energy windows available on a given camera design. Three-window techniques cannot be used on a two-window gamma camera.

A third limitation to the GM method is the classical uncertainty in the outline of any given organ of interest. This issue goes back to the origins of nuclear medicine and is associated with the epithet "unclear medicine." Nuclear images are inherently diffuse with blurred edges make outlining the tissue problematic. Notice that finding a true outline or contour is difficult even in the simplest conditions such as a single organ along the line of sight into the body. Because of the general poor spatial resolution of nuclear gamma camera images (greater than 1 cm at the surface and multiple cm at depth), any edge cannot be seen as a sharp line on the imaging device. The observer has to assign given pixels to one of two categories—inside or outside the organ—by allocating a region of interest around the probable tissue. A number of techniques have been applied to the outlining issue, and no optimal method has yet been established.

There are at least three techniques to generate an organ or tumor ROI. The most primitive is to have the reader simply define organ pixels based on experience and anatomic expectations. This method is admittedly subjective but is probably the most common in current usage. To become more objective, the reader may invoke a computer system. Among computer-driver techniques are selection of a contour at a fixed fraction of the maximum activity in the organ. Selection of the 50% contour is a typical example of this type. Of course, other contours are possible and may be chosen by other investigators. If a totally computer-driven choice is desired, the activity in the tissue may be assayed via finding solutions of the LaPlace equation

$$\nabla_{x,y}^2 A(t) = 0 \tag{5.8}$$

Here, the user is looking for the slowest-changing slope area around the organ's structure at time t.

Accuracy of the GM method may be measured using a variety of either animal experiments with sacrifice (Eary, Appelbaum, et al. 1989) or humanoid phantoms (van Rensburg, Lotter, et al. 1988). Results have generally been on the order of ±30% in such experiments (cf. Table 5.1). It must be pointed out that the error can be considerably

TABLE 5.1 Evaluations of the Accuracy of the Geometric Mean (GM) Method

EXPERIMENT	DETAILS	RESULT	REFERENCE
Animal (dogs) assay with ^{131}I	Dogs sacrificed at various time points	± 30 % for liver and other major organs	Eary et al. (1989)
Phantom with ^{111}In	Pseudo-organs imaged and compared with activity known	± 35% for major tissues	Van Rensburg et al. (1988)
Phantom with ^{111}In	Anthropomorphic phantom imaged	± 30% for liver, spleen	Liu et al. (1996)

larger in some circumstances. One of these difficult situations is the possibility of mistaking two separate nuclear medicine sources for two projections of a single source. In tumor imaging, this situation often is an issue since the lesion may be observable only from one side of the patient. In colorectal cancer, the hot region in the anterior view may indeed be that of the tumor, whereas the corresponding posterior area of increased activity is simply a view of the spleen. Using one or more gamma camera detectors without recourse to anatomic imaging information makes such misinterpretations difficult to prevent. Access to anatomic imaging, as an adjunct to camera information, is becoming an important additional aspect in quantitative activity measurements. Several methods, in fact, explicitly depend on using this type of geometric data to simplify the activity evaluations. One of these is the CT-assisted matrix inversion (CAMI) method (Liu, Williams, et al. 1996).

5.3.3 CAMI Imaging to Determine $A(t)$

It is interesting to consider the more general physical situation that led to Equation 5.2 and attempt to couple nuclear and anatomic imaging. Imagine a geometric ray passing through the patient and intersecting various radioactive organs of interest. Here, we explicitly drop the assumption that only a single organ along this ray contains activity and allow a spatial sequence of such tissues. In effect, we have a linear array of different tissue activity densities starting at the front (or back) or the patient and extending to the opposite side. In this strategy, the fundamental nuclear image set is either one of the pair of whole-body images used to search for metastatic sites. Obtaining this pair is standard procedure in scanning a cancer patient so that no additional nuclear imaging is required.

Explicitly, we set the counts recorded by the camera along this ray to be equal to an integral over these sources in sequence:

$$N/\Delta t = \varepsilon \int_{0}^{L} a(y)BF(y)\exp(-\mu y)dy \tag{5.9}$$

Here, we generalize our notation so that the unknown activity's linear density (Bq/cm) is symbolized by $a(y)$. Thus, the product $a(y)dy$ is a differential amount of activity (dA) at a distance y from the surface of the patient. Notice that a backscatter factor (BF) is explicitly contained in the calculation. Yet Equation 5.9 may be repeated for many possible rays going through the anatomy. If we assume that BF can be separately measured at the relevant photon energy for different thicknesses (y), the set $a(y)$ remains the only unknown in the equality. As such, Equation 5.9 is not much of an improvement on Equation 5.2; in fact, it explicitly shows the limitations of the GM method.

If, as an additional step, we now introduce the corresponding CT or other anatomic data set so that the sequence of tissues along the ray may be defined as to location (y) and type, then Equation 5.9 is soluble. The constant activity densities factor out of the integral. Each such linear activity density represents one most probable value for each tissue or organ in the patient. In most RP studies there are no more than four or five such tissues.

Yet the number of geometric rays may be made quite large—on the order of hundreds—so that the mathematical problem is greatly overdetermined. For example, while passing a ray through the abdomen, one would have a_M, a_L, and a_K, respectively, for activity densities in muscle, liver, and kidney. Each of these parts of $a(y)$ would have known geometric starting and stopping values (from CT or MRI) where their particular tissues begin and end. Solutions for a_i are obtained using matrix inversion to solve Equation 5.9 for these unknowns. This method is defined as CAMI (Liu et al. 1996). When invented, there were no hybrid scanners, so the images from the two imaging modalities (gamma camera and CT scanner) had to be mathematically fused initially. With hybrid imagers such as single-photon emission computer tomography (SPECT)–CT, this first step in the process is eliminated to make the CAMI method more easily implemented.

Because of the matrix inversion, the final result with CAMI is a set of linear tissue-specific activity densities. These values are not the ultimately desired unknowns, which are the total activities in the source tissues. To obtain $A_i(t)$ for tissue i, one integrates over all rays that pass through that organ to find its total volume (V_i) and then multiplies that result from the best-fit value for $a_i(t)$:

$$a_i(t)V_i = a_i(t) \sum_{j=1}^{j=N} \ell_{i,j} = A_i(t) \tag{5.10}$$

where $\ell_{i,j}$ is the length of the j-th ray within the i-th organ. These length terms are implicitly of the same cross-sectional area so that the sum over lengths is also a sum over volumes. This cross-sectional area may be as small as 1 pixel or large enough so that statistical certainty may be enhanced. Notice that the term in the summation sign is simply the anatomic volume of interest for the i-th organ. This volume has been obtained by explicitly using CT or MRI information, which, in turn, defines the size and location of the organ or interest. While CT images have been emphasized for this quantitation, MRI results are also possible. The CAMI method is optimal if the two images (nuclear and anatomic) are locked into spatial register by the image acquisition software so that fusion postscanning is not required.

As noted already, finding the true volume of the source organ has always been a concern in purely nuclear imaging. Although a number of strategies to generate an ROI are known, it is better if anatomic data are used for volume measurement in the first place. Thus, CAMI avoids the ROI issue by simply using the CT or MRI anatomic data to define a given tissue's location and size. As Chapter 4 demonstrated, this size, or more particularly the associated mass, is also vital to the estimation of radiation dose. Keep in mind that the medical oncologist will be using this size, as a function of time, to track the effects of various therapies being done on the patient. We should notice that, in general, sizes of nuclear images, except for perhaps the fluorodeoxy glucose (FDG) result, do not have very much impact in the patient's progress notes. Medical records instead cite only the positive (or perhaps negative) outcome of the nuclear scan and the radiologist's uptake value relative to an adjacent major organ (e.g., a reading of +1).

The CAMI method also avoids the requirement of having a pair (or group) of opposed images or any series of projections. The algorithm may be implemented using a single projection. It is important to remember that the cancer patient will, by definition, have whole-body nuclear images taken throughout the clinical imaging protocol to determine extent of disease. One of these images at each time point would suffice for implementing the CAMI method if CT (or MRI) data were also available simultaneously. Of course, a user can also try to improve the statistical certainty of the result by adding other projections and keeping the same limited set of unknowns. If both anterior and posterior whole-body images are available, they may be included in the activity calculation.

Accuracy of CAMI was investigated using a humanoid phantom with several radioactive ^{111}In sources. Errors are seen to vary between 5 and 15%, with an average value near 7% (Liu et al. 1996). The result is generally better than the GM approach and avoids the possible ambiguity issues of two objects appearing as one in the GM. Notice that the CAMI technique explicitly avoids the problem of superimposed sources by simply considering them in the first place using the anatomic information available. CAMI calculations are generally done assuming that all rays passing through a given organ pass through the same activity density—that is, that a_i is a constant in a given tissue. This assumption may be dropped if there is evidence that heterogeneity occurs as, for example, in the cortex, medulla, and pelvis of the kidney. Another context of differential amounts of activity is the variegated accumulation of the RP at a tumor sites. The topic of differential uptake in lesions is described in more detail in Chapter 6.

The CAMI method, since it relies on a single projection, is unlikely to be the best possible method to estimate $A(t)$ for a given organ or tissue. It seems reasonable that multiple projections would provide better estimates since there will be some angles at which the separation of sources is more optimal than in the two standard planar views (anterior and posterior whole-body images). Again, however, there would be a requisite geometric set of data so that the reconstruction of the unknown activities could be done most efficiently. That optimal method is therefore presumably based on 3-D nuclear imaging. Two such techniques are available: SPECT and PET.

5.3.4 Quantitative SPECT Imaging to Determine *A(t)*

In SPECT, the pair of gamma camera heads is usually rotated around the patient in a 360^0 arc to gather a number (e.g., 60) single projections. Rotations may be discrete ("step-and-shoot") or continuous. These images are then processed to produce a set of standard cross-sections of the patient's radioactivity distribution. However, all resultant sections given to the radiologist—coronal, sagittal, and transverse—have arbitrary ranges of the image parameter. No absolute SPECT values are available. A person reading a standard SPECT tomographic section is placed in the same position as the reader of a planar gamma camera image as previously described. One can describe a voxel value only by reference to some other voxel in a known organ in the same section (e.g., the liver). This result is also similar to an MRI image wherein any recorded value is defined simply as a local "signal." This nonquantitative limitation has long been an obstruction to practical determination of activity at depth in the patient.

To make SPECT a quantitative imaging method, four separate issues must be considered in the reconstruction algorithms. Some researchers assert that the four must be done in a certain sequence. Three of these four were alluded to in the earlier description of camera methods for determining activity at depth: patient attenuation, patient scatter, and collimator response. One additional factor of explicit interest in quantitative SPECT (QSPECT) is an activity correction for small sources. In this last item, structures are considered small if they have dimensions about twice the resolution of the gamma camera or less. Effects of reduced source count loss are, in fact, even observable at emitter volume dimensions on the order of several times the spatial resolution. Generally, these last corrections are termed count recovery coefficients (CRCs). Essentially, the poor spatial resolution of the gamma camera directly leads to a blurring out of the "island" source into the "sea" of background activity. We will see that this aspect of QSPECT is similar to results in PET imaging wherein determination of the CRC is also a key issue for small lesions being evaluated for activity. If the physicist is interested in making good estimates of tumor-absorbed doses, such corrections, which can approach factors of two, are important.

Koral, Dewaraja, et al. (2003) and Dewaraja, Wilderman, et al. (2005) documented their progress with QSPECT at the University of Michigan over the past 15 years. One of their important discoveries was a use of a power-law relationship between the spherical source's size and the recovery coefficient value. Using a depth-dependent detector modeling term, nonuniform correction for attenuation and a three-window scatter correction, the authors were able to find errors of < 12% for ^{131}I in a Monte Carlo simulation of the clinical situation. In this case, the Zubal phantom (Zubal, Harrell, et al. 1994) was used to represent the hypothetical patient for a set of 60 projections. The ordered-subset expectation maximization (OSEM) reconstruction method was followed (Hudson and Larkin 1994).

Use of the three-window technique (Macey et al. 1995) was required due to the high-energy photon (637 keV) produced in the decay of ^{131}I. One window is set above the 360 keV peak of interest to subtract the gamma rays scattered "down" into this energy bin. The third energy window is then set below the 360 keV bin to subtract out the photons that were originally at 360 keV but that were Compton-scattered into a lower-energy form.

In a comparable experiment based on imaging a physical phantom, errors on the order of 10% or less were reported by Macey and the DeNardos (DeNardo, Macey, et al. 1989). This particular QSPECT analysis was based on the Alderson phantom loaded with ^{99m}Tc. In this case, attenuation correction was done postreconstruction. A linear-attenuation coefficient (μ) value was derived as a function of depth into the physical phantom. The authors commented that there was some difficulty in implementing the attenuation correction due to lack of a sharp boundary of the nuclear image. This problem was described earlier as endemic to nuclear imaging whether of the camera or PET type. No anatomic data were included in the University of California–Davis analysis as it was entirely done using nuclear camera images.

Gamma camera imaging is the dominant method for ionizing radiation quantitation in the nuclear medicine clinic. It is not the only modality, and a second choice for following the animal or patient is possible with positron emission tomography. This technique has a long history of being quantitative because of the simplicity of dual-photon detection process. In fact, a specific parameter has been invented to help with the evaluation of the amount of activity in a given voxel of a PET tomographic image.

5.3.5 PET Image Quantitation and the SUV Value

By analogy with the earlier camera work, PET imaging personnel have developed several indicators of uptake in lesions and normal tissues. The most common of these parameters is the Standard Uptake Values (SUV). Generally, these indicators are modeled on the uptake concept described in Chapters 1 and 2 so that corrections for both injected activity and whole body mass are included in the definitions. Explicitly, one sets:

$$SUV(t,\text{conventional},\%\text{ID/ml}) = \frac{N(\text{voxel})\varepsilon}{A_0\exp(-\lambda t)M}$$

5.11

where N is the number of counts recorded per unit volume per unit time in a voxel. In the definition, ε is the PET scanner efficiency in MBq/count/s, $A_0\exp(-\lambda t)$ is the initial total activity injected decay-corrected to the time of scanning and M is the total body mass. This last term is sometimes modified due to the tendency of modern patients to be overweight. In that case, the body mass is replaced by the lean body mass determined by finding the fraction of M that is not attributable to adipose tissues. Other denominators may also be used such as the total body surface area of the patient. This last datum cannot readily be measured but is generated by use of a standard formula based on patient total body mass and height. Use of these latter SUV values is less common in the literature. Finally, we should mention that [18]FDG is the present-day standard for SUV measurement of tumor viability and, to some degree, size.

5.3.6 Diagnostic Use of the Standard Uptake Value Parameter

Standard uptake values (SUVs) using [18]FDG are often used by the radiologist to determine the presence and viability of tumor at sites within the body. If the SUV is found to be above a lower cutoff, the radiologist will call the site as positive for malignancy. Likewise, the effectiveness of any given therapy may be followed by tabulating the variation of SUV over time and treatment modality. This functional method is preferred to following the variation in the physical size of the lesion using anatomic imaging. A CT or MRI evaluation, for example, may be demonstrating only the sum of both active lesion and necrotic residual tissue volumes. This total size may not, in fact, show much variation with therapy.

Confusion may arise due to what is the best SUV for diagnosis: that is, does one find an average value over the entire mass or simply the maximum value inside the lesion? Given a technique, the oncologist may then follow the therapy of a patient by observing the change of the SUV over time. This last process is very important and depends on consistency in the methodology in determining SUV over the months or even years of the treatment process.

Positron tomography is seen to be more directly quantitative than voxel activity values obtainable via CAMI or QSPECT. The appeal of the positron technique is that the PET user does not need to provide corrections to the data sets and that the resultant images are, if calibrated using standard sources, in units of activity per ml. To keep the system consistent, every PET facility is required to provide quarterly calibration checks in both the 2-D and 3-D (no collimator) modes. Here, a known ^{18}F or other relevant positron source is first assayed in a dose calibrator and then imaged in the detector system. Records, including trend lines, are kept of these results. A careful use of time recording and clock synchronization is needed during this calibration procedure as well as during the patient study since the SUV definition requires the activity in the denomination of Equation 5.11 to be that at the midpoint of the time interval of patient measurement. If the half-life of the emitter is short, correction for the radioactive decay can be difficult.

Even though SUVs are, by far, the standard quantitative output of a PET system, nuclear radiologists have discovered a number of practical caveats (Wahl, Jacene, et al. 2009). Several of these have already been mentioned herein, including use of lean body mass or body surface area in lieu of total body mass in the denominator of the definition. Other variables include the time of patient measurement of the standard uptake value. In this regard, there is not a lot of latitude using ^{18}FDG since the half-life of ^{18}F is only 110 m or slightly less than 2 hours. Usually the imaging is done approximately 1 hour after injection of the ^{18}FDG molecule. Some investigators indicate that they use later times—up to 2 hours postintravenous (post-IV) injection. Clearly, whatever the value, the time delay must be kept constant so that the results are reproducible.

Another major issue with use of the SUV is the method to determine the region of interest in the patient. As described previously for gamma camera projections or tomographic images, multiple methods can also be applied to outlining a possible lesion in the PET system. Among the major strategies are subjective outline, the percentage of the maximum method, and use of a computer-driven algorithm. For consistency of the clinical assay, the chosen method must be applied coherently by one or more observers. Yet, even given any of these, radiologists still face a further challenge: where and how, in the selected ROI, do they generate the SUV? Some clinical readers prefer the maximum value within the region. Finding the largest voxel signal value is direct and simple to generate from the digital image. It may be, of course, that this maximum refers to one or a few voxels in size and thus becomes subject to criticism. Statistical issues arise as to the significance of a very strong signal from a limited segment of the tissue of interest. At the other extreme, the entire ROI can be used to find a local spatial average of the SUV. Such mean results naturally dilute the "hotter" voxels and may lead to an underestimate of the severity of the disease. After all, if we select a large enough ROI, almost any tumor would be lost in the background counts of blood and other normal tissues.

An ongoing complication with SUV determination is the blood glucose level of the patient at the time of PET scanning. This situation arises since FDG is a labeled sugar and competition occurs at the tumor sites between this particular RP and unlabeled glucose in the blood. While a diabetic patient may provide wide variation in the blood sugar value and is an obvious example, there is the more general problem of controlling any given patient for glucose level during a sequence of PET scanning procedures

TABLE 5.2 Limits on the use of the [18]FDG SUV for Detecting and Following the Course of Therapy of Malignancies

FACTOR INFLUENCING THE SUV	POSSIBLE EFFECT	RESULTANT UNCERTAINTY IN THE SUV VALUE IF UNCONTROLLED
Blood glucose level	As glucose level increases, the apparent SUV decreases due to competition between sugar molecules.	±15%
ROI method	True lesion size is subjective; must be standardized in the clinical method.	±55% is reported
Body mass/area	Volume of FDG distribution in a patient	±20% due to fat vs muscle
Timing of measurement	Synchronization of hot lab versus imaging area clock	±10%
Method of reconstruction	Change of SUV due to number of iterations; change due to method of reconstruction	±100% is possible

Source: Boellaard R., *J. Nucl. Med.* 50(Supp 1), Table 1, 2009. With permission of the Society of Nuclear Medicine.

during therapy. Generally, the patient is fasting for some hours (4h is typical) before the imaging, but that duration without food is not a guarantee that the same glucose level will occur at the time of scanning. Although we cannot reduce the blood sugar level arbitrarily, the clinical staff must attempt to control for it by measuring and maintaining the level from one imaging time to the next in a given patient. Likewise, for the consistency of the image assay between patients, the blood level must be controlled across all individuals coming into the nuclear imaging facility for PET FDG studies. Table 5.2 gives a summary of the dominant issues in determining the SUV value for [18]FDG and the uncertainties each may cause in the eventual calculation (Boellaard 2009).

While [18]FDG is almost a unique agent in present-day PET imaging of tumor sites, this domination may not hold indefinitely. As we noted in Chapters 1 and 2, multiple types of targeting mechanism may be applied in the future. Liposomes, antibodies, engineered small proteins, segments of RNA and DNA, and the various nanostructures may all play a role in the future of cancer detection. Yet there is still the issue of a lack of appropriate radioactive labels, as described earlier, for PET imaging. There are, as listed in Chapter 1, relatively few positron-emitting radionuclides available for labeling to the multitude of future agents. Most of these potential labels are not very long-lived, so RP materials that target relatively slowly to the target molecule cannot be followed over extended times. As we noted with the imaging figure of merit (IFOM) discussion of Chapter 3, very short-lived labels may not provide much information on the targeting process since they simply die away before the agent gets to the structure, function, or molecule of interest in the animal or patient.

5.3.7 Other PET Radionuclides and Image Quantitation

While ^{18}F may be a very popular radionuclide at present, several other positron emitters can be used as potential radiolabels for the variety of novel agents being developed. Among the most important are ^{86}Y and ^{124}I. In the former case, this isotope of yttrium (12% positron decay) can be used as a label to measure the amount of ^{90}Y-labeled compound that is present in an animal or patient. Recall that ^{90}Y is a pure beta emitter, so only its brake radiation passes outside the body of the patient. The decay scheme of ^{86}Y is more complex than ^{18}F and contains a number of single-photon emissions. There are photons seen at 627 keV (33%) and 1.08 MeV (83%), among other energies. An additional limitation is that the half-life of ^{86}Y is only 14.7 h, so long-term imaging with this radiolabel is not possible.

The case for ^{124}I (23% positron decay) as a potentially important label can be made even more strongly than that for ^{86}Y. As noted in Chapter 3, there is always a place for an iodine isotope in the tagging of novel agents. This follows from the single-step technique of attaching the iodine to a tyrosine molecule in the protein or other structure of interest. No chelator is required, and the labeling can be done with the necessarily limited amounts of research material available from the engineering group. This last fact is very important in testing novel agents and their variants. Radiopharmacists will prefer to use iodination to get biodistribution data as soon as possible after production of a new agent. As in the case of ^{86}Y, the decay spectrum of ^{124}I is relatively complicated— at least compared with ^{18}F. For example, there are single-photon emissions at 603 keV (63%) and 1.69 MeV (11%). The half-life of ^{124}I is 100 h, so this label may be used in tagging longer-circulating agents in the blood of the animal or patient.

5.4 BONE MARROW $A(t)$ VALUES

Before leaving the topic of activity quantitation in vivo, we must consider one of the most perplexing problems of therapy using internal emitters: what is the activity in the red marrow? In this case, imaging the source organ is a difficult task due to the existence of blood-forming tissues in a variety of locations including the mandible, humerus, femur, and spine. As the patient is treated with chemotherapy or radiation, or simply ages, the amount of viable marrow may decrease in these various regions in a differential way. There is also a possibility that tumors may present within the marrow space in certain conditions such as lymphoma.

One of the most difficult tasks for determination of radiation dose is the amount of energy given to the patient's red marrow. Marrow is the most sensitive tissue with regard to absorbed dose as it differs by about one order of magnitude from other normal organ values for T5/5 or T50/5 in the external beam literature (Emami, Lyman, et al. 1991). These dose values refer to toxicity to 5% of the patients (T5/5) or 50% of the patients (T50/5) in 5 years, respectively. While other tissues have values of about 20 Gy or higher, red marrow is considered to have a sensitive response to ionizing radiation at values as low as 1.5 to 2.0 Gy. This sensitivity implies that red marrow doses must be calculated with some urgency and that appropriate care in the treatment planning of patients for internal emitter therapy is necessary. These calculations are done with a diagnostic level of radioactivity in the treatment planning phase of the study.

In the early 1990s, a fundamental algorithm was developed by the American Association of Physicists in Medicine (AAPM) assuming that blood circulation through the marrow acted as an effective source organ. Siegel, Wessels, et al. (1990) then used the $\tilde{A}$ of the blood as a surrogate for the $\tilde{A}$ of the red marrow with an efficiency factor (f) of approximately 0.2 to 0.4. Explicitly, the task group set

$$AUC(RM) = f \bullet AUC(blood) \frac{1500}{5000} \tag{5.12}$$

where the *ad hoc* factor (f) is indicated, and 5,000 and 1,500 refer, respectively, to the mass (in grams) of whole blood and red marrow in a nominal adult patient. The equality, if one transposes the RM mass value to the left-hand side (LHS), states that the concentration of activity (u) in the RM is proportional to that in the whole blood. Sgouros (1993) was later able to show that the theoretical best value for this constant was approximately 0.3. The latter value is typically used in day-to-day RM absorbed dose estimates. Several questions may be raised about these various assumptions.

In the previous analysis, the presence of activity in the marrow has been explicitly neglected. Finding activity that is truly in the marrow is not an easy task. Generally, areas in the lumbar vertebrae are used as sources assignable to the RM itself (Sgouros, Stabin, et al. 2000). Many observers decline to include any red marrow targeting since it may be difficult to show in an image. If demonstrable, such sources would have to be added to that of the blood so that at least two $\tilde{A}$ components would then be included in the MIRD-type calculation for the patient's absorbed dose estimate. Thus, the estimate would involve both blood and red marrow as source organ contributors to the RM $\tilde{A}$ value.

Both organ mass values assumed in Equation 5.12 may be questioned for an individual patient. Probably most important is the uncertainty in the 1,500 gram value assumed for the RM. At present, the actual red marrow mass of a given person is difficult to measure and is one of the fundamental unknowns in estimating red marrow absorbed dose. It is well documented that as patients age or undergo chemotherapy, the red marrow mass will decrease correspondingly. Such decrements are generally irreversible. Besides these variations, females usually have lower values than males of the same age. Because of the uncertainty in the mass, there is a corresponding uncertainty in the estimate of RM absorbed dose. This ambiguity is probably one of the primary causes of the apparent disparity between estimated RM absorbed dose and clinical effects observed in the blood, as Chapters 8 and 9 will show.

The mass of the whole blood is a second uncertainty in Equation 5.12. This parameter can be estimated by taking blood samples near the time of injection ($t = 0$). The plasma volume may then be calculated if the RP is restricted to the plasma space of the patient. Given a corresponding hematocrit, and this volume, a total blood volume may be estimated for the patient. If the agent moves outside the plasma, however, standard clinical formulas can be used to predict total blood volume; these involve the patient weight and height.

5.5 COMBINATIONS OF METHODS FOR PRACTICAL ACTIVITY MEASUREMENTS

Having seen problematic issues in RM activity measurement, it is important to generalize to the situation in other tissues. It is unlikely that any single-activity quantitation method will suffice for the dose estimation process within a clinical protocol involving a novel RP. Blood curve data must be taken as they are generally essential due to the issues of pharmaceutical handling by the body and eventual red marrow dose estimates. Chromatographic column assays of the blood samples may be made as well. These chemical–physical analyses permit the observer to see if the label, for example, is coming off the RP or if the RP itself is labile due to enzyme or other action within the patient. One looks at the various masses and associated labels (if any) in the fractions coming off the column. It may be observed in an intact antibody study, for example, that there is a radioactive massive protein (MW = 300 kDa), implying that a radiolabeled antibody has combined with its circulating antigen to form an antibody–antigen complex in the blood. If substantial numbers of such complexes are found, the targeting to the tumor may have been interrupted by too much circulating antigen.

Similar probe and column analyses may be made of the concentration of the urine and feces of the patient. If the total volume of both outputs is measured, the analyst may determine the total activity excreted via these routes. Likewise, the column analyses can be invoked to provide information on the chemical makeup of the materials coming through the kidneys and gut.

Probe counting of the whole body of the patient is also quite commonly used to determine whole-body clearance of the novel agent. Here, the scientist must keep the geometry fixed to allow repeat measurements at multiple time points. It is much easier to find the total activity in the body via the probe than by integration over the various camera images—even the whole-body images—since these may omit some parts of the anatomy of the patient. Extremities and head are the two most common omissions.

Use of the GM image pair is the present standard for the requisite sequence of four to seven clinical images taken during a typical RP study. Whole-body images may be used as well as localized results (vignettes) of the body zones that seem of greatest interest. Geometric mean results are then computed for each imaging time point, and the results are tabulated. As Koral and associates (2003) showed, it is best if a QSPECT image is used to more definitively quantitate the GM results at some important time during the high accumulation phase of the study. In intact antibodies, this time is approximately 48 h postinjection of the RP.

It is difficult to justify a multiple sequence of QSPECT images during an RP protocol since the time required may become excessive. The whole-body image pair, which typically takes between 15 and 30 minutes to acquire, is needed by the reading radiologist at each time point for a complete report of the cancer patient. It is unlikely that the average patient would be willing to lie in place for an additional 20 minutes at each time point to permit nuclear tomography. A single QSPECT result, however, may be necessary to improve the absolute accuracy of the GM mean result as described already. This has been the technique advocated by the Michigan group (Koral et al. 2003).

Following a patient for a sequence of extended imaging periods during a PET–CT study is likewise a problematic strategy. The primary limitation here centers around the lack of a satisfactory label for the RP. As noted in Chapter 1, the number of useful positron emitters is strictly limited—particularly if the biological lifetime of the pharmaceutical in the blood and other tissues is much beyond 10 hours. To put this in another way, the IFOM (Chapter 3) of the labeled agent may be so poor as to preclude imaging for the most interesting and important parts of the study. Table 5.3 gives a summary of

TABLE 5.3 Six Methods Used to Quantitate Activity In Vivo

METHOD	ADVANTAGES	DISADVANTAGES
Purely Nuclear Methods		
Single probe external to the body of the patient	Simple to use; good for whole-body results	No image. Sensitive to exact position. Only used for surface structure or whole body.
Well counting of blood or other tissue samples	Gives exact results for samples taken from the patient	One-time snapshot. Sample may be nonrepresentative of the patient's organ of interest. Ethical, surgical questions.
Geometric mean imaging with gamma camera	Accurate for organ activity if the total patient thickness is known; transmission study may be used	Overlying and underlying tissues may conflict with the assay. There may not be a suitable image pair. Outline of organs is difficult.
Hybrid: Nuclear/Anatomic Methods		
CAMI technique	Superimposed organs can be assayed. Organ size and position must be available via the CT (or MRI) imaging.	Requires CT (or MRI) anatomic data to invert matrix. Sensitive to scatter, collimator design, and resolution.
QSPECT done using SPECT–CT imager	Gives accurate activities for cross sections of organs. Not sensitive to superimposed organs. Organ size and position are from anatomic imager.	Requires consideration of four issues relating to attenuation, scatter, collimator sensitivity, and count recovery for small lesions at sizes comparable to spatial resolution.
PET–CT imager	Similar to QSPECT. Attenuation correction is much easier due to back-to-back 511 keV photons. Method is much more sensitive than gamma camera imaging.	Similar to QSPECT. Presently, issues of SUV measurement for [18]FDG. Not enough suitable radionuclides.

the various standard methods to assay activity in vivo along with the advantages and disadvantages of each technique.

5.6 SUMMARY

Determination of the amount of activity inside a living animal or patient has been and continues to be an issue in the development of nuclear medicine. It can be considered the fundamental problem of the clinical field. Typically, the various temporal measurements are accomplished by a number of different methods. At least six strategies are available. Techniques such as single probes for whole-body imaging and well counters for tissue samples provide the ideal situation since the uncertainties of activity estimation are minimal and probably given by Poisson statistics. Yet such samples, aside from blood, are relatively rare and, in any event, are restricted in scope. Surgical samples are particularly prone to being of very limited value. The most common technique of activity quantitation with the gamma camera is the geometric mean method. Here, the uncertainty is on the order of ±30% for animal and phantom studies. Overlap of organs in the image field is probably the single greatest limitation of the GM method. Additional difficulties with the technique arise when the anterior and posterior hot source images are not of the same source. All three of these methods are purely nuclear in character and do not require complementary anatomic imaging to provide patient geometry.

Activity estimates made with a synthesis of both nuclear and anatomic (CT or MRI) data are relatively recent developments and account for the other three methods of activity estimation listed here. Implementation of CAMI is the simplest in that only planar nuclear images are required along with the CT information. Uncertainties with CAMI are ±10% for stacked organs (collinear along a mathematical ray) in an anthropomorphic phantom. Here, fusion is required between the nuclear 2-D projections and the scout view of the anatomic scanner. If one goes to complete 3-D nuclear data sets, associated activity uncertainties may be as small as ±5% for QSPECT or PET imaging. Thus, the use of hybrid imaging systems (QSPECT–CT and PET–CT) is recommended so that image fusion may be performed with minimal errors in registration of the nuclear and anatomic formats. In the case of PET–CT, multiple factors enter into the seemingly simple technology to provide consistent SUV values. As noted, some of these contributions, if not controlled, may induce errors of 50% or more. Consistent reconstruction methods and other technical variables are required to achieve high accuracy in this modality.

Even with hybrid imagers, several issues remain. One is the ambiguity of the lesion size—hence total activity—due to the poorly resolved edge in any nuclear image. Multiple methods occur for outlining such structures, and no best method is at hand. The SUV situation for PET imaging is a good example of this difficulty of implementing quantitative measurements in the nuclear medicine imaging center. Also, there is the associated matter of count recovery coefficient corrections for small sources—that is, those whose size is comparable to or smaller than approximately twice the nuclear system imaging resolution. These corrections may be on the order of 40 to 50% for lesions 10 ml or less in volume. Since this true size may not be determinable by the nuclear investigator, the actual size is presumably best given by the anatomic data set.

In practical terms, quantitative biodistribution studies generally require several techniques to be used simultaneously. Blood, some tissue samples, and excreta are best followed using probe technology. Because of the need of the whole-body images for the cancer patient, a sequence of geometric mean images is needed from $t = 0$ to the end of the study. As a final complement to these traditional studies, QSPECT (or PET) should be added at some time points to normalized the GM results. Hybrid imagers having CT (or MRI) capability along with the gamma or PET imagers provide the best way to establish the absolute amount of activity at depth in the patient. With all of these analytic hardware and software pieces in place, one anticipates an uncertainty approaching ±5% in activity measurements in vivo.

REFERENCES

Boellaard, R. 2009. Standards for PET image acquisition and quantitative data analysis. *J Nucl Med* **50**(Suppl 1): 11S–20S.

DeNardo, G. L., D. J. Macey, et al. 1989. Quantitative SPECT of uptake of monoclonal antibodies. *Semin Nucl Med* **19**(1): 22–32.

Dewaraja, Y. K., S. J. Wilderman, et al. 2005. Accurate dosimetry in 131I radionuclide therapy using patient-specific, 3-dimensional methods for SPECT reconstruction and absorbed dose calculation. *J Nucl Med* **46**(5): 840–9.

Eary, J. F., F. L. Appelbaum, et al. 1989. Preliminary validation of the opposing view method for quantitative gamma camera imaging. *Med Phys* **16**(3): 382–7.

Emami, B., J. Lyman, et al. 1991. Tolerance of normal tissue to therapeutic irradiation. *Int J Radiat Oncol Biol Phys* **21**(1): 109–22.

Hine, G. J. and J. A. Sorenson. 1967. *Instrumentation in nuclear medicine*. New York: Academic Press.

Hudson, H. M. and R. S. Larkin. 1994. Accelerated image reconstruction using ordered subsets of projection data. *IEEE Trans Med Imaging* **13**(4): 601–9.

Koral, K. F., Y. Dewaraja, et al. 2003. Update on hybrid conjugate-view SPECT tumor dosimetry and response in 131I-tositumomab therapy of previously untreated lymphoma patients. *J Nucl Med* **44**(3): 457–64.

Liu, A., L. E. Williams, et al. 1996. A CT assisted method for absolute quantitation of internal radioactivity. *Med Phys* **23**(11): 1919–28.

Macey, D. J., E. J. Grant, et al. 1995. Improved conjugate view quantitation of I-131 by subtraction of scatter and septal penetration events with a triple energy window method. *Med Phys* **22**(10): 1637–43.

Sgouros, G. 1993. Bone marrow dosimetry for radioimmunotherapy: theoretical considerations. *J Nucl Med* **34**(4): 689–94.

Sgouros, G., M. Stabin, et al. 2000. Red marrow dosimetry for radiolabeled antibodies that bind to marrow, bone, or blood components. *Med Phys* **27**(9): 2150–64.

Siegel, J. A., B. A. Wessels, et al. 1990. Bone marrow dosimetry and toxicity for radioimmunotherapy. *Antibody Immunoconj Radiopharm* **3**: 213–233.

Siegel, J. A., S. R. Thomas, et al. 1999. MIRD pamphlet no. 16: Techniques for quantitative radio-pharmaceutical biodistribution data acquisition and analysis for use in human radiation dose estimates. *J Nucl Med* **40**(2): 37S–61S.

Thomas, S. R., H. R. Maxon, et al. 1976. In vivo quantitation of lesion radioactivity using external counting methods. *Med Phys* **3**(4): 253–5.

Van Rensburg, A. J., M. G. Lotter, et al. 1988. An evaluation of four methods of 111In planar image quantification. *Med Phys* **15**(6): 853–61.

Wahl, R. L., H. Jacene, et al. 2009. From RECIST to PERCIST: evolving considerations for PET response criteria in solid tumors. *J Nucl Med* **50**(Suppl 1): 122S–50S.

Zubal, I. G., C. R. Harrell, et al. 1994. Computerized three-dimensional segmented human anatomy. *Med Phys* 21(2): 299–302.

Modeling and Temporal Integration

6

6.1 REASONS FOR MODELING

As seen in Chapter 4, estimation of absorbed dose requires integration of each organ's time–activity curve out to infinity (or at least to many times the physical half-life of the radiolabel). Prior to such integration, a mathematical method is required to allow interpolation and extrapolation of necessarily sparse animal and human data sets such as $a(t)$, $u(t)$, or $A(t)$. A number of different types of mathematical pictures exist, and several software packages are available from universities and other resources. This chapter describes various model types and several of these resources. Examples of some of the model styles are included. It is important to mention that any model, no matter how well justified conceptually, can be criticized for being limited or overly specific. Thus, the particular choice may be difficult to make.

One important feature of every mathematical representation is that the analyst must carefully choose a range of interest—that is, precisely how much of the real world the model is proposed to explain. When Johannes Kepler developed his elliptic model of planetary orbits, for example, he kept his analysis restricted to Tycho Brahe's well-documented motion of Mars. Mercury has greater ellipticity, but its observational data were sparse due to the difficulty in observing a near-solar object. Generally, models are probably best known for their limits rather than their extent. The modeler selects a certain set of data or even a particular subset before proceeding to the analysis. Judicious choices at these initial entry points are extremely important for the ultimate success of the process. Some models pertain only to a particular group of organs and others only to a limited mass sequence or time interval. Such choices as to the logical limits of the model are every bit as important as the model itself. In fact, these two aspects go hand in hand and can never be separated in the final report.

Models can have very practical and important applications. In radiopharmaceutical (RP) analyses, integration of organ curves is a primary—but not the sole—motivation for modeling. Other reasons for functional representations of data sets are possible. Some of these applications are given in Table 6.1. Mathematical representations are generally

TABLE 6.1　Reasons for Mathematical Models of Biodistribution Data Sets

REASON FOR MODELING	TYPES OF MODEL	TYPICAL IMPLEMENTATION
Integration of time–activity curves	Multicompartmental Multiple exponentials Splines Power law Polynomials	Integration of $A(t)$ from $t = 0$ to infinity for one of more organs. Estimate errors in the integrations.
Interpolation of time–activity curves	As above	Finding best time for imaging or clearing of radiotherapy compound
Using surrogate gamma labels for eventual pure beta emitters	As above	Converting from ^{111}In label to ^{90}Y label for data set
Find errors in experimental data sets	Multicompartmental	Model parameters show unusual or suspect data values obtained by the experimenter.
Categorizing individual animals or patients	Multicompartmental	Model parameters show distribution into possible subpopulations.
Recognizing abnormal regions in an image	Functional image	Regional cerebral blood flow (rCBF)

useful in all scientific and engineering practices, so the reader can appreciate the process as almost a necessity. Modeling, in the world of RP, is not a theory but is rather a phenomenological representation of radiopharmaceutical information. It is difficult to remember, use, and compare data sets if they cannot be summarized in some kind of mathematical shorthand format. The issue in RP studies is exactly what sort of model is appropriate and over what interval this mathematical form can be fitted to the data.

6.1.1　Correction for Radiodecay

An ambiguity occurs prior to beginning any specific RP modeling approach for kinetic analyses. A general analytic question arises as to what points are to be analyzed: data as taken such as $a(t)$, or those corrected for decay such as $u(t)$? Traditionally, experimenters have analyzed decay-corrected numbers (i.e., as published) and have placed these into the modeling format. It turns out, however, that the data as taken are more appropriate because of both mathematical and statistical reasons.

One mathematical basis for including decay explicitly in the model is that the form of the analysis may contain a variety of nonhomogeneous or nonlinear terms. Among these are time delays, constants, time derivatives, and a dependence on unknowns raised to a power greater than unity (Williams, Odom-Maryon, et al. 1995). A relevant example for RP analyses is any term involving products of two unknowns such as in antibody–antigen binding. Unphysical results occur if corrected data are entered into the modeling equalities in such cases. Finally, the statistician will point out that decay-corrected results are not appropriate for direct usage as they give inflated magnitudes. This increase occurs since correcting for decay makes the pharmaceutical amount at later time points of comparable magnitude to the value at the early points. Yet the activity values at later times are usually numerically much smaller—due to physical decay among other factors. This result is particularly the case for radiolabels with short half-lives such as those used in positron imaging. By first correcting for decay, the analyst is artificially increasing the amount of pharmaceutical present in every compartment. Admittedly, weighting the later time points could mitigate this problem, but it is better to handle the situation correctly in the first place.

6.2 TWO FORMATS FOR MODELING

Two types of mathematical representation appear in the RP literature: compartmental (Jacquez 1985) and noncompartmental modeling (Norwich 1997; Veng-Pedersen 2001). The former is based on the assumption that the organ or tissue being sampled has uniform concentration of the pharmaceutical and that it connects with various other uniformly filled compartments. This assumption leads to a set of ordinary differential equations that govern movement out of, and into, each organ. In many nuclear medicine situations, uniformity is a necessary assumption since little information is available about the distribution within the tissue itself. For example, if one samples the patient's venous blood at a brachial vein, there are probably no data on the blood activity anywhere else in the body. Use of the heart image is problematic since it will contain both heart wall and content. Likewise, major organs such as the liver, spleen, and kidneys will often appear to be uniform—at least in planar images.

If the observer is not able to obtain simultaneous multiorgan data or is not amenable to assuming uniformity of the agent in the organ or tissue, the alternative choice is the noncompartmental model. This is a purely mathematical form that, again, is up to the user. The most common such picture is the multiexponential representation of each organ's activity *vs* time curve. Many other logical possibilities are available including power-law representations, wavelets, Fourier series, and convolutions. Most of these functional forms can be justified by prior literature, mathematical elegance, or other intellectual support.

6.3 COMPARTMENT MODELS

More commonly, in compartment models each organ or perhaps a part thereof is assigned a mathematical volume or "box" in a diagram. For example, in the lowest-order example of Figure 6.1, two compartments interact with each other but not with any other entity.

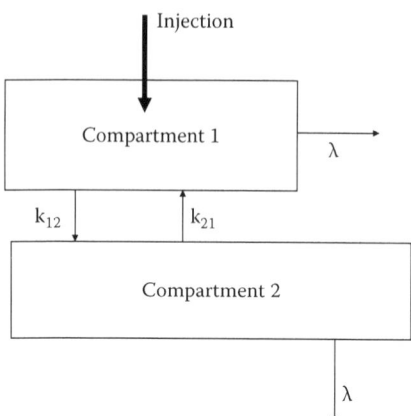

FIGURE 6.1 An illustrative two-compartment closed model with decay. Injection of a spike bolus at $t = 0$ would represent the boundary condition.

In general, if it is thought that compartment i interacts with compartment j, then two arrows are drawn: one going from i to j and one in the reverse direction. These represent first-order rate constants governing the movement of material between the two boxes. Usually, the targeting agent is injected into the blood, so a so-called mammillary approach is often used with blood appearing in a central role and interacting with all measured organs. A logical alternative is the catenary picture whereby two or more compartments pass along an agent in a serial fashion with the RP going sequentially from the first organ to the second organ and so forth. In the general situation in vivo, it is likely that both mammillary and catenary features will be found in any data set.

In the compartmental picture, decay of the label is considered as one means of exiting every compartment in the model. Figure 6.2 illustrates a multicompartment model that has been constructed at the City of Hope (Odom-Maryon, Williams, et al. 1997) to represent clinical results with the intact antibody T84.66 having a ^{111}In label.

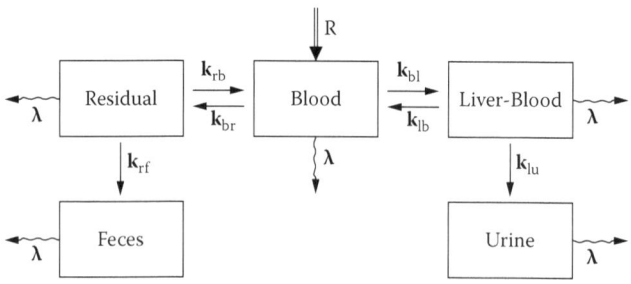

FIGURE 6.2 Five-compartment model with excretion and decay. This representation was used in the analysis of 20 patients receiving cT84.66, an antibody against CEA. (From Odom-Maryon, T. L. et al., *J. Nucl. Med.*, 38, 12, 1997. With permission of the Society of Nuclear Medicine.)

In both examples, radiodecay (with decay constant λ) is included as an exit route out of each compartment. In other words, the analyst is using data as taken by the observers and not the data as corrected for decay.

In the compartmental approach, the analyst uses a set of differential equations to describe the movement of activity. In Figure 6.1, we may write these equalities as

$$dA_1/dt = -k_{12}A_1 + k_{21}A_2 - \lambda A_1$$

$$dA_2/dt = -k_{21}A_2 + k_{12}A_1 - \lambda A_2$$

(6.1)

with explicit use of radiodecay in each compartment. If one adds the two equations, it is seen that the sum is

$$d(A_1 + A_2)/dt = -\lambda(A_1 + A_2)$$

(6.2)

In other words, the total of the two activities $(A_1 + A_2)$ decays at the physical rate. A similar conservation law will hold if we expand our list of compartments unless one (or more) of them has a route outside the body of the animal or patient. An example of this would be a renal or gut segment that exhausts to the exterior, as shown in Figure 6.2. Notice that we are using the activity in the compartment $[A(t)]$ in the previous example. Use of $a(t)$ in a similar analysis would be more problematic since conservation of material is possible only for the quantity $a(t) \cdot m$, where m is the mass of the organ or tumor.

Compartmental analyses eventually become appropriate as the spatial level of sampling or imaging approaches the physical size of the system. In this limit, the observer must necessarily assume that the activity being observed is uniform in the voxel of interest. In gamma camera imaging, this limit is on the order of 1 centimeter at the camera face and increases with depth into the patient, whereas it will be approximately 5 mm for a positron emission tomography (PET) system throughout the anatomy.

6.4 NONCOMPARTMENT MODELS

6.4.1 Multiple-Exponential Functions

A very common data representation is a set of one or more exponential functions fitted to a single organ or tissue. This single-tissue form of analysis is common in both radiopharmaceutical and unlabeled drug contexts. In the latter case, it is generally the method of choice to describe blood curves.

Radiodecay is the best-known example of an exponential format with the number of radioactive nuclei in an isolated system decreasing as

$$A(t) = A_0 \exp(-\lambda t)$$

(6.3)

Here, A_0 is the initial $(t = 0)$ amount of activity (MBq), and λ is the physical decay constant. Taking the logarithm (ln) of both sides of Equation 6.3 reveals that the resultant equation is a linear model and, using least-squares methods, that the best-fit parameters (A_0 and λ) are unique. The latter simplification is not the case in other multiexponential situations so that resultant solutions may depend on such diverse items as the starting points in parameter space and the precise algorithm used in the analysis. Nonuniqueness of the solution implies that the analyst attempts the least-squares fitting process more than once for at least some of the data sets. In such searches, different starting points for the fitting procedure are used, and the results of each "best fit" are compared.

Several traditional models have been used in pharmacology to represent both nonradioactive and RP materials in a single tissue, such as the blood. One parameter of great interest to the pharmacologist is the initial volume (V_i) of distribution of the injected tracer species. Here, we determine the maximum concentration of the tracer in the blood (C_{max}) and divide that value into the total amount of injected material (A_0)

$$V_i = A_0/C_{max} \tag{6.4a}$$

This volume may turn out to be equal to total plasma volume in the case of an intact antibody. With agents able to move out of the bloodstream, V_i may be much larger, up to a limit of the total extra cellular space in the test animal or patient. An addition volume (V_{ss}) may be estimated if the concentration approaches an asymptotic (steady-state) limit (C_{ss}):

$$V_{ss} = A_0/C_{ss} \tag{6.4b}$$

An associated single simple parameter for modeling the agent is the average blood clearance rate. In this computation, the analyst sets the rate (CL) equal to the ratio of the amount of the injected material to the integral of that injection out to very long times

$$CL = A_0 \bigg/ \int_0^\infty a(t)dt \tag{6.5}$$

If the analyst cannot follow the agent out to sufficiently long times or, for whatever reason, wants to use only a limited integration interval, the denominator integration is appropriately truncated. Since the $A(t)$ of the blood is usually a monotonically decreasing function, truncation will lead to a larger clearance value.

To check the coherence of the proposed CL variable, one may substitute a simple situation such as Equation 6.3 into Equation 6.5. The outcome is that the clearance rate is simply λ—a reassuring result. Note that in this case we generalize beyond Equation 6.3 being a representation of radioactive decay and allow the equality to represent a mono-exponential clearing in general. Figure 6.3 illustrates the use of these simple parameters in the analysis of a blood curve.

As noted, Equation 6.3 is a functional form that is monotone decreasing (as λ is > 0), so this result is not a good representation of a curve that starts from zero, rises to a maximum, and ultimately stays constant or decreases at large time values. Such functions are the usual forms seen for nonblood time–activity curves in animal and clinical studies. Such curves cannot be represented by a single-exponential function, and larger numbers ($n \geq 2$) of exponential terms are required.

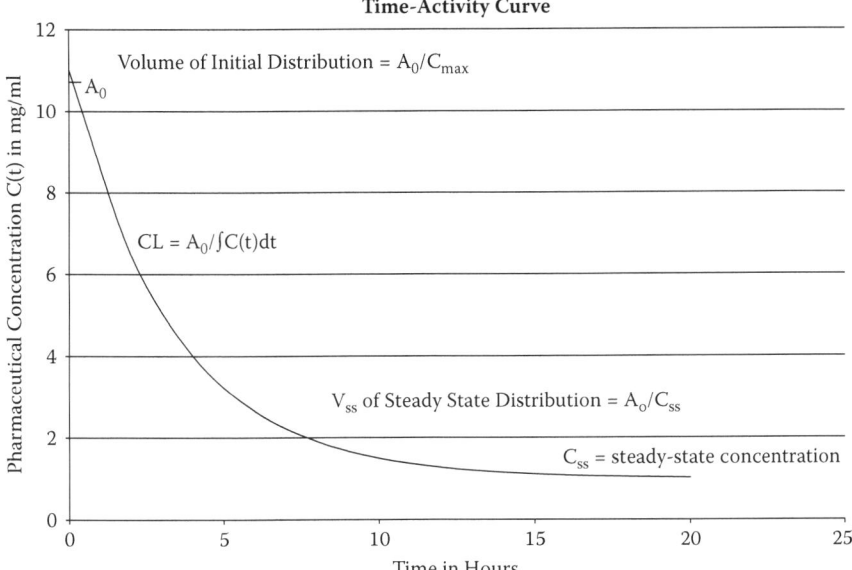

FIGURE 6.3 Concepts used in simple noncompartmental analysis of a blood curve are shown using a concentration versus time function [C(t)]. Clearance rate (CL) as well as volumes of distribution are shown. An initial (V_i) as well as a steady-state volume (V_{ss}) are included. These fitted parameters are used in Chapter 11 for allometric comparisons of human and animal protein blood clearance data.

Up-and-down curves imply at least a double-exponential representation of the organ or tissue via

$$A(t) = A_1 \exp(-k_1 t) + A_2 \exp(-k_2 t) \tag{6.6}$$

where the k_i are strictly positive rate constants in units of inverse time. Amplitudes, given by the A_i parameters, show the relative influence of the two clearance terms. Unlike the rate values, amplitudes may be required to be negative to fit a given data set. In the two-exponential (and higher-order) situation, the solution to the fitting procedure is not unique—unlike the case of a single exponential as in Equation 6.3 or the models of $u_T(m)$ in Chapter 3. Instead, the algorithm must search in the four-dimensional space (A_1, A_2, k_1, k_2) to find a "best fit" to the data. Generally, this is done using some form of least-squares technique. In other words, the solution lies at a nonlocal minimum in the sum of the squares of the residuals where a residual is defined as the difference between a given data point and the fitted curve of Equation 6.6. A more extended discussion of the statistical nature of this hunting process is described later in this chapter. Such solutions require the use of computers. An example of the two-exponential fitting process is in Chapter 3 for the five cT84.66 antibodies developed at City of Hope by Anna Wu and Paul Yazaki (Williams, Wu, et al. 2001).

Many investigators do not go beyond the two-exponential analysis. A primary reason for this limit is the sparseness of a typical data set. In such cases of only two k_i

values, the researcher may refer to so-called α and β half-lives. These two times are given by

$$T_{1/2\alpha} = \ln(2)/k_1$$
$$T_{1/2\beta} = \ln(2)/k_2 \tag{6.7}$$

where the alpha half-life is, by definition, the shorter of the two. We and other authors have tabulated these two parameters in reporting on the behavior of a biexponential function. It is important in such tables also to give the relative weighting of the two half-lives (A_1 and A_2, respectively); otherwise, the reader does not get a clear impression of which of the two processes is dominant. It may turn out, for example, that one of the two amplitudes is much larger than the other such that the curve is essentially a single exponential with a slight perturbation.

The previous argument can be generalized to a set of three (or more) exponentials. This expansion may become necessary when the data are recorded out to longer and longer times. It is expected that more data points—particularly later in time—will inherently require more exponentials. Solutions in such situations may become more difficult to obtain because of the increasing number of free parameters. To simplify this scenario, the experimenter may elect to define or constrain one or more amplitude variables *a priori* and thereby will limit the computer time required for a best-fit solution. For example, in the two-exponential case for blood, it may be possible to set $A_1 + A_2 = A_0$ to represent the fact that the sum of the two terms is the initial value of total injected activity into that tissue. It is, however, generally difficult to have a well-mixed distribution of activity in the blood immediately postinjection. Thus, this constraint may not be valid. Measurements at this time are problematic and also suffer from the mixing problem. A simpler case arises for most other tissues (major organs such as the liver, spleen, kidneys, and lungs) since their RP activity at $t = 0$ should be nil such that $A_1 + A_2 = 0$. This constraint is easier to justify.

Another method to simplify the analytic process is to determine one (or perhaps more) of the k_i by a traditional process called "curve stripping." Here, the observer looks at the longest times initially and attempts to fit a single rate constant and amplitude over a suitably short interval. The resultant unique single exponential term, which is of the form of Equation 6.3, is then subtracted from the measured organ curve over the entire time range and a reduced set of exponentials is fitted, in turn, to an earlier time interval. This iterative method was developed in the precomputer era and can be used to determine starting values for one or more of the requisite rate constants and amplitudes in computer searches for best fits to the data.

Using multiple exponential terms is generally done on an organ-by-organ basis. In other words, the analyst finds the best statistical fit to data one organ at a time. Thus, any resultant optimal rate constants refer only to that single tissue. While efficient in some ways, such results completely neglect the interaction between tissues. To determine such interorgan parameters, the analyst may prefer to go to a compartmental picture in the first place as described above.

At this point, we should point out that multiexponentials are also a result of compartmental modeling. If one were to use a boxes-and-arrows approach as in Figures 6.1 and 6.2, the results would be a set of multiexponential functions for each compartment. In fact, if the solution were obtained by matrix inversion, each compartment would have the same set of exponential functions. The only difference between one tissue and another would then be their respective amplitude values. Such resultant rate constants are then termed eigenvalues (Gr. Eigen: characteristic) for that specific system of equations. In general, the kinetic parameters would be algebraic mixtures of the compartment-to-compartment transfer rate constants (k_{ij}) in the original compartmental model. Occurrence of multiexponentials in both compartmental and noncompartmental models has been one of the factors that has led to the popularity of this form of analysis in many RP assays.

When using a series of exponentials, an analyst has the added opportunity to refer to various phases of a RP study. By observing the blood curve, for example, one could refer to the rapidly changing beginning portion ($T_{1/2\alpha}$) as the early phase, whereas the slowest-changing portion might be called the late segment of the study. Physiological processes involved in these segments can be alluded to but are somewhat difficult to prove.

Interpolation and integration of exponential functions are relatively easy. In particular, the integral for a multiple-exponential is simply the sum

$$\text{AUC} = \sum_{i=1}^{i=N} A_i / k_i \tag{6.8}$$

This simplification is one of the most attractive features of this type of analytic procedure. Convergence is clear, and the process of integration is one line of code. The indeterminate value of N in Equation 6.8 does cause some uncertainty, as one can be unsure if enough exponential terms have been included. A number of analysts would prefer a simpler functional form. One such model is the power law, described in the following section. Before we treat that representation, one item of interest to the modeler is the limit of a large number of exponential functions.

As mentioned already, as organ–activity curves are taken out to longer and longer times, more and more exponential functions are often required. The logical question arises as to what would happen if the number of such functions were allowed to increase without limit—that is, an infinite set of exp(–kt) terms. The fundamental result is that this sum, if put into an integral format rather than a summation, may be represented by the definition of the gamma function (Γ) whereby

$$\Gamma(b) = t^b \int_0^\infty k^{b-1} \exp(-kt) \, dk \tag{6.9}$$

where both time and the exponent of time (b) are constrained to be positive. Note that the rate constant in the exponential is now a continuous variable and is integrated over. By taking the time variable to the left-hand side (LHS) of the equality, one finds that

$$t^{-b} = \frac{\int_0^\infty k^{b-1} \exp(-kt)dk}{\Gamma(b)} \tag{6.10}$$

The LHS of Equation 6.10 is termed a power-law representation of a time–activity function. The multiple rate constants and the exponential terms have been taken out of the formulation due to the integration procedure.

6.4.2 Power-Law Modeling

A number of modelers have discovered that power-law equalities are relatively good representations of blood, plasma, or other data sets. Here, the assumed function is

$$A(t) = A_0 t^{-b} \tag{6.11}$$

where A_0 and b are fitted constants with $A_0 > 0$ and $b > 0$. Care must be exercised when approaching the value $t = 0$ in the analysis due to the net negative exponent. Temporal offsets are often used so that time may be constrained to be strictly positive and, if fact, above a certain minimum value. This value could be the time required for adequate mixing of the unlabeled agent or RP into the blood system as previously described. Earlier than that, the observer has no simple measured entity available, and partial differentials, depending on space and time, would need to be invoked. As in radiodecay, Equation 6.11 may be converted into a linear form by taking the logarithm of both sides so that a unique solution is available via least-square analysis. Notice that, as in Equation 6.3, there are only two parameters in the linear fitting process. This economy of variables has been one of the most attractive features of the power-law model.

Other than blood, it is unlikely that a tissue could be fitted over its entire course with Equation 6.11 since most organ results must initially rise to a maximum and then decrease. However, one may be tempted to try fitting $A(t)$ or $u(t)$ values at sufficiently long times after the maximum is achieved—that is, use the equality of Equation 6.11 over the monotonically decreasing part of the curve.

Historically, power laws have had a wide application in many of the sciences—particularly astronomy and allometry. The latter topic is discussed in Chapter 11. Recently, power-law application to spatial objects has led to the concept of fractals (Mandelbrot 1967). For example, the coastlines of countries are said to be of a fractal type. A characteristic of such spatial functions is that the scale of the image cannot be determined by looking at a picture of a country's seashore. The coast "looks the same" over a wide range of dimensional—ranging from cm to km or even hundreds of km. In time analyses, the analogous result is that a temporal curve appears to have the same dependence on time during all time intervals. One cannot, therefore, by looking at a power-law time–activity curve, determine if the temporal interval is early, intermediate, or late in the organ's clearance of the tracer. Notice that this ambiguity is in direct contrast to a multiexponential representation where the rate of organ decrease (or increase) depends on the absolute time of the observation. In other words, using exponential fitting, the analyst essentially defines the various phases of the RP motion through the system of interest. With power-law representation, this association is totally lost.

Power-law representations cannot persist over unlimited temporal or spatial intervals. For example, fractal relationships must eventually break down as the magnitude of the abscissa scale approaches smaller and smaller values. In the distance realm, the coastlines would appear to change as we begin to see clusters or grains of sand. In a time context, we have already seen that the very early blood activity values may be impossible to represent via a power-law function due to mixing issues immediately postinjection.

Similar discrepancies occur at relatively extended distances and long times. This large-scale breakdown occurs because we lose sight of the object of interest as the dimensions of the problem are enlarged. If we allow the spatial dimensions to become on the order of the size of a country, coastlines become absorbed into the pixel size and cannot be separated out for analysis. At very long times, activity also becomes indistinguishable from the inevitable background counting and is thus simply lost to the observer.

Interpolation of power-law time–activity curves is very simple. Integration may present a challenge in that the integral will be unbounded if $b \geq -1$ upon substitution of the integration limit $t = \infty$. It is also unbounded if $b < -1$ upon substituting the limit $t = 0$. There seems to be little room in the analysis except to exclude these two time points out of the process, that is, to integrate from a finite time out to another finite time. Alternatively, as described already, one may simply state that extended time intervals are not appropriate for integration of power-law curves. Hence, one would integrate only over the interval that has been initially explored by the investigator. Other time intervals are then accomplished by extrapolation, for example, by using decay of the radioactive label of the RP.

6.4.3 Tumor Uptake as a Function of Tumor Mass

In the power-law model, it is possible that the model will hold in variables that do not involve, at least explicitly, time or space. One important such example (cf. Chapter 3) occurs in the dependence of tumor uptake [$u(t)$] upon tumor mass (m)

$$u_T(m) = u_0 m^b \tag{6.12}$$

where $u_T(t)$ is tumor uptake in %ID/g, and m is the mass of the lesion in g. A similar analysis holds for the $a(t)$ parameter except that the exponent b increases by one.

The result of Equation 6.12 was originally shown for both liposomal and antibody tracers in a number of murine tumor models at 48 h postinjection (Williams, Duda, et al. 1988). These early results are described in Chapter 3 where Table 3.1 gives u_0 and b parameters for the best fits to these various mouse data sets. In general, the results were consistent with an exponent (b) between -0.33 and -0.50. Such narrow limits of the exponential results were one of the more rewarding results of this study.

Having use of this power-law relationship is important since the absorber dose rate (Chapter 4) is directly proportion to $u(t)$ for a short-ranged, charged particle. If targeted radiotherapy using alpha or beta radiation is being considered, the resultant dose for a smaller mass will be much higher than for a larger tumor of the same type. In the case of $b = -0.5$, the dose rate increases by approximately threefold as the mass decreases by a factor of 10. Thus, if a mass law of the form of Equation 6.12 can be shown, smaller lesions become more amenable to radiation therapy. Because of spatial resolution limitations in nuclear medicine, it is unlikely that very small lesions, on the order of mm size or smaller, will be visible in either animal or clinical imaging contexts.

A geometric picture (Figure 3.1) may be invoked to justify this mass-dependent tumor power law. In Chapter 3, we consider either spherical or cylindrical masses being perfused from capillaries having access from one side of the lesion. If we assume that only a limited distance into the tumor can be accessed, via diffusion, in the time of the experiment, the uptake can be expressed as a ratio of area of the tumor to its mass (or volume). This result is essentially that of the power law or fractal. It must be emphasized, as in the case of temporal analyses, that as m approaches zero this mass-related power law must also fail. If we look at the two tumor models in the figures, we realize that as their size decreases to that of the tracer diffusion distance (0.1–0.2 mm) the uptake approaches a constant value, and the untargeted zone at the inside of the lesion does not exist. This distance value (e.g., a 100 μm diameter sphere) is a mass of approximately 1 microgram. We would anticipate that the mass law must have a cutoff at approximately this mass level or smaller.

It is likely that establishing comparable results for patient tumor data would be much more problematic than for laboratory animals due to several factors. First, when we select a group of consecutive patients, there will be wide interpatient variations in the amount of molecular (or other) target available in the tumors. With murine data, this factor is not important since the tumor being grown will be the same cellular type and have similar antigen expression in all of the animals in the experiment. Second, there will be variability in the timing of data acquisitions due to the variation in the time the patient is taken into surgery. In mouse experiments, this time is generally fixed at a specific time: 48 h in the data sets shown in Chapter 3. The comparison of animal and clinical results for the $u_T(m)$ relationships is more extensively discussed in Chapter 11.

Another mitigating effect against $u_T(m)$ power-law dependence involves the use of radioiodine as the radiolabel. Dehalogenation begins to remove activity from the lesion as soon as deposition begins. Generally, it is difficult to prove, with $p < .05$, that such iodinated tracers demonstrate an inverse relationship with mass. As noted already, however, at least one clinical trial involving an iodine label does show the power-law representation (see Chapter 11).

6.4.4 Sigmoidal Functions

In this segment devoted to modeling, we should describe one type of function typically used in multiple scientific fields including agronomy, economics, and radiation therapy: the provision of an S-shaped population curve. Such representations are important in fitting data on cell death, for example, as well as tumor regression with radiation sources (Withers and Suwinski 1998). Chapter 9 gives an application in which the investigators use one such function to allocate the relative amounts of lymphoma tumor size decrease to a pair of simultaneous therapy agents: the protein by itself as well as the radiation dose delivered by the protein.

To understand the use of sigmoidal functions, of which there are multiple mathematical forms (Lyles, Poindexter, et al. 2008), it is best to remember that the biological system is prone to saturation. The radiobiologist can kill only 100% of the cells in the culture by irradiation using a ^{60}Co source. Likewise, if the absorbed dose approaches zero, there can only be 0% cell death—not an increase in the number of viable tumor cells. Similarly, tumor size can regress to a zero dimension at only high enough dose values at which point there is no further required treatment. Thus, the biological system under study is inherently limited at either end of the independent variable's range.

Most work with sigmoidal functions has involved time as the independent variable with the logistic curve being an example

$$dn/dt = a_1 n(t)[a_2 - n(t)] \tag{6.13}$$

Here, $n(t)$ represents members of a population (e.g., a number of cells in a tissue-culture experiment). Multiple other formats have been used historically, and a summary is available (Lyles et al. 2008). Chapter 9 returns to this topic, where absorbed dose—not time—becomes the independent variable in the sigmoidal equation.

6.5 BASIS FUNCTIONS

General application of a sequence of exponents or power laws to multiple agents or long time intervals is difficult. As we have seen, both short times and long times present issues of fitting the data due to either a lack of uniform distribution ($t = 0$) or uncertain extrapolation beyond the last recorded data point. Engineers, working in either spatial or temporal spaces, are often inclined to forgo a given functional representation and instead use sets of mutually independent functions to try to achieve a general solution. These are termed sets of basis functions; they presumable may be used to describe individual curves taken during almost any experimental process. One of the most traditional of these is a group of sine or cosine functions in a Fourier approach. The fast Fourier transform (FFT) is generally used to generate this set of basis functions. While more often applied to spatial analyses, this approach can also be applied to dynamic curves. It would clearly be of use in fitting functions going through a maximum during the observation interval. The method does require that the function be periodic; this is clearly not rigorous for $A(t)$ or $a(t)$ curves, which essentially continue out to indefinitely large times.

One area of application of sinusoidal functions is in the modeling of circadian rhythms that occur in production of certain hormones and other chemicals in the body of the animal or patient. These functions are often on a 24 h basis conditioned by the earth's rotation. There are other clocks (Zeitgebers in German) internal to mammals. If the species is isolated away from the sky (e.g., in a cave), these secondary timers will begin to function and will provide the periodic stimulus necessary to keep the physiological systems functioning. Generally, internal clocks are seen to run slightly slower than the earth's rotation.

A second possible basis representation is the application of so-called wavelets (Kaiser 1994) that permit a small group of square-wave or other relatively rapidly changing functions to represent time–activity curves. These are particularly useful in cases where the organ activity moves relatively rapidly from a minimum to a maximum or the converse situation (i.e., sharp upswings or downturns). If one were to apply Fourier analyses to such examples, the number of sinusoidal component functions would have to be very large to generate such relatively sharp edges.

In a basis set analysis, the user halts the fitting process when sufficiently many functions are available to represent the given curve to within a predefined precision. A question may be raised on how good this fitting is and when to stop the modeling process. As an alternative, simpler functions can be applied over short intervals of the time–activity curves.

6.6 DATA REPRESENTATION WITH TRAPEZOIDS AND SPLINES

If the user declines to use exponential, power-law, or basis functions in a modeling context, there remains the use of simple functional sets to locally describe a time–activity curve by giving a general method to directly represent the area between adjacent time points. If one assumes a linear representation of the activity between measured points, the local (i-th) area under the curve (AUC) is

$$\text{AUC}_i = 1/2(A_{i+1} - A_i)\Delta t_i + A_i \Delta t_i = 1/2(A_i + A_{i+1})\Delta t_i \qquad (6.14)$$

where A is a biodistribution activity variable. Notice that no computer-based fitting is required since one uses all data directly. One feature of this analytic process is that the time interval is generally not constant in RP biodistributions. Thus, the time step must be carried as a variable through the calculation. Implicit in trapezoidal analysis is that a linear interpolation may be made between adjacent data points (A_{i+1} and A_i) in the organ data set. Total AUC is found by adding segmental values out to some sufficiently long times. Methods to take into account the curve beyond the last recorded data point are necessary. A conservative approach is to simply extrapolate from that point using the physical decay of Equation 6.3.

Cubic splines are becoming one of the common methods (Wahba 1990), whereby the experimenter fits two adjacent points and their first and second derivatives in $A(t)$ for each organ. The process is continued along the curve, and the resulting splines are represented by a four-parameter function. Like the trapezoidal case and unlike exponential functions, these cubics are simply representations of the curves. While not an elegant mathematical form, the functions are useful in that they may be fitted to lie atop the data precisely. Unlike trapezoids, curvature is now allowed, and the functional forms seem much more natural to the observer. Splines have continuous first and second derivatives at junction points between fitted segments. Neither of these requirements is satisfied by trapezoids.

With splines, interpolation is performed locally with the four regionally fitted parameters. Similarly, integration is done in closed form using a segmented approach such that each successive spline is integrated and their overall sum recorded much as with trapezoids. The terminal data point can be problematic in that some assumption must be made about how the organ or tumor curve will continue beyond the last measurement. A linear form is often assumed for nondecay agents; in the case of RP materials, the terminal form may simply be the monexponential of Equation 6.3.

Fitting organ or tumor data with such linear functions or cubic polynomials may be satisfying at first glance, but other methods that attempt to probe more deeply into the data are often employed. We have omitted consideration of blood curve effects up to this point. One method to take such effects into account is use of blood curve deconvolution.

6.7 DECONVOLUTION AS A MODELING STRATEGY

When observing a time–activity curve in a normal organ or in a tumor, it is tempting to state that the result depends only on the targeting agent (RP) and the measured tissue. Yet there is at least one other variable in this situation: the blood curve that determines input into the structure of interest. A blood curve may be affected by influences independent of the agent–tissue combination. Consider a blood curve being artificially maintained at a constant value by an intravenous (IV) infusion. The resultant organ uptake curve should initially show a monotonic increase with time. This result would probably continue until the tissue receptor became saturated whereupon the curve would approach a constant value. If, on the other hand, a blood curve would be rapidly driven to zero because of excretion of the agent via the renal system, for example, most target tissues (other than kidneys and bladder) would show only a minimal uptake. Thus, the organ or tumor activity, to some degree, must reflect the arterial blood curve of the agent. In general, any such blood curve is determined by multiple interactions of the tracer during circulation within a variety of tissues in the body of the animal or patient. There is also the possibility of intervention into the shape of the blood curve by external means.

Because of the blood curve's effect on every organ, it comes as no revelation that many observers are interested in eliminating this dependence. In other words, in determining what the organ or tumor response would have been had there been a bolus input of the RP at the tissue's arterial side. To achieve this idealized objective, the analyst uses the method of deconvolution.

Before discussing this type of analysis, we need mention that direct injection into the arterial side of an organ or tumor is not a simple, benign, or pleasant task. Such interventions are inherently difficult and may present a distinct morbidity. In the case of rodents, microscopic surgical manipulations on the high-pressure system are called for. Even with accessible arterials, it is problematic that an experimenter or clinician would attempt direct arterial input. Few neurologists, for example, relish attempting an injection into one or the other external carotid artery. This would particularly be the case for a known (or possible) stroke patient or other individual of clinical interest. Instead, after the first few circulation times in the animal or patient's body, it is usually assumed that the venous sampling conventionally being done is equivalent to measuring the arterial input for the tracer of interest. Thus, one may state that deconvolution is a mathematical manipulation to eliminate the necessity—and associated difficulty—of arterial injection into a specific tissue.

In a deconvolution strategy, the analyst makes the assumption that any tumor (or normal organ) uptake curve is represented by the integral equation

$$u(t) = \int_0^t b(t-\tau)h(\tau)d\tau \tag{6.15}$$

where $u(t)$ is the measured uptake, $b(t)$ the measured venous blood curve (in the same %ID/g units), and $h(t)$ is the unknown impulse response function. The $h(t)$ function gives the resultant organ curve if the experimenter could have injected a bolus into the organ's arterial supply. By comparing impulse response functions, the investigator may compare the response of a given tumor (or other) tissue with a variety of tracers. This has been done for the five cognate cT84.66 antibodies targeting the LS174T tumor as described earlier (Williams et al. 2001) and is shown in Figure 3.10.

Recall, from Chapter 3, that the diabody has the optimal form of the five cognates in that its resultant $h(t)$ function is wider and higher than that of any of the other four members of this set. If all five cognates could be injected as a bolus into the similarly sized tumors, diabodies would then be present longer at the tumor site and show correspondingly reduced loss from the tumor bed. Such an extended residence allows maximal therapy effect since the radioactive label is present for longer periods of time. A similar argument would hold for a chemotherapeutic or immunotherapeutic agent based on the diabody. Presumably this result reflects the low molecular weight (MW) (hence ease of diffusion) and divalent binding of this member of the cT84.66 cognate group. By far the worst-performing member is the single-chain antibody, which has a relatively poor $h(t)$ result due to its monovalent binding to the carcinoembryonic antigen (CEA) target molecules. One could use results such as these to argue for the use of the diabody and against the use of single-chain antibody in a subsequent clinical trial.

There is some interest in the constancy of $h(t)$ for tumors across a variety of tracer–tumor combinations. As noted in Chapters 2 and 3, it is difficult to directly compare one antibody with another since they generally have distinct target molecules, different tumor models, and disparate binding affinities. To effect such a comparison, we proposed using the reduced impulse response function whereby each $h(t)$ is divided with its integral from zero to infinity. We have been able to show, in the case of two different diabodies, that the reduced forms were essentially the same (Williams et al. 2001). In that example, we used the cT84.66 diabody from City of Hope and the analogous molecule developed by Adams, Schier, et al. (1998) from Fox Chase. This equivalence, which eliminates the specifics of their molecular targets, implies that both these diabodies interact over time with their antigen-producing tumors in essentially the same way.

Given a tracer's organ or tumor impulse response function, the experimenter can predict the subsequent response of that organ for any possible form of blood curve. This is one of the most important applications of deconvolution analysis. For example, radiation oncologists may elect to truncate the blood curve of a therapeutic agent when the patient's red marrow absorbed radiation dose estimate begins to approach possibly dangerous levels. This has been termed extracorporal absorption (ECA) and has actually been employed in treating lymphoma patients at Lund University (Linden, Kurkus, et al. 2005). In that case, truncation is achieved using a biotinylated antibody injection along with an avidin-containing column connected to the patient's blood system.

One interesting mathematical result of deconvolution analysis is that truncation of the blood curve does **not** alter the original ratio of the two areas under the curve

$$\text{AUC}_T/\text{AUC}_B = \int_0^\infty u_T(t)\exp(-\lambda t)dt \Bigg/ \int_0^\infty u_B(t)\exp(-\lambda t)dt = R \qquad (6.16)$$

where the subscripts T and B refer to tumor and blood, respectively. Note the presence of R, the traditional therapeutic figure of merit from Chapter 3, on the right-hand side (RHS). A similar argument may be made for normal tissues. The constant indicated is characteristic of the particular agent and tumor target being studied. This ratio could be used to indicate relative effectiveness of therapeutic agents. Note that we assume an instantaneous decrease of the blood curve to zero at some random point in time. This result of Equation 6.16 may be known to electrical engineers but came as a surprise to our group when it was first discovered via following the AUC ratio at different truncation time points (Williams, Liu, et al. 2000). One thus concludes that truncation does not alter the traditional therapeutic ratio but only permits the patient to receive a maximum tumor dose without undue hazard to any normal tissue. A similar conclusion presumably holds for chemotherapeutics when subjected to blood curve analysis of the agent. In other words, one would predict that the chemical agent has a fixed ratio of AUCs just as does the RP and that truncation does not affect the value.

6.8 STATISTICAL MATTERS

Statistical considerations enter into modeling from the beginning. Multiple questions arise during fitting that require answers that can come only from a discussion of relative probabilities. At the lowest level, an analyst may wish to know if the compartmental fit is significant at some level of confidence. Yet other issues may arise such as if a simpler model exists that represents the data equally well—or better. The Akaike criterion

TABLE 6.2   Statistical Issues Arising in the Compartmental Modeling

STATISTICAL QUESTION	GENERAL METHOD	RESULT
Finding best-fit parameters	Steepest descent	Optimal set of parameters
Choosing one model from many possible nested models	Akaike criteria	Optimal model for the situation at hand. For example, is a single exponential a better fit than a double exponential?
Using the data most efficiently	Weighting the individual points per their reliability	Adjunct to steepest descent method
Comparing one biodistribution with another	Looking at the fitted parameters of the model	Are the two parameter sets equal?
Finding variation in computed quantities using fitted parameters	Bootstrapping, Monte Carlo (MC), or differential forms	Distribution of a computed quantity such as area under the curve (AUC)

(Akaike, Parzen, et al. 1998) is often employed in this situation so that a minimalist model may be determined from a nested set of compartmental models.

In addition is the issue of weighting the residuals in the method of steepest descent. If weighting is employed, what should the relative values be for a given data set? Often, these are equated to the inverse of the data variances for the individual points. However, other weights may be used.

Another topic of significance in statistical thinking is the nature of animal and patient parameter distributions. It is often assumed that these fitted quantities have a scatterplot that is a bell-shaped curve. Yet such a Gaussian result is unlikely to be the case since the distribution of k_{ij}, for example, must be skewed toward positive values and away from the origin ($k_{ij} = 0$). In some of the hunting algorithms used in the method of steepest descent, the origin is excluded more strongly than in others. This asymmetry is a consequence of the fact that if k_{ij} is negative then the corresponding exponential term goes to infinity with increasing time values. Such solutions are distinctly nonphysical. A general list of some of the statistical questions arising in modeling is contained in Table 6.2.

6.9 METHODS TO ESTIMATE ERRORS IN CALCULATED PARAMETERS SUCH AS AUC

Uncertainties in the directly computed quantities, such as the quartet $\{A_1, A_2, k_1, k_2\}$, are given by the fitting method. Yet one issue that is generally not considered in modeling analysis is estimation of errors in any subsequently calculated parameters (F). One example of this would be the error in the AUC values needed in absorbed dose estimates. In any theoretical calculation of physical results in either scientific or engineering contexts, there must be included an estimate of associated uncertainties. Several methods are in current use for AUC in this regard. These attempts to find variations in area under the curve are themselves estimates and do not necessarily give similar predictions for area uncertainties.

6.9.1 Bootstrapping

In this strategy (Iwi, Millard, et al. 1999), the analyst uses a given model and subsets of the total data and watches how the resultant AUC—or other variable of interest—changes. A mean and standard deviation will eventually be generated for the computed parameter. For example, one might have 10 animals at each time point and a total of five time points. Instead of using all 50 results in one RP analysis, the statistician may instead take five animals randomly at each time for each of the five time points. A random number generator is used to select the relevant animals at each point. This hypothetical 25-animal analysis is then repeated for other groups of five animals, and the results are tabulated. A large number of subsets are generally possible without repeating the animals chosen. Finally, there is a resultant mean or median AUC as well as an estimate of its variation.

Bootstrapping can be criticized in that the analytic picture is given *a priori* so that if the model is not particularly good in representing the data these resultant estimates of error may be inaccurate. In its favor, the method strictly uses only data obtained by the experimenter and follows the variation that the set of raw information allows in the computation of a given quantity of interest.

6.9.2 Monte Carlo Methods

In Monte Carlo (MC) analysis, the investigator uses the fitted parameter set and its associated covariance matrix to estimate statistics for the AUC or other computed quantity. It must be emphasized that the covariance matrix is a consequence of finding all of the model parameters in a single hunting episode using all of the data. Unlike classical physics experiments whereby so-called external variables can be held constant, the RP analyst must usually contend with a situation whereby all the unknowns are determined simultaneously. This complexity means that every model parameter generally has a nonzero covariance term with every other parameter. In other words, the resultant fit is a set of numerical values tied together in a bundle that no amount of computation analysis can untangle. No parameter is free of any other, and the entire set of these values must be handled as a unit (Kaplan, Williams, et al. 1997).

A random number generator is required to move within this parameter space. While the program used in the fitting may show only standard deviations, the actual distribution of fitted parameters is stored and will be followed in the MC simulations. Since all the fitted parameters are determined simultaneously, the covariance matrix permits only variation within these predetermined limits. Such variations are not necessarily along a Gaussian curve since the fitted values may demonstrate other histograms—particularly for rate constants as described before. Log-normal distributions are one common alternative possibility.

Multiple MC sequential runs (10^4 to 10^6) are taken to determine the variation possible in the various computed parameters. Results can be displayed as a histogram. One difficulty that often occurs is that some MC runs may lead to very large area values. These areas may appear as the largest number available to the computer on which the MC strategy is implemented. If even only one such outcome (e.g., 10^{20} in relevant units) occurs, the average AUC becomes essentially unbounded. Thus, investigators will have to limit the values selected for their tabulations by excluding large outliers. The appearance of very large areas, for example, may indicate that at least some curve integrals may not be stable—probably due to slow falloff of the activity variable with time. If the data being analyzed are as taken (i.e., not corrected for decay), the appearance of infinite areas is less likely since the activities at long times are being downwardly modulated by decay of the label. If, however, one must analyze the pharmaceutical data, such as in a chemotherapy context, the issue of unbounded areas may indeed be problematic and needs to be carefully considered in the presentation of the results. In this instance, the experimenter should be careful to follow the agent for as long a period as is possible to try to find a maximum of uptake—and a subsequent decrease—in the tissues of interest.

Again, as in bootstrapping, the mathematical picture is a given constraint and, as such, may be incorrect. Many repetitions of the MC protocol are time-intensive and may have to be run offline or at night. Parallel processors can help with the logistics of the analysis. In any event, appearances of very large values in the results are always disconcerting and have to be excluded from the summary.

6.9.3 Differential Methods to Estimate AUC Errors

A third method is possible to estimate uncertainties in computed quantities such as the AUC. Imagine a generic variable F that is computed by the analyst using fitted parameters from the model. In this technique, the analyst takes various possible partial derivatives of the derived function F for a given tissue. These functional derivatives are then evaluated at the point of best fit in the parameter space of the analysis. By doing a Taylor's series expansion about the point of optimal fit, the resultant variance for the computed function becomes

$$Var(F) = \sum_{i=1}^{N} (\partial F / \partial x_i)_{min} \sum_{j=1}^{N} (\partial F / \partial x_j)_{min} COV(x_i, x_j) \qquad (6.17)$$

In this scalar result, $F(x_i)$ is defined to be a function of N variables (x_i). Notice that all possible algebraic derivatives are required, as is the covariance matrix of those variables (COV). Analytic derivatives are obtained by differentiating the functional form such as a multiexponential or power-law fitting. For a two-exponential function, there are four variables, and $N = 4$ in the summation. Notice that the result is a form of double dot product as the covariance matrix is projected onto a single number, the variance of the computed quantity. If this had been a classical physics experiment where there are intentionally no cross-terms in *COV*, the only item in the matrix would be the diagonal elements that represent the errors in each of the independent variables.

The various methods shown here for estimating the uncertainty in a calculated quantity such as *F* will not necessarily give the same result. Differences of the estimates comes directly from the variety of techniques; it is an example of the "blind men feeling the elephant" situation in that as the techniques differ greatly their results may be quite disparate. For example, the vivid excursions observed in some cases of the MC approach may cause the average uncertainty for that method to be much larger than that seen with either the bootstrap or differential techniques. As noted, the latter pair may be criticized separately for their respective limitations. In the approach of Equation 6.17, for example, the user can be criticized for relying on the lowest-order terms in a Taylor's series expansion of the function around its mean value. Higher-order terms may yield larger error estimates. Since these are not considered, their significance is unknown.

6.10 PARTIAL DIFFERENTIAL EQUATIONS AS A MORE GENERAL MODELING FORMAT

Until now, we have analyzed the RP in terms of either time or space. A more complete format to attempt description of the RP is a set of partial differential equations (PDEs) that give simultaneous spatial and temporal variations of activity within the tissues of interest. A relatively simple example for a spherical tumor (Fujimori, Covell, et al. 1990) is

$$
\partial c_i / \partial t = D\left[\left(\partial^2 c_i / \partial r^2\right)_{r=r_i} - \frac{2}{R-r_i}\left(\partial c_i / \partial r\right)_{r=r_i}\right] - k_f c_i a_i + k_b comp_i \tag{6.18a}
$$

$$
da_i / dt = -vk_f c_i a_i + vk_b comp_i \tag{6.18b}
$$

where c_i is the antibody molar concentration, and a_i is the corresponding molar antigen concentration. Molar concentration of the complex of antibody–antigen is then *comp_i*. Each of these values is given at radius r_i, and R is the outer radius of the spherical lesion. Rate constants are indicated by the k parameters, and v is the valence of the antibody. In this analysis, convection is neglected, and only the diffusion and binding aspects are explicitly included. One outcome of the PDE approach of Equation 6.18 is the theoretical prediction of the *surface barrier* restriction of antibody access to tumors. Said restrictions should equally well apply to other antigen-specific agents that access the lesion from the blood supply side. This barrier is a verbal construct that describes how the protein agents are limited to relatively shallow depths within the tumor due to binding with the antigen molecules. Although the term has a somewhat negative connotation, surface effects are not unexpected and simply reflect the specific binding of antibody to antigen within the tumor space. As noted already and in Chapter 3, the mass law for tumor uptake as a function of tumor mass implies that the RP is preferentially deposited on the perfused outside of the spherical lesion. Such surface deposition at the tumor–capillary juncture also implies an absorbed dose differential between the surface and interior of the typical tumor.

Experimental data contributed by Jain and co-workers (Jain 1990) at Carnegie Mellon University and later at Harvard University show that convection terms should generally be added to the aforementioned analysis. In experiments done on tumors implanted in the ears of rabbits, this group discovered a significant convection term— implying material flow out of the tumor and back into the blood supply. Such flows, apparently due to the relatively ineffective lymphatic systems of the lesion, further reduce the flux of antibodies (the tracer being studied) into the tumor volume.

Because of the number of free parameters, it is difficult to actually carry out a computation of the type outlined in Equation 6.18. Many of these variables must, of necessity, be estimated from other experiments—often involving in vitro testing. Measurement at the microscopic level required by the analysis is a further limitation. One cannot expect to determine the spatial distribution of activity from the blood supply side into the tumor volume using even microPET devices. As noted in Chapter 2, the spatial resolution of gamma cameras and even PET scanners is on the order of 1 cm and, in the case of cameras, becomes worse at depth. Thus, predictions of antibody targeting are more theoretical and may be used to compare hypothetical agents prior to experiment. In that sense, these results are analogous to the use of imaging figure of merit (IFOM) in Chapter 3—a tool to differentiate one targeting vehicle from another.

6.11 SOME STANDARD SOFTWARE PACKAGES FOR MODELING

Multiple software packages are available for modeling. Many of these are free of charge and are accessible from the Internet. In most cases, these resources provide many standard compartmental or noncompartmental models already written into code. For example, single, double, and even triple exponential functions are often available to be applied to an organ curve analysis. In addition, opportunities exist to write differential or algebraic equations that can be used to model general data sets of interest. Among the latter packages are ADAPT II, SAAM II, and R. The following sections describe these in the order listed.

6.11.1 ADAPT II

ADAPT II is a FORTRAN-based package (D'Argenio and Schumitzky 1979) available from the Department of Bioengineering at University of Southern California. The software is distributed free of charge. It may be run on a desktop computer equipped with an appropriate compiler or on a VAX system. Its fundamental mode of operation requires the user to write differential equations (DEs) of the type shown in Equation 6.1. These may be of any form including nonlinear models that may be appropriate for antibody–antigen binding. If the user is willing to accept other representations of data, many preprogrammed models are already available on the package as compiled files. ADAPT II has three internal modules from which users may choose: SIM, SAMPLE, and ID.

The SIM module allows users to test various assumed models by letting the chosen differential or algebraic equations evolve over time and noting the shape of any resultant curves. SIM may be a logical first choice in biodistribution analyses in that it would be usefully implemented before any model is chosen—based purely on the shape of the curves. The problem is that there are an unlimited number of models such that users are unable to go over all their features in any finite time. It is probably best if only a very limited set of models is chosen and then tested. Some reference to the RP literature is helpful.

Given a mathematical model that has been preselected for whatever reason, users may next opt for the SAMPLE module, which gives optimal times for sampling the data for each of the input data streams. This capability is unique to ADAPT II and can be of some use in planning an experimental trial. One must, of course, be aware that the sampling times depend on the selection of a mathematical picture in the first place. This decision could be difficult to make if no data or results from other observers are at hand. Application of SAMPLE is probably best done if a model representation of the data has already been established by earlier work with similar targeting agents.

Finally, the ID module is probably the most important of the three as it leads to fitting data with a model or set of equations users have previously written. It is independent of SIM and SAMPLE. The hunting algorithm is a modified Nelder–Mead strategy that explicitly excludes negative rate constant (k) values. The algorithm automatically shifts between "stiff" and "nonstiff" methodologies. A stiff rate constant set are those wherein one constant is much larger than another. Included in the tabulation of results is the covariance matrix that shows the correlation among the fitted parameters. This matrix is useful for estimating errors in quantities computed from the fitted parameter set, such as the AUC and various temporal moments. This process, using MC methods, was previously described.

Assuming use of ID, the user is given the best-fit solution of the modeling equations. Included graphical outputs show both the data points and the fitted curves for each organ being modeled.

6.11.2 SAAM II

Available from the University of Washington, SAAM II (Foster, Boston, et al. 1989) has features similar to ADAPT II. It has an initial purchase cost as well as a yearly update fee. These charges allow the user to ask directly for telephone and Internet help in applications of SAAM II. Both compartmental and equation-based analyses are possible where users are given this choice via a set of pull-down menus. One of SAAM II's most attractive aspects is its graphic user interface (GUI). Here, the analyst can directly draw the boxes and arrows onto the computer screen to facilitate design of a compartmental model of the type shown in Figures 6.1 and 6.2. Likewise, bolus inputs and clearance routes may be indicated directly on the screen by the user. For beginning users of modeling software, this facility can prove extremely useful.

Integration of time–activity curves is accomplished in SAAM II (as in ADAPT II) by writing the desired integral as part of a differential equation. Thus, if we wish to integrate a function $A(t)$ out to the end of some time interval (t), we write a new unknown function (y) as the integral

$$dy/dt = A(t) \qquad (6.19)$$

and require that the program implicitly find y which is equivalent to AUC $[A(t)]$. This integral is done on a temporal, step-by-step process via Equation 6.19. Convergence of the result is proved if the value of y approaches a constant. If we wish to integrate every organ's AUC, similar equalities must be added—one for each tissue. If we have five organs that need integration, five equalities of the type shown must be added to the DE equation set in SAAM II or ADAPT II. This essentially doubles the number of equalities in the analysis with a corresponding increase in computation time. Because of this extensive time commitment, the user may prefer to integrate the function analytically, if at all possible, as shown in Equation 6.8.

6.11.3 The R Development

So far, we have discussed only two (of many) representative modeling packages that offer fitting algorithms and present their results as tabular and graphical outputs with covariance matrices and confidence intervals. It may be that users wish to pursue the raw data in a deeper way initially before even beginning the modeling. In that case, the R software package (Maindonald and Braun 2007) may be of interest. This is freeware that offers extensive support, via the Internet, for many applications in the statistical context. Coding is done with command lines using a unique language that must be learned by the user. This aspect differs from the use of the widely known FORTRAN programming required in ADAPT II or SAAM II.

As an example of many features that R will support is testing data for Gaussian variation. As noted before, it is a common assumption of many analysts that their data are well described by bell-shaped distributions. Also available are cubic spline and wavelet fitting functions uncommon in other software packages. One of the strongest features of R is its graphics output capabilities. These displays and printouts are much more flexible that the common analytic package and can be used to present complicated data sets and analyses to audiences. Users wishing to make presentations of their results, and even of the raw data, may find this facility helpful.

The R package does not, however, have a GUI or another relatively easy way (at present) to communicate its results to or from databases such as Excel. Some effort would be required to get biodistribution information into R or, conversely, back to the database as results from any given R analysis. This limitation implies considerable time in the translation of information between the two analytic tools.

6.12 SUMMARY

Because of the need to integrate time–activity curves in the dose–estimation process, there is an associated need for interpolation and extrapolation of $a(t)$, $A(t)$, or $u(t)$ for each organ. Modeling provides this method; it may be elaborate or very simple, but it is required of the dose estimator. Other advantages of having a mathematical descriptor of each tissue are also listed and go back to a tradition in scientific and engineering analyses.

Multiple methods can be applied to fitting experimental data. In RP analyses, the dichotomy is between compartmental models involving the entire physiological system and representations relating only to single organs or even subsets thereof. In the former case, the various organs are mathematically interconnected so that a set of rate constants and associated amplitudes are found in what may be an elaborate fitting procedure. Noncompartmental methods generally represent only one organ at a time. While expedient, such limited results leave the analyst less than satisfied as to the overall behavior of the RP in the living animal or patient. For example, the rapidly decreasing blood curve is a fact, but the organ or organs responsible for that decrease are not manifest if one uses a single-organ picture of the process. By invoking a multicompartmental analysis, it may be clearly seen where that agent is being excreted.

Several standard formats are known in the literature for RP curve representation. Probably the most popular is the use of a sum of two or more exponential functions. These analytic forms are well known and permit immediate integration and easy interpolation. A less common format is the power law whereby the time–activity curves are represented by time to an inverse power. An additional example of this format is given for tumor uptake of tracers as a function of tumor mass at a fixed time point. Power laws are directly associated with a fractal picture of nature. Here, by looking at the function, the observer cannot tell what temporal (or spatial) interval is being investigated. This is in direct contradiction to the exponential situation whereby the concept of early and late half-times is popular in the literature. Power laws, however, have difficulty dealing with the values of the function when the independent parameter (e.g., time in organ modeling) goes to zero. Special care is needed to exclude this value from the analysis. A similar result holds for the tumor uptake versus tumor mass curves; these measurements cannot be logically taken down to a mass of zero grams. Thus, some of the dynamic range of the independent variable cannot be accomplished via the power law, and a breakdown of the format is inherent in the physical world. Fractal representations are not conceptually possible as continuing indefinitely in either time or space.

Actual fitting of these pharmacokinetic data requires software. Several available packages are listed and, in a brief fashion, reviewed. Generally, the package algorithm must hunt in a multiparameter space, so the solution to the analysis is not unique. Also, unlike classical physics experiments, the fitted parameters are determined as a group, so the covariance among them is nonzero. It is this interrelationship that must be interrogated to find the uncertainties in the resultant model variables. The analyst is thus left with a least-squares solution, which is a bundle of fitted parameters and not a single best value for each variable in the problem. This type of "solution by committee" is an example of the analytic complexity of biological systems.

REFERENCES

Adams, G. P., R. Schier, et al. 1998. Prolonged in vivo tumour retention of a human diabody targeting the extracellular domain of human HER2/neu. *Br J Cancer* **77**(9): 1405–12.

Akaike, H., E. Parzen, et al. 1998. *Selected papers of Hirotugu Akaike*. New York: Springer.

D'Argenio, D. Z. and A. Schumitzky. 1979. A program package for simulation and parameter estimation in pharmacokinetic systems. *Comput Programs Biomed* **9**(2): 115–34.

Foster, D. M., R. C. Boston, et al. 1989. A resource facility for kinetic analysis: modeling using the SAAM computer programs. *Health Phys* **57**(Suppl 1): 457–66.

Fujimori, K., D. G. Covell, et al. 1990. A modeling analysis of monoclonal antibody percolation through tumors: a binding-site barrier. *J Nucl Med* **31**(7): 1191–8.

Iwi, G., R. K. Millard, et al. 1999. Bootstrap resampling: a powerful method of assessing confidence intervals for doses from experimental data. *Phys Med Biol* **44**(4): N55–62.

Jacquez, J. A. 1985. *Compartmental analysis in biology and medicine*. Ann Arbor: University of Michigan Press.

Jain, R. K. 1990. Physiological barriers to delivery of monoclonal antibodies and other macromolecules in tumors. *Cancer Res* **50**(3 Suppl): 814s–819s.

Kaiser, G. 1994. *A friendly guide to wavelets*. Boston: Birkhäuser.

Kaplan, D. D., L. E. Williams, et al. 1997. Estimating residence times and their associated errors in patient absorbed-dose calculation. *J Nucl Med Technol* **25**(4): 264–8.

Linden, O., J. Kurkus, et al. 2005. A novel platform for radioimmunotherapy: extracorporeal depletion of biotinylated and 90Y-labeled rituximab in patients with refractory B-cell lymphoma. *Cancer Biother Radiopharm* **20**(4): 457–66.

Lyles, R. H., C. Poindexter, et al. 2008. Nonlinear model-based estimates of IC(50) for studies involving continuous therapeutic dose-response data. *Contemp Clin Trials* **29**(6): 878–86.

Maindonald, J. H. and J. Braun. 2007. *Data analysis and graphics using R: an example-based approach*. Cambridge, UK: Cambridge University Press.

Mandelbrot, B. 1967. How long is the coast of Britain? Statistical self-similarity and fractional dimension. *Science* **156**(3775): 636.

Norwich, K. H. 1997. Noncompartmental models of whole-body clearance of tracers: a review. *Ann Biomed Eng* **25**(3): 421–39.

Odom-Maryon, T. L., L. E. Williams, et al. 1997. Pharmacokinetic modeling and absorbed dose estimation for chimeric anti-CEA antibody in humans. *J Nucl Med* **38**(12): 1959–66.

Veng-Pedersen, P. 2001. Noncompartmentally-based pharmacokinetic modeling. *Adv Drug Deliv Rev* **48**(2–3): 265–300.

Wahba, G. 1990. *Spline models for observational data*. Philadelphia: Society for Industrial and Applied Mathematics.

Williams, L. E., R. B. Duda, et al. 1988. Tumor uptake as a function of tumor mass: a mathematic model. *J Nucl Med* **29**(1): 103–9.

Williams, L. E., A. Liu, et al. 2000. Truncation of blood curves to enhance imaging and therapy with monoclonal antibodies. *Med Phys* **27**(5): 988–94.

Williams, L. E., T. L. Odom-Maryon, et al. 1995. On the correction for radioactive decay in pharmacokinetic modeling. *Med Phys* **22**(10): 1619–26.

Williams, L. E., A. M. Wu, et al. 2001. Numerical selection of optimal tumor imaging agents with application to engineered antibodies. *Cancer Biother Radiopharm* **16**(1): 25–35.

Withers, H. R. and R. Suwinski. 1998. Radiation dose response for subclinical metastases. *Semin Radiat Oncol* **8**(3): 224–8.

Functions Used to Determine Absorbed Dose Given Activity Integrals

7

7.1 INTRODUCTION

Three types of functions are used in conjunction with the measured activity integrals ($\tilde{A}$ or $\tilde{a}$) to estimate radiation doses. This chapter is devoted to the measurement and estimation of these functions. We will describe the point-source function (PSF), voxel source kernel (VSK), and the standard $\mathbf{S}$ matrix of the Medical Internal Radiation Dose (MIRD) approach. Each of these parameters has its applications in theoretical, animal, and clinical dose calculations. Additionally, some inert materials require dose estimates such as inside dosimeter volumes. While measurement was possible for early PSF representations, we will see that Monte Carlo (MC) methods have become standard for the calculation of all three formats of the various dose-estimating functions.

Other than certain simple geometries, such as at the location of dosimeters, most absorbed dose applications involve use of an appropriate $\mathbf{S}$ matrix. Standard forms of $\mathbf{S}$ are available in the literature and online. These results, derived from MC calculations on anthropomorphic phantoms, may be used in regulatory situations and to compare one radiopharmaceutical (RP) with another. In the latter case, one would evaluate absorbed dose estimates in the relevant phantom when both agents have the same clinical application. One could, in principle, compare two anti-carcinoembryonic antigen (CEA) agents that are intended for imaging breast cancer patients with respect to their relative dose values using the adult female phantom geometry. Regulatory agencies may, indeed, obtain such comparisons and would need the assurance that the same phantom geometry had been used for each dose estimation calculation.

While phantom-associated **S** matrices may have application in regulatory affairs and in scientific comparisons, there will always be a need for patient-specific **S** values. For particle emitters that have essentially no associated photon-absorbed dose, some standard corrections can be made to phantom **S** matrices to yield values appropriate for an individual undergoing treatment. This conversion requires that anatomic information on patient organ masses is available—preferably from nonnuclear imaging options. Hybrid scanners such as single-photon emission computer tomography (SPECT)–computed tomography (CT) or positron emission tomography (PET)–CT make this possibility into a much more practical matter. On occasion, the mass may change over the course of treatment.

Humans are not the sole species of interest in absorbed dose estimation. Investigators doing work on a given tumor type would want to see what level of radiation dose would induce growth delay and would reduce lesion sizes in small rodents and other animal models. Primary among these models is the nude mouse. There is the additional concern of normal organ toxicity. Because of the small size of these test animals, it may be that the range of particulate radiation is sufficient to go from one organ to an adjacent tissue. For example, there could be a finite hepatic absorbed dose caused by a ^{90}Y RP label delivered to the liver from sources in the kidneys in a nude mouse. By the same token, radiation dose would be lost from the source organs due to penetration of the high-energy beta particles out of that tissue before they had lost all of their energy. Thus, until we have reliable **S** matrices for the mouse, such computations cannot be done. Some recent work on this topic is included herein.

Given a dose function, the dose computation for a given target tissue involves a summation over all source tissues that contain activity. For point-source and voxel source functions, these summations are in the form of convolutions as indicated in Chapter 4. When an **S** matrix is involved in the estimation, the summation process becomes a matrix product of **S** times the vector of integrated source activities. Considered in a more general light, such summations are another form of a convolution except that multiple targets are involved in the matrix product whereas only one target organ is included in a typical point-source or voxel source computation. In the following, we describe all three of these dose functions and their uses in absorbed dose calculations.

7.2 POINT-SOURCE FUNCTION

Probably the earliest format for absorbed dose calculation is the point-source function. Originally, these functions were obtained in air at a variable distance from the small source. In that case, the inverse-square law dominated the mathematical form. Eventually, tissue-equivalent material was interposed between source and detector to simulate passage of the ionizing radiation through living tissue. For extended sources, the estimator assumes that knowledge of $\tilde{A}$ is known on a point-by-point basis from a tissue biopsy or imaging method. We must admit that the latter situation is not likely due to the restricted spatial resolution of SPECT and PET systems. Alternatively, the input $\tilde{A}$ distribution may be of a purely theoretical nature such as a worst-case scenario where a majority of the activity is deposited in a particular set of source points. The dose is then computed at a number of locations in the organs of interest assuming that the medium is uniform and of a tissue-like nature. Use of a point-source function is given in Chapter 4 for the correction to the absorbed dose reading in an implanted thermoluminescent dosimeter (TLD). Such calculations may be compared to the measurements done with the dosimeters when placed in a radioactive medium. This was a purely theoretical computation but one that was needed in comparing TLD readings with the actual absorbed dose estimates at that location.

Point-source functions have been implemented using two strategies. The older technique is to simply measure the exposure at various distances from a known point-source of the radionuclide of interest. For example, Berger (1971) was able to determine the dose rate at various distances (r) from sources of ^{90}Y and ^{131}I and represented these as relatively simple functions. These relationships are monotonically decreasing with r from the point-source with some evidence of a shoulder at short intervals. Although mathematical forms may be used, a simple table of results versus distance is often sufficient for applications. In a software routine, the user may have the PSF read directly from the table using interpolation between the measured data points.

After development of Monte Carlo methods in the 1960s, several standard MC algorithms were employed to determine a form for the PSF. Here, the user considers a set of points that represent absorbed dose rate as a function of distance from the source. By allowing enough emitted-particle histories to be followed, the investigator may achieve sufficient precision ($\pm 2\%$ is generally acceptable) to enable use of this empirical function for any radionuclide of interest. As in the case of measured dose rates, interpolation can be used to find the rates at values of r not explicitly in the MC calculations. This interpolation is taken directly from tabulated values and may be of a linear format or rely on one of the functions mentioned in Chapter 6 such as a set of exponentials or a cubic spline.

7.3 VOXEL SOURCE KERNEL

Experimental knowledge of the activity and its integrals over time at a point-by-point location in source tissues is clearly not an easy objective to achieve. Generally, one would anticipate such measurements only with biopsy data. Thus, the PSF has limited application other than in a theoretical situation such as a radiation safety problem. Designing of shielding of a source within a nuclear medicine department would be one example of this usage. For practical situations inside living tissue, a similar function, but based on larger, voxel-sized sources, has been defined. This function is termed the voxel source kernel. Two examples have been extensively described in the literature (Bolch, Bouchet, et al. 1999; Liu, Williams, et al. 1998). Both have been derived using MC methods inside a uniform tissue-equivalent medium that extends to infinity. Typically, the voxel size has been on the order of several mm on an edge, so it is small compared with gamma camera or positron detector spatial resolution.

In the VSK strategy, the user generates an absorbed dose from a given voxel to spatially adjacent voxels. Generally, VSK results are done in the form of tables with the absorbed dose to adjacent voxels in three dimensions being indicated. Each table is specific to a given radionuclide. Symmetry considerations make these summaries easy to read, and they can be inserted directly into an absorbed dose program for calculations given the voxelized $\tilde{A}$ data set. As in the case of a point-source kernel, applications are somewhat limited due to two reasons that are intrinsic to these simple techniques.

An initial limitation is lack of data on a voxel-by-voxel basis in the animal or patient. As was pointed out in Chapter 5, there is no standard method to determine activity at depth in total organs of living creatures—let alone in segments of these organs. To achieve voxel-sized A (or, via integration, $\tilde{A}$) values, the CT-assisted matrix inversion (CAMI) method, quality SPECT (QSPECT), or PET would be required. Other techniques such as geometric mean (GM) or direct measurement will generally not permit activity values to be determined at each voxel in source organs. A second limitation on point- or voxel source methods is that they are measured or derived assuming a uniform soft-tissue-equivalent phantom that extends to large distances. An animal or patient satisfies neither of these two simplifying aspects, so a method is needed that allows interstitial items such as bones and lungs and also involves photon and particle interactions within a finite-sized phantom.

The VSK developed at City of Hope (Liu et al. 1998) was used to compute patient-specific organ estimates for eight organs in approximately 1 minute on an early (circa 1995) workstation. Input information on the local activity concentration was determined earlier using the CAMI 3-D method described in Chapter 5. Voxel size in the VSK was 4 mm × 4 mm × 10 mm, with the last dimension being along the long (z) axis of the patient's body. It was difficult to justify any smaller size for image-derived activity data since the gamma camera is not able to resolve structures of this magnitude even at the surface of the collimator. Table 7.1 gives an abbreviated listing of voxel kernels for ^{90}Y using the MCNP-4A Monte Carlo program from Los Alamos.

The MIRD Committee has also published voxel source kernel dose functions for ^{90}Y and four other radionuclides: ^{32}P, ^{89}Sr ^{99m}Tc, and ^{131}I (Bolch et al. 1999). In their MC analyses using the EGS4 code, the voxels were generally cubic, being either 3 mm or 6 mm on an edge. This reference contains a number of tabulations for the cubic voxel examples of each of the five radionuclides. Other simulations were carried out in rectangular geometry for ^{131}I. For example, the authors generated a VSK for autoradiography having voxel sizes given by 50 μm × 50 μm × 1.0 mm.

TABLE 7.1　Partial Listing of VSK Values for ^{90}Y

Y VOXEL ↓	X VOXEL →			
	0	1	2	3
0 ($z = 0$)	0.39 mGy/Bq-s	0.71×10^{-1}	0.32×10^{-2}	0.19×10^{-4}
1 ($z = 0$)	0.71×10^{-1}	0.24×10^{-1}	0.13×10^{-2}	0.78×10^{-5}
0 ($z = 1$)	0.23×10^{-1}	0.70×10^{-2}	0.39×10^{-3}	0.31×10^{-5}

Note: Assuming a tissue density of 1 g/cm³.
Source: Liu, A. et al., *Nucl. Med. Biol.*, 25, 4, 1998. With permission of Elsevier.

Nonuniformities in the tissue being considered are always of concern to the absorbed dose estimator. This aspect of the computation is of interest in both PSF and VSK approaches. In the MC computer codes previously mentioned, it is possible that one or more voxels in the array be made into something other than soft tissue. A bone voxel or a lung voxel, for example, may be introduced spatially into the simulation. As Kwok, Prestwich, et al. (1985) showed, there can be large increases in the absorbed dose at interfaces between soft tissue and bone. This result is largely a backscatter phenomenon from the higher density, higher atomic number material in the bone matrix. In such designated nonuniform analyses, the computation becomes a particular result. If one were to generate a set of such specialized voxels to represent an animal or patient, the computational time would extend from the simpler analysis that assumes that all voxels have soft-tissue composition. This complexity is a fundamental reason for using whole-organ geometries and the corresponding S matrix.

7.4 S MATRIX CONSIDERATIONS

7.4.1 Methodology of the S Matrix

By far the most common strategy to go from $\tilde{A}$ to absorbed dose (D) or from A to dD/dt is by using the S matrix. As in the case of modern PSF or VSK functions, the S values are computed using Monte Carlo methods. In this instance, the geometry is not a uniform, soft-tissue-equivalent medium, and the extent of the medium is distinctly finite. Anthropomorphic phantoms of various sizes have been proposed for these computations, and the results are generally made available from the MIRD Committee of the Society of Nuclear Medicine. These phantoms have internal skeletons and also lung-equivalent tissue inserts so that they are inherently more appropriate than the PSF or VSK forms already mentioned. Organ masses are defined by average values obtained at biopsy or autopsy tabulations. Organ shapes are given by mathematical functions that define the tissue of interest. For example, the thyroid is given by a pair of ellipsoids, and the thyroid isthmus is neglected. The phantom selection is now extensive. Cristy and Eckerman (1987) defined a number of phantoms in their "family." Three prenatal cases, including 3 mo, 6 mo, and 9 mo fetuses are included. A newborn, 5-year-old, and 10-year-olds are now available. Finally, there are a 15-year-old and a pair of adult forms appropriate for both men and women. Several of these are given in Figure 4.1.

Monte Carlo calculations being done in a reasonable time require simplifying assumptions. In the traditional MIRD phantom-based approach, activity in every source organ is assumed to be uniformly distributed. An obvious reason for this assumption is that there are an infinite number of nonuniform distributions but only one uniform result. Associated with this simplification is the clinical result that most investigators obtain only a single $A(t)$ value for each organ. Limited spatial resolution is the usual reason for this constraint; that is, there is no $A(x,y,z,t)$ to be considered. Similarly, a single absorbed dose to the entire target organ is then calculated using the matrix product of S

and $\tilde{A}$. Thus, the typical **S** matrix calculation involves both uniform distribution of activity in each source and an average absorbed dose to every target.

Other simplifications occur in these Monte Carlo analyses. For example, to expedite calculations of **S**, both kidneys are traditionally lumped into a single renal source tissue. Likewise, the two lungs are made into one organ. By symmetry of the problem, these particular tissues are also summed together to form one target organ made from the two kidneys and one lung target organ made from the two lungs, respectively. This assumption can sometimes prove useful when only one of the pair can be observed by nuclear imaging in a given patient.

Yet human anatomy is not bilaterally symmetric. The two kidneys have different positions in the body; the left is isolated from other major organs, whereas the right is closely associated with the liver. Likewise, the relatively smaller left lung is wrapped around the heart and major vessels, and the right lung is free of such contact. It may be anticipated, in a real-world situation, that each member of these two paired organ systems might have different estimated absorbed doses. If, for example, there is extensive uptake of a RP by the liver, one would expect an elevated right kidney absorbed dose compared with that received by the left kidney. Major blood pool or heart uptake—or extended blood circulation times—could contribute enhanced absorbed dose to the left lung compared with the right lung. These differences are not calculable via the standard MIRD phantom results and require some other approach. In the future, separation of the renal and pulmonary tissues into their respective component organs should become part of the standard dose estimation method.

7.4.2 S Matrix Symmetry

The nature of the **S** array is of some interest, and it has not been extensively discussed in the literature. Theoretically, **S** is a square matrix with the number of rows, which correspond to the number of target organs, equal to the number of columns, which correspond to the source organs. In other words, there is an entry for every possible organ (approximately 30) in the selected phantom's body. In clinical practice, however, **S** immediately becomes rectangular since the number of clearly identifiable sources in a patient is usually less than 10 and is often under 5. This limitation follows from the lack of demonstrable accumulation data in each of the multiple anatomically possible sources. For example, in antibody work, there is usually activity seen in heart content, liver, spleen, and perhaps kidneys and bladder. The latter pair is particularly expected if the label is a radioactive form of iodine. A radiologist would be hard-pressed to find activity in the adrenals, bone marrow, stomach, intestines, and thyroid in the case of a radiometal-labeled antibody, for example.

The small number of demonstrable source organs, to some degree, is a reflection of some of the classical problems ("the tools of ignorance") of nuclear medicine imaging alluded to in Chapter 2. In either camera or PET applications, it may be difficult to resolve small organs or accumulations if their size is twice the resolution of the system or less. We have described the use of recovery coefficients in this context of small sources in Chapter 5. If the investigator has recourse to a hybrid, high-resolution imager, such as PET–CT, then the situation becomes simpler in that any nuclear image voxel may be correlated with an anatomic source organ. Many of these tissues will show essentially the same activity per unit volume, so the activity is best assigned to a general background organ such as the residual body of the patient or animal.

It may be proven mathematically that S is symmetric about its diagonal in the case of a large Monte Carlo universe—that is, a large enough uniform phantom (Loevinger and Berman 1968). The result is

$$S_{i,j} = S_{j,i} \tag{7.1}$$

Such a result is not literally true for typical matrices calculated for finite MIRD phantoms having skeletal as well as lung inserts. It is interesting to see how close to symmetry the results of such MC calculations actually are. Let us consider off-diagonal S matrix elements for a sequence of commonly used radionuclides and the adult male phantom as made available via the OLINDA software package (Stabin, Sparks, et al. 2005). As shown in Table 7.2, the symmetry shows the greatest disparity for the kidney–stomach combination of source–target organs. This is because there is an asymmetry in this situation since the stomach as a source is actually the content of that organ, whereas the stomach as a target is the wall of that organ. Thus, one can expect some differences as we cross the diagonal of the S matrix. As we increase the dominant energy of photon emission, this difference decreases. The latter result is demonstrated by reviewing a sequence of commonly used radionuclides from ^{99m}Tc through ^{131}I and to ^{18}F. Here, dominant photon energies are 140, 360, and 511 keV, respectively, as listed in Chapter 1. As we get to 511 keV for ^{18}F, the stomach–kidney disparity disappears completely. Physically, this result shows that the higher photon energies pass more readily through materials so that the assumptions made in the symmetry argument become progressively more reliable. In more conventional situations such as the combination of kidneys and liver, the symmetry of S is exact to three significant figures as shown. This agreement holds for all three radionuclides shown in Table 7.2.

TABLE 7.2 Symmetry of the Human **S** Matrix for Selected Radionuclides

ORGAN COMBINATION	**S** ELEMENT ^{99m}Tc mGy/MBq-s	**S** ELEMENT ^{131}I mGy/MBq-s	**S** ELEMENT ^{18}F mGy/MBq-s
Kidneys to stomach (wall)	2.53 E-07	6.98 E-07	1.74 E-06
Stomach (contents) to kidneys	2.73 E-07	7.22 E-07	1.74 E-06
Difference	+7.9%	+3.5%	0%
Kidneys to liver	2.93 E-07	8.19 E-07	2.06 E-06
Liver to kidneys	2.93 E-07	8.19 E-07	2.06 E-06
Difference	0%	0%	0%
Kidneys to ovaries	7.02 E-08	2.17 E-07	5.70 E-07
Ovaries to kidneys	7.02 E-08	2.17 E-07	5.70 E-07
Difference	0%	0%	0%

Source: Stabin, M. G. et al., *J. Nucl. Med.*, *46*, 6, 2005.

7.4.3 Target Organ Mass Dependence of S for Particles

Let us momentarily move from a photon orientation (considering a label appropriate for imaging) to a particulate emitter (intended for therapy) in this discussion. One anticipates that the dominant absorbed dose terms should now be those along the diagonal of **S**—the so-called self-dose. In this analysis, we will restrict ourselves to two well-known therapy radionuclides: ^{131}I and ^{90}Y. Both of these emitters have been extensively applied to nuclear medicine therapy as is shown in Chapters 8 and 9. Use of ^{131}I goes back to the development of reactors in the middle 1950s and is still the radionuclide of choice (as Na^{131}I) for thyroid therapy. Let us first consider ^{131}I diagonal elements available from the OLINDA literature (Stabin et al. 2005). A selection of self-dose terms is given in Table 7.3 along with the mass (m) of the corresponding tissue as defined in the adult male phantom. Notice that the variation in the product of $S_{i,i}$ times the organ mass (m) does not vary greatly over a tissue mass range of 10 to 1,900 grams. This stability occurs even in this case of a mixed photon-plus-beta emitter.

If we shift our consideration to the pure beta emitter ^{90}Y, these variations become less but do not disappear. In Table 7.4, we find that the product of diagonal **S** elements and organ mass has a more limited variation than that determined for ^{131}I. In particular, for the eight organs chosen, the coefficient of variation (CV) is now 4.5% instead of the 9.8% seen for ^{131}I. Notice also that the product essentially has a limit of 0.150 mGy-g/MBq-s for larger organs in the yttrium-90 analysis. In the limiting case, the organ is massive enough so that even the high-energy betas from ^{90}Y cannot escape from the tissue and all available energy is deposited in the source (= target) organ. The corresponding value for ^{131}I is estimated as 4.95 E-02 mGy-g/MBq-s using the whole body result from Table 7.3.

It is also important to compare the tables (7.3 and 7.4) to see which of the two radionuclides is the more attractive as a possible therapy label. By comparing the mass*S products for the two radionuclides, 0.150 mG-g/MBq-s (^{90}Y) and 0.0495 mGy-g/MBq-s

TABLE 7.3 Diagonal Elements of Human **S** Matrix for ^{131}I

ORGAN	$S_{i,i}$ mGy/MBq-s	ORGAN MASS (m) (grams)	$m\,S_{i,i}$ mGy-g/MBq-s
Adrenals	1.99 E-03	16.3	3.24 E-02
Brain	2.91 E-05	1420	4.13 E-02
Kidneys	1.18 E-04	299	3.53 E-02
Liver	2.15 E-05	1910	4.11 E-02
Lungs	3.40 E-05	1000	3.40 E-02
Spleen	1.96 E-04	183	3.59 E-02
Testes	8.60 E-04	39.1	3.36 E-02
Thyroid	1.59 E-03	20.7	3.29 E-02
Whole body	7.16 E-07	70 E 03	5.01 E-02

Source: Stabin, M. G. et al., *J. Nucl. Med., 46,* 6, 2005.
Notes: Excluding the whole body, the average results are 3.58 E-02 ± 0.35 E-02 for eight values in the last column. This analysis leads to a CV of 9.8%.

TABLE 7.4 Diagonal Elements of the Human **S** Matrix for ^{90}Y

ORGAN	$S_{i,i}$ mGy/MBq-s	ORGAN MASS (m) (grams)	m $S_{i,i}$ mGyg/MBq-s
Adrenals	8.26 E-03	16.3	0.135
Brain	1.05 E-04	1420	0.149
Kidneys	4.81 E-04	299	0.143
Liver	7.83 E-05	1910	0.149
Lungs	1.50 E-04	1000	0.149
Spleen	7.79 E-04	183	0.150
Testes	3.54 E-03	39.1	0.138
Thyroid	6.57 E-03	20.7	0.136

Source: Stabin, M. G. et al., *J. Nucl. Med., 46,* 6, 2005.
Notes: Mean results for the eight values in the last column are 0.144 ± 0.0064. This analysis results in a CV of 4.5%.

(^{131}I), a ratio of 2.90 is obtained. Therefore, using ^{90}Y, the investigator can achieve nearly a threefold increase in the self-dose value for **S** compared with using ^{131}I at the same activity levels. This ratio is further corroborated in Chapter 8 where we compare two anti-CEA antibodies per their respective absorbed doses per injected unit of activity. To make such a simple comparison, we assume that the RP being used would deposit radioactivity independently of the radiolabel. As noted in Chapter 3, this assumption is not necessarily valid due to loss of label (dehalogenation) with iodinated radiopharmaceuticals. The self-dose enhancement is essentially the ratio of average beta-ray energies for the two radionuclides. Although tumors were not included in these tabular comparisons, lesion absorbed dose enhancement should also be comparable to the ratio given by the normal organs. Keep in mind that such elevated normal organ absorbed doses may correspondingly restrict the amount of activity allowed with a radioactive yttrium-90 label. On the other hand, use of ^{131}I may not be so readily accomplished in a logistical sense as ^{90}Y due to the photon emissions of the former. These gamma and x-rays will, in general, lead to increased radiation protection and possible patient hospitalization requirements.

7.4.4 Target Organ Mass Dependence of S for Photons

If we consider photon emitters of ionizing radiation, there is a pair of operational relationships that may help the dose estimator. One rule holds in the case when the two organs are the same tissue. The second rule is invoked in the logical alternative when the two tissues are disjoint. One of the earliest expositions of these two dependencies is the MIRD 11 publication (Snyder 1975).

If the source and target are the same, the **S** dependence is usually assumed to be roughly proportional to the target mass to the negative 2/3 power. Recall that the analogous rule for particles, such as beta rays, is that **S** is proportional to target mass to the negative first power as described already. This photon result is not a theoretical assertion based on geometric arguments but is an outcome from multiple MC analyses of different target masses. The originators hoped, by giving a rough variation with mass, to allow correction to the doses predicted using a standard MIRD phantom for a specific patient. Using a set of 12 humanoid German (GSF) phantoms, one group has assayed the variation of specific absorbed fractions [φ/m(target)] at fixed photon energies from 10 keV to 5 MeV. Above 100 keV, their results were consistent with the MIRD 11 assertion using the thyroid as an example of a tissue having a single location. But variations of up to 18% away from the $m^{-2/3}$ rule appeared for a distributed tissue as exemplified by the red marrow (Petoussi-Henss, Bolch, et al. 2007). Keep in mind that, in both these examples, the source and target were the same organ. A second study, using MC analyses of a specific phantom, analyzed the **S** dependence on the total body (TB) mass when the TB was used as source and target organ (self-dose). Calculations were carried out for the photon component of three radionuclides: ^{131}I, ^{137}Cs, and ^{186}Re. Mass dependencies reported were, in the same order, $m^{-0.75}$, $m^{-0.77}$, and $m^{-0.70}$ (Siegel and Stabin 2007). It was clear that all three were close to the $m^{-0.67}$ value given in MIRD 11, although none was exactly equal to that conjectured amount.

A logical second case for study is the situation whereby source and target are different organs. Here, the traditional rule has been that no target-mass correction is needed for these off-diagonal **S** matrix elements for photons (Snyder 1975). Heuristically, this result has been argued as being due to a form of compensation: as the target gets larger, the fraction of photons absorbed proportionally increases, yet that result (φ) is divided by the same mass such that no net effect on **S** takes place. Using the GSF phantoms (Petoussi-Hens et al. 2007), it was found that the specific absorbed fraction [φ/m(target)] did not vary significantly (error < 5%) from target mass independence for a variety of source organs and a red marrow target. Thus, the conjecture appeared to have been verified.

7.5 APPLICATIONS OF S MATRICES

7.5.1 Standard (Phantom) S Values

As shown already, pure beta emitters exhibit diagonal **S** elements that are essentially inversely proportional to organ mass. Off-diagonal elements are generally set to zero, although brake radiation (cf. Chapter 4) would need to be included in more exact considerations. Similar results hold for alpha emitters although these radionuclides generally have multiple other emissions including photons and beta particles. Two obvious issues arise for mass values that are to be permitted in such calculations. First, it may be important—even necessary—that **S** matrices be used that are representative of one (or another) phantom. This application is available from a number of software platforms including the OLINDA (Stabin et al. 2005) program at Vanderbilt University. As previously noted, some 10 phantoms are presently available, and other, more anthropomorphic forms are currently being implemented. So regulatory agencies may understand new RP applications, these standard phantoms are appropriate for use in documents submitted to these agencies. Only by using this method can regulators understand the context and magnitude of absorbed does estimates in the application under review. Use of standard phantoms is thus a required aspect of the IND or other RP approval submission process.

A second reason for using standard phantoms in absorbed dose calculations is the possibility of comparing one RP with another. This analysis can be done in-house prior to the submission process or by the regulatory bodies themselves. For example, a pharmaceutical manufacturer may have generated a revised form of a previously developed radioactive therapeutic drug directed against prostate cancer. To see if the absorbed doses have been changed significantly from the original form of the agent, the analyst would look at the dose profiles, organ by organ, in a relevant adult male phantom. By comparing the two profiles, the changes can be clearly seen only if the same relevant phantom is common to both analyses. This application of standard **S** values may be termed the scientific reason for phantom usage in absorbed dose estimations. Presumably, such comparisons are of interest and are routinely made by manufacturers as well as by regulatory agencies such as the U.S. Food and Drug Administration (FDA). There are essentially no published reports of this type, but the situation should improve in the future due to the great amount of nanoengineering as described in Chapter 1.

7.5.2 An Aside: Changes in $\tilde{A}$ Needed in Phantom Studies

As noted in Chapter 4, one should not take time–activity data from patients or animals and substitute these directly into the MIRD formalism. Such substitutions are not logically acceptable because the subject being imaged will not have the distribution of organ masses (and total body mass) that is characteristic of any member of the phantom "family." The standard correction is

$$A(\text{phantom}) = A(pt)\frac{m(\text{phantom})/M(\text{phantom})}{m(pt)/M(pt)} \tag{7.2}$$

where $A(pt)$ is the measured patient's organ activity at a given time point, m is the organ mass of interest, and M is the total patient mass. Although we use the word *patient*, keep in mind that the actual person involved might be a healthy volunteer. To the lowest order, this manipulation corrects for perfusion differences between the test subject and the phantom selected. To apply Equation 7.2 requires knowledge of both $m(pt)$ and $M(pt)$. Whereas the latter is standard clinical information, the former is not. Thus, corrections of the type shown in Equation 7.2 may not be achievable.

In many historical examples from the MIRD literature, such specific organ mass information was simply not available, so the Equation 7.2 corrections were not possible. Here, patient or volunteer anatomic data could not be obtained because of lack of CT or MRI imaging during the nuclear scanning studies. Very early MIRD documents were published before the invention of these anatomic imaging devices. Even if the CT or MRI equipment were available, there would also be ethical (e.g., radiation dose) or financial constraints (e.g., imaging costs) on their usage in a dose estimation protocol. Therefore, these early results are not, strictly speaking, accurate for the phantom. Chapter 8 describes some of these estimates.

It can be asserted that organ masses are obtainable by nuclear medicine imaging alone. As noted in Chapter 5, this strategy, while applied frequently in the past, does not necessarily give rise to correct sizes due to the difficulty of finding organ outlines. Therefore, the standard for sizing has become the anatomic image set. These images are also of importance in the medical oncologist's or radiation oncologist's assessment of a patient during therapy. Thus, the anatomic set should always be available for the clinical follow-up of the cancer patient. There is also the possibility that these tissue sizes will vary during therapy. While tumors are one obvious example of mass variation, other changes in normal organs may have tumor content such as the spleen in the case of non-Hodgkin's lymphoma (NHL). Such mass variations can be followed by anatomic imaging.

Equation 7.2 is generally applied in all animal-derived data sets. In most calculations of this type, the animals are sacrificed at various time points following the relevant research protocol as outlined in Chapter 2. Activity in each biopsied organ as well as the tissue's mass are directly measured. Total animal mass is also assayed so that all necessary variables needed in the equality are available. Because of this information, phantom estimates based on animal data are internally consistent and may be used with greater confidence than typical phantom estimates based on patient or volunteer imaging data.

7.5.3 Elaboration of Standard S Matrices for Kidney

It has been a standard assumption involved in generation of the various organ-sized **S** matrices that there is no internal structure within either the source or target soft tissues. Here, we explicitly neglect any pathological variations in the organ due to disease or tumor sites; that is, we deal only with normal tissues. In part, this simplification follows from the physical appearance of the organ and, to an important degree, from the uniformity of the tissue as seen via nuclear imaging. The renal system is a clear exception to this rule.

The kidneys are important in the handling of many RPs or, in any event, the radioactive debris left behind in the blood by metabolism of the radiopharmaceutical. Excretion of radioactive iodine and iodinated metabolic products left after dehalogenation is one example of this processing. As we will see in Chapter 9, there is now evidence of renal toxicity, caused by targeted radionuclide therapy (TRT) involving the use of small proteins. The MIRD Committee has published a new geometric model of the renal system (Bouchet, Bolch, et al. 2003) such that there are now four segmented volumes instead of the traditional single organ: cortical, papillary, medullary, and pelvic zones. These are defined mathematically by various partial and complete ellipsoids. Total renal adult mass is still restricted by the International Commission on Radiological Protection (ICRP) value of 288 grams. Six phantoms were developed: newborn, 1-year-old, 5-year-old, 10-year-old, 15-year-old, and adult. Across the phantoms, the rough breakdown of the various regions by fractional volume is 70% (cortex), 25% (medulla), 4% (pelvis), and 1% (papillae).

In Bouchet et al.'s (2003) report, the MIRD team tabulated new **S** matrices for the six different phantoms and 26 radionuclides including ^{90}Y and ^{131}I. These matrices have four source segments (cortex, medulla, pelvis, and papillae) and five targets including the aforementioned segments as well as the total kidneys. Notice that, as in the traditional MIRD format, no difference was allocated between left and right kidney, so the two kidneys are taken as a single system. In addition, there were tables of absorbed fractions (φ) for photon and electron energies from 10 keV to 4.0 MeV. In the former case, the authors analyzed various situations where the time in the segmented parts of the kidneys was allowed to vary and compared those results with what would have been calculated had the overall residence time been assumed in the original homogeneous kidney model. For ^{90}Y, if one let the radionuclide pass through the renal pelvis and medulla very rapidly (0% residence times in each), the absorbed cortex dose was a factor of 1.29 higher compared with that calculated assuming an unsegmented kidney analysis. Likewise, if the fractional residence time in the pelvis (25%) and medulla (50%) were relatively large, resultant cortical absorbed dose was reduced to only 0.48 of the intact kidney value. Similar ratios arose if ^{131}I was the radioactive label where the ratios were 1.27 and 0.38, respectively. The authors concluded that absorbed dose estimates, if anatomic activity information were available, would vary greatly from the values that have been used in traditional **S** matrices where the kidneys have no internal structure.

7.6 MODIFICATION OF S FOR PATIENT-SPECIFIC ABSORBED DOSE ESTIMATES

Since the mid 1990s, there has been growing interest in generating a set of absorbed doses that pertains to a specific individual. Here, the $\tilde{A}$ value will be correct as taken since we are concerned with a specific patient. Yet the phantom's **S** value will not be appropriate since the patient does not necessarily have the organ masses or geometry of any standard phantom. Motivation for this work lies in the requirement of determining the total activity to be given to a patient for possible radioimmunotherapy and other forms of internal emitter therapy. In this case, at least the diagonal elements of **S** must be modified to suit the individual. Such results are a form of treatment planning similar to that done in external beam therapy. It has been documented, for example, that major organs in patients may show considerable mass variation from the phantom values assigned to a person of that sex and age.

To the lowest order, the **S** correction for a beta or alpha emitter would be

$$S_{i,i}(pt) = S_{i,i}(\text{phantom}) \frac{m_i(\text{phantom})}{m_i(pt)} \tag{7.3}$$

where the correction term is the ratio of organ masses: m_i (phantom)/$m_i(pt)$. Notice that the correction for photons is entirely lacking in this assertion. As already seen, this photon correction will depend on whether the dose is self-delivered or from another tissue. In the former case, the variation goes as $m(pt)^{-2/3}$; in the latter, there is no correction needed. Generally, even if these changes are incorporated, they will be small compared with that given for the beta-derived self-dose.

The magnitude of the Equation 7.3 correction is also generally larger than that seen for the activity value. In one survey of liver and spleen sizes of patients undergoing radioimmunotherapy (RIT), the City of Hope group found corrections of up to a factor of three for these tissues. In a renal analysis (Wong, Shibata, et al. 2003), the four women had a kidney phantom-to-patient mass ration of 1.01 ± 0.22, whereas the 10 male patients had a corresponding mean and standard deviation of 0.84 ± 0.26. Range values for the renal mass correction were 0.69 to 1.2 and 0.58 to 1.45 for the female and male patients, respectively (Williams, Liu, et al. 2002).

7.7 INVERTING THE S MATRIX TO MEASURE ACTIVITY

An additional detail regarding the **S** array is the possibility of inverting this matrix to permit the determination of $\tilde{A}$ values given an externally measured dose or dose rate. This strategy is a reversal of the traditional application in that we now wish to determine an unknown set of $\tilde{A}$ values at a number of possible locations within the body. Here, the activity levels become the unknowns, whereas absorbed dose (D) and **S** are measured and calculated, respectively. We recall the difficulty in determining activity at depth, $A(t)$, as outlined in Chapter 5. If one assumes the equality of Equation 7.4, there arises a possibility that we can compute the integral of the activity in the source organ using a dose evaluated by an external device. In this formalism, dose is assumed to be measured by standard means at several selected locations on the skin surface. Recall that the equation for dose given the integral of activity is

$$D = \mathbf{S}\tilde{A} \tag{7.4}$$

If we know the relevant **S** array and D is measured at several points on the phantom (or patient) surface, then one can assert

$$\mathbf{S}^{-1} D = \tilde{A} \tag{7.5}$$

In this consideration, we do not use standard values as published but instead must calculate a new set of **S** values that relate surface regions of the phantom (the dosimeter "targets") to source organs within the phantom. Potential surface regions of interest would be at the umbilicus, xiphoid process, and other traditional landmark areas on the skin. One may consider each site to be an appropriately sized volume at the phantom or patient surface—for example, a 6 cc cylindrical volume representing an ion chamber or a volume 1 cc or smaller for a TLD.

In an early attempt to realize the previously given situation, our group calculated point-to-point **S** matrix values for locations within and on the skin of the RANDO phantom. Our interest was in eventually using the method to determine the actual amount of ^{90}Y present in the patient during targeted radionuclide therapy. Recall that ^{90}Y is a pure beta emitter, so the only detectable emission at the skin surface would be brake radiation. Generally in this situation the therapist has had to use a surrogate

[111]In-labeled agent of the same type to measure activity in the various source organs over time. While that procedure may be useful, it would be better to actually measure the yttrium *in situ* directly for the patient. Some simplification to the clinical protocol would also be possible.

It is noteworthy that the original MIRD **S** values did not include any target positions external to the phantom—for example, on the aforementioned surface locations. Had such values been available, one could have tested the internal consistency of the dose estimation calculations by measuring absorbed doses (or dose rates) at one or another point on the corresponding patient's skin. Notice that, unlike the implanted TLD dosimeters mentioned in Chapter 4, these measurements could be done in air without invasion of the tissues or the loss of signal due to aqueous media surrounding the detector. One would hope that such sites will be made available in the next generation of phantoms.

7.8 VARIATION OF TARGET MASS DURING THERAPY

So far, we have assumed that the denominator of **S** is fixed, that is, that the target organ mass is constant during the irradiation. That this assumption may not always be valid was shown by DeNardo, DeNardo, and coworkers (1998) at the University of California–Davis (UC–Davis) with their results for NHL lesions being treated with the [131]I- Lym-1 antibody. A general description of this therapy is included in Chapter 8 where other radiolabels are also considered. While using the ^{131}I label for Lym-1, some lymph node lesions were observed to decrease in size over the 1- to 2-week period during which the patient was being followed for activity measurements in the nuclear medicine clinic.

The UC–Davis group (Hartmann Siantar, DeNardo, et al. 2003) analyzed lymph node mass variation during NHL therapy. Using a technique described more completely in Chapter 11, the investigators were able to measure external nodal size directly using calipers assuming an ellipsoidal geometry for the lesion. In seven patients, a total of 37 nodes could be measured in this fashion. Nodal masses were individually fitted as a function of time with a single exponential function analogous to the description of radioactive decay

$$\text{mass} = m_0 \exp(-\mu t) \tag{7.6}$$

This term was substituted as the denominator in the standard dose estimation formula

$$\text{Dose(node)} = A_0 \int_0^\infty X(m_0)\exp(-\lambda_{\text{eff}}t)\big/m_0\exp(-\mu t) \tag{7.7}$$

where $X(m_0)$ is the product of **S** and lesion mass at the initial mass value (m_0). The analysts assumed that the clearance of activity from the lesion could be described by a single effective decay constant (λ_{eff}). In 32 lesions, the authors performed the integral of Equation 7.7; that is, the summation converged to a finite value. In the remaining five nodes, it was found that the rate of decrease of the tumor size was so great that the dose rate was still increasing at the last time point at which the mass was measured. Mathematically, the problem occurred in these five nodes since the μ value exceeded the effective decay constant λ_{eff}. Thus, the node was shrinking faster than the activity was clearing from the lesions. To look at this situation as we did with the inversion of **S**, if one could place a hypothetical dosimeter into one of these nodes, it would be seen that the dose rate would be a monotonically increasing function of time out to the last observation. The authors had to forgo further analyses for this subset of the NHL tumor sites since extrapolation of the dose would lead to an infinite result.

Taking the 32 analyzable nodes into their summary, the authors found that 70% of these NHL sites required an absorbed dose mass correction greater than a factor of 1.2. Furthermore, 50% of the nodes required a dose correction that exceeded 2.0. In this analysis, a dose correction meant that the absorbed dose as determined by Equation 7.7 was larger than the result that would have been obtained with the standard MIRD format assuming a constant lesion mass equal to the initial value m_0.

As noted in Chapter 4, the defined dependence on the inverse of the target tissue mass makes calculating absorbed dose sometimes difficult. In particular, the denominator in a dose estimate may become very problematic if the mass may be seen to become vanishingly small during a course of therapy. Use of an exponential form for this behavior, as given by Equation 7.6 for metastatic lymph nodes, leads directly to this issue. By shrinking the target mass rapidly enough, the dose rate can become increasing with time so that a finite total absorbed dose result is difficult to establish in the limited amount of observation time postinjection. If one had a more extended interval to measure the nodes, it will probably eventuate that the limiting mass was not zero but some finite value. This final magnitude could be the original (predisease) size of that node or perhaps the sum of that value plus the necrotic remnants of the lymphoma. In either case, Equation 7.6 could be suitably modified to approach this desired endpoint as a limit. For example, the estimator could add a constant term (m_{res}) to Equation 7.6, so the limit as t → ∞ is simply this measured residual mass value.

A similar report on tumor regression during the first weeks of TRT has been published by the group at Lund University (Hindorf, Linden, et al. 2003). In this example of three NHL patients, the authors used ^{90}Y-epratuzumab, an antibody to a B-cell surface antigen. Unlike the UC–Davis report, the group at Lund University used CT scan information to measure tumor sizes as a function of time. These anatomic measurements at depth were required since the investigators did not have surface lymph nodes available for assessment. A single lesion in each patient was found to regress in size during the several-week course of therapy. In one case, the absorbed dose enhancement due to tumor regression was a factor of 1.7. Lund investigators did point out that some of the mass reduction might be attributed to the antitumor properties of the unlabeled epratuzumab anti-CD22 antibody (Leonard, Coleman, et al. 2004). Thus, there may be more than a single causative factor in reducing lesion size. Chapter 9 describes similar conclusions regarding the anti-CD20 antibody tositumomab. In that context of NHL therapy, a sigmoidal form (probit function) was used to determine what fraction of tumor size regression could be attributed to the immune aspects of the therapy. As we go to more combinatory therapies, such breakdowns may be very important in assessing the effectiveness of a given aspect of the combination.

7.9 MURINE S VALUES ESTIMATED USING MONTE CARLO TECHNIQUES

Because of the necessity of perfecting TRT in an animal model prior to clinical trials, there is an ongoing need for realistic murine **S** matrices. Our group at City of Hope, working with our colleagues from Battelle Institute, was one of the first to attempt this computation. Initially, 10 nude mice were sacrificed to find the size and separation of the major organs. As is common in MC work, an ellipsoidal function was used to approximate the spatial form of a majority of the organ systems: liver, spleen, kidneys, heart, lungs, and a tumor planted on the flank (Hui, Fisher, et al. 1994). Femur and enclosed marrow were approximated by mathematical cylinders whose size was dictated by the anatomic data set. A total of 13 tissues were considered as sources in the analysis. Fourteen target tissues were included with the one additional target being the carcass. The EGS4 Monte Carlo program was used to propagate the ^{90}Y beta radiation spectrum from source organs to target organs throughout the body of the animal. Recall that the range of betas from yttrium-90 is 11 mm so that considerable cross-organ absorbed dose values were seen in murine organs whose sizes were in the range of 1 mm to 1 cm in greatest dimension. For example, the heart was an ellipsoid 8.5 mm × 5.5 mm × 5.0 mm in the representation. Figure 7.1 shows an idealized cross-section of a source organ with

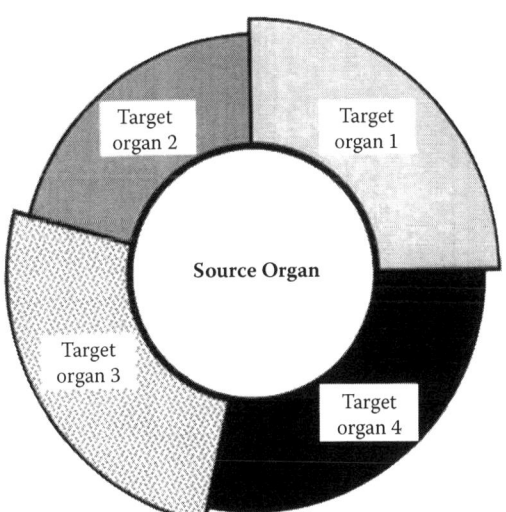

FIGURE 7.1 Organ arrangements, shown in transaxial view, in the Hui et al. nude mouse model. Notice the adjacent tissues alongside the source organ at the center. (Reprinted from Hui, T. E. et al., *Cancer*, 73, 3, 1994. With permission of John Wiley and Sons.)

TABLE 7.5   Murine Cross-Organ and Diagonal φ Values for [90]Y

| TARGET ORGAN ↓ | SOURCE ORGANS → | | | | | |
	LIVER	SPLEEN	KIDNEYS	LUNGS	HEART	STOMACH
Liver	0.67	0	0.11	0.053	0.055	0.12
Spleen	0	0.34	0.040	0	0	0.079
Kidneys	0.017	0.132	0.46	0	0	0.013
Lungs	0.06	0	0	0.29	0.27	0.02
Heart	0.007	0	0	0.14	0.45	0
Stomach	0.017	0.33	0.054	0.018	0	0.61

Source: Hui, T. E. et al., *Cancer*, 73, 3, 1994. Reprinted with permission of John Wiley and Sons.

four adjacent tissues. The fraction of source surface area subtended by each secondary organ was determined using the anatomic results of the dissections.

Absorbed fractions (φ) were computed for each of the source organs. Table 7.5 gives a subset of the results. Many organs had a major fraction of their radiation dose contributed by tissues outside of themselves; that is, the self-dose was a minority of the total dose estimated. Lungs are one example of this type; the heart contribution to the lung was almost as large as the lung-to-lung contribution. For the liver, there were two major outside sources of beta radiation from [90]Y: the kidneys and the stomach. Each of these would contribute approximately 12% of the total hepatic dose if the same amounts of activity were in each of the three source tissues.

It was clear that the situation in the murine model was considerably different from that found using a similar analysis clinically. In the patient case, all absorbed fractions would be unity for self-doses and zero everywhere else. An equivalent statement is that **S** would be strictly diagonal for clinical applications. For the mouse, wherein the body dimensions are comparable to the range of betas from ^{90}Y, the **S** matrix has major off-diagonal elements for a majority of the murine organs.

Since the original results derived using anatomic data along with EGS4, other groups have also done murine computations with other geometric assumptions and different MC algorithms. Flynn, Green, et al. (2001) at Royal Free and University College London derived results for ^{131}I as well as [90]Y. This group included a differential kidney model to separate cortex from medulla in the mouse. Sgouros and coworkers (Kolbert, Watson, et al. 2003) analyzed a more limited number of organs but for more radionuclides. In this example the Memorial Sloan-Kettering Cancer Center (MSKCC) group considered only the liver, spleen, and two separated kidneys but for an extensive list of radionuclides including ^{90}Y, ^{131}I, [153]Sm, ^{32}P, and [188]Re. A point-source function was the basis of the effort; convolution with a uniform activity distribution in the various source organs led to absorbed fractions. The geometry of the MSKCC work was based on the MRI image of a single mouse. Most recently, the group at Lund University (Hindorf, Ljungberg, et al. 2004) looked into the murine calculations for a number of radionuclides including ^{99m}Tc, ^{111}In, ^{131}I, and [90]Y. In the Swedish work, the authors investigated the spatial dependence of their computer model of the mouse by varying the separation

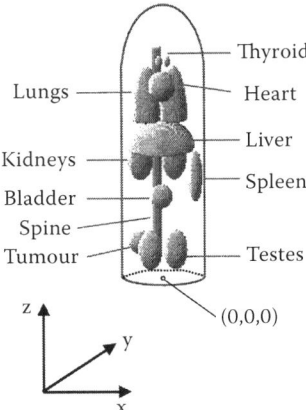

FIGURE 7.2   Lund U mouse phantom for S matrix generation. Ellipsoids were used for most of the organ representations in the Monte Carlo analyses. (From Hindorf, C., Ljungberg, M., and Strand, S. E., *J. Nucl. Med.*, 45, 11, 2004. With permission of the Society of Nuclear Medicine.)

of the right kidney from the left kidney and from the liver. It was found that the absorbed fraction of ^{90}Y emissions was particularly sensitive to such variations. For example, by moving the right kidney 2.3 mm vertically (along the body length or z dimension) in the model, the right kidney to liver **S** value decreased to less than 20% of its original value. This analysis showed that the absorbed fractions, and equivalently the **S** matrices, of mice are very dependent on the actual dimensions imposed upon the problem. An image of the Lund University mouse phantom is given in Figure 7.2.

Table 7.6 contains a summary of these four reports for ^{90}Y. Results shown in the table for the diagonal elements of **S** demonstrate that there was great similarity in the outcome of these separate analyses. In almost all cases, the absorbed fractions were within

TABLE 7.6   Comparison of ^{90}Y Absorbed Fractions (φ) for Murine Phantoms

ORGAN	CITY OF HOPE[a]	U.C. LONDON[b]	MSKCC[c]	LUND U[d]
Liver	0.67	0.69	0.69	0.64
Spleen	0.34	0.37	0.34	0.38
Kidneys	0.46	Not available	0.52 (average)	0.48
Lungs	0.29	0.31	Not done	0.22
Heart	0.45	0.49	Not done	0.47
Whole body	0.88	Not done	Not done	0.86

Source: Hindorf, C., Ljungberg, M., and Strand, S. E., *J. Nucl. Med.*, 45, 11, 2004. Reprinted with permission of the Society of Nuclear Medicine.
Note: Diagonal values only.
[a] Hui, Fisher, et al. (1994).
[b] Flynn, Green, et al. (2001).
[c] Kolbert, Watson, et al. (2003).
[d] Hindorf, Ljungberg, et al. (2004).

10% of the other values. One exception occurred in the lung calculations where the Lund University results were smaller than the other two estimates.

Off-diagonal (cross-organ) $\mathbf{S}$ terms were not as readily comparable across the reports. Generally, agreement in this case was not very good. For example, MSKCC reported a φ(liver → right kidney) and φ(liver → left kidney) of 3.8×10^{-3} and 7.4×10^{-6}, respectively. At the City of Hope, the average renal absorbed fraction for a source in the liver was found to be significantly larger: 7×10^{-2}. As noted by the Lund University group, these calculations are extremely sensitive to the spatial separation of source and target murine tissues. Changes of separation of 1 or 2 mm are very significant when the 90th percentile distance for energy loss by ^{90}Y beta particles is only 5.1 mm.

7.10 SUMMARY

Three dose functions (one is an array) have been described for converting activity integrals ($\tilde{A}$) into absorbed dose estimates. Two of these entities—the point-source function and the voxel source kernel—must be convolved with the spatial activity (cf. Chapter 4) distribution to give dose values. Both functions are usually developed using Monte Carlo methods, which include two simplifying assumptions: that the specimen being analyzed is uniformly made of soft tissue and without boundaries. Thus, effects due to lungs, bones, or the edge of the phantom (or animal) are neglected.

The $\mathbf{S}$ matrix is a third option for absorbed dose estimation. This array has become the standard tool for computing dose in various tissues in either animals or patients. As in the previous two dose functions, $\mathbf{S}$ is generated using MC software—but assuming a geometry appropriate for one or another anthropomorphic or animal phantom. This phantom is not a physical object but rather a mathematical construct that exists inside the computer doing the Monte Carlo histories. Because the phantom has lung and bone-equivalent inserts and is of finite extent, the two limitations listed for the PSF and VSK functions do not occur. Yet there is a price to pay. Any resultant matrix is specific for the phantom with which it was generated. One strategy around this limitation is to perform the Monte Carlo analyses with a set of various mathematical constructs to represent patients of various ages and of both sexes. Animal versions of $\mathbf{S}$ are also possible in the same way for varieties of rodents.

An experimenter is thus in a quandary as to what phantom-derived $\mathbf{S}$ matrix to apply or how to modify a given $\mathbf{S}$ matrix to suit an individual undergoing TRT. In clinical therapy considerations, the dominant terms in the matrix are diagonal ones (i.e., the self-dose component of the process). Since dose goes as the inverse of target mass, one can modify these diagonal elements of $\mathbf{S}$ to become the correct values for a given patient by multiplying elements by the ratio of phantom organ mass divided by patient organ mass for that tissue. Again, specific information for the patient's organ masses is needed from anatomical imaging. Either CT or MRI scanning is appropriate and the hybrid scanner is an ideal way to bring the two images (nuclear and anatomic) into fusion. For small animal therapy, the cross-organ terms (off-diagonal) terms can be as important as the diagonal results and need to be considered so that this type of simple correction would not suffice.

From the previous discussion, it is clear that Monte Carlo methods have been essential in developing dose functions. Limitations on the general use of such standard functions follow from the phantom assumptions made in their generation. Thus, one is tempted to suggest that a true MC analysis be performed on individual patients for targeted radiation therapy (Yoriyaz, Stabin, et al. 2001). Likewise, the animal being used as an experimental model or being treated could have its own **S** matrix formulated. As computer speeds improve, this possibility may become the method of choice and the use of predetermined **S** matrices would fade away for clinical therapy. As noted already, this individualized method would require adequate geometric images of the subject— either patient or mouse. It may be the last requirement that is the rub. To determine sizes of major organ systems, the analysts will have to segment the various anatomic slices obtained from CT or MRI. This segmentation process may take considerable time if every voxel in the image needs to be given an anatomic location—that is, an organ to which it belongs. A realistic compromise would be to do the Monte Carlo analysis only for the patient's major organ systems and tumor sites. This limited computation would be reminiscent of an external beam treatment plan in which only certain slices of the patient's scan are segmented prior to being entered into the planning system software.

Phantom usage would not disappear in the world of patient-specific MC dose estimation. Dose estimators would still need to use phantoms as standard mathematical models to provide the geometry for evaluating novel agents. This is the one aspect of the fundamental issue of developmental and regulatory RP comparison that was described in Chapters 1 through 3. While interpatient analyses would always be of interest, there must be a method to compare agents prior to the start of clinical trials. Only with suitable phantoms and their associated **S** matrices could such referencing be done by using Equation 7.2. At the time of writing, the FDA has established that the OLINDA phantom set and computation are standard for absorbed dose estimations involved in submission of RP applications to the agency.

REFERENCES

Berger, M. J. 1971. Distribution of absorbed dose around point sources of electrons and beta particles in water and other media. *J Nucl Med* Suppl 5: 5–23.

Bolch, W. E., L. G. Bouchet, et al. 1999. MIRD pamphlet No. 17: the dosimetry of nonuniform activity distributions—radionuclide S values at the voxel level. Medical Internal Radiation Dose Committee. *J Nucl Med* **40**(1): 11S–36S.

Bouchet, L. G., W. E. Bolch, et al. 2003. MIRD Pamphlet No 19: absorbed fractions and radionuclide S values for six age-dependent multiregion models of the kidney. *J Nucl Med* **44**(7): 1113–47.

Cristy, M. and K. F. Eckerman. 1987. Specific absorbed dose fractions of energy at various ages from internal photon sources. I methods. *Oak Ridge Natonal Laboratory Report ORNL/TM-8381/VI.*

DeNardo, G. L., S. J. DeNardo, et al. 1998. Low-dose, fractionated radioimmunotherapy for B-cell malignancies using 131I-Lym-1 antibody. *Cancer Biother Radiopharm* **13**(4): 239–54.

Flynn, A. A., A. J. Green, et al. 2001. A mouse model for calculating the absorbed beta-particle dose from (131)I- and (90)Y-labeled immunoconjugates, including a method for dealing with heterogeneity in kidney and tumor. *Radiat Res* **156**(1): 28–35.

Hartmann Siantar, C. L., G. L. DeNardo, et al. 2003. Impact of nodal regression on radiation dose for lymphoma patients after radioimmunotherapy. *J Nucl Med* **44**(8): 1322–9.

Hindorf, C., O. Linden, et al. 2003. Change in tumor-absorbed dose due to decrease in mass during fractionated radioimmunotherapy in lymphoma patients. *Clin Cancer Res* **9**(10 Pt 2): 4003S–6S.

Hindorf, C., M. Ljungberg, et al. 2004. Evaluation of parameters influencing S values in mouse dosimetry. *J Nucl Med* **45**(11): 1960–5.

Hui, T. E., D. R. Fisher, et al. 1994. A mouse model for calculating cross-organ beta doses from yttrium-90-labeled immunoconjugates. *Cancer* **73**(3 Suppl): 951–7.

Kolbert, K. S., T. Watson, et al. 2003. Murine S factors for liver, spleen, and kidney. *J Nucl Med* **44**(5): 784–91.

Kwok, C. S., W. V. Prestwich, et al. 1985. Calculation of radiation doses for nonuniformity distributed beta and gamma radionuclides in soft tissue. *Med Phys* **12**(4): 405–12.

Leonard, J. P., M. Coleman, et al. 2004. Epratuzumab, a humanized anti-CD22 antibody, in aggressive non-Hodgkin's lymphoma: phase I/II clinical trial results. *Clin Cancer Res* **10**(16): 5327–34.

Liu, A., L. E. Williams, et al. 1998. Monte Carlo-assisted voxel source kernel method (MAVSK) for internal beta dosimetry. *Nucl Med Biol* **25**(4): 423–33.

Loevinger, R. and M. Berman. 1968. A formalism for calculation of absorbed dose from radionuclides. *Phys Med Biol* **13**(2): 205–17.

Petoussi-Henss, N., W. E. Bolch, et al. 2007. Patient-specific scaling of reference S-values for cross-organ radionuclide S-values: what is appropriate? *Radiat Prot Dosimetry* **127**(1–4): 192–6.

Siegel, J. A. and M. G. Stabin. 2007. Mass scaling of S values for blood-based estimation of red marrow absorbed dose: the quest for an appropriate method. *J Nucl Med* **48**(2): 253–6.

Snyder, W. S. 1975. *S, absorbed dose per unit cumulated activity for selected radionuclides and organs*. New York: Society of Nuclear Medicine.

Stabin, M. G., R. B. Sparks, et al. 2005. OLINDA/EXM: the second-generation personal computer software for internal dose assessment in nuclear medicine. *J Nucl Med* **46**(6): 1023–7.

Williams, L. E., A. Liu, et al. 2002. The two types of correction of absorbed dose estimates for internal emitters. *Cancer* **94**(4 Suppl): 1231–4.

Wong, J. Y., S. Shibata, et al. 2003. A Phase I trial of 90Y-anti-carcinoembryonic antigen chimeric T84.66 radioimmunotherapy with 5-fluorouracil in patients with metastatic colorectal cancer. *Clin Cancer Res* **9**(16 Pt 1): 5842–52.

Yoriyaz, H., M. G. Stabin, et al. 2001. Monte Carlo MCNP-4B-based absorbed dose distribution estimates for patient-specific dosimetry. *J Nucl Med* **42**(4): 662–9.

Absorbed Dose Estimates without Clinical Correlations

8

8.1 INTRODUCTION

From the 1960s to the present day, there have been a number of estimates of absorbed dose in the nuclear medicine literature. Most of these calculations refer to diagnostic clinical studies, although a few of the reports concern animals and patients receiving therapeutic amounts of various agents. This chapter emphasizes therapeutic dose estimates; most of the latter are examples of radioimmunotherapy (RIT). Specifically, we discuss applications in which there is either no manifest positive correlation of clinical outcome with estimated absorbed dose or no attempt to obtain such an analysis. We defer to Chapter 9 those relatively few literature studies in which there is a correlation between absorbed organ dose and effect. The latter examples, although decidedly in the minority, are very significant to the development of improved dose estimation techniques. Hence, we emphasize those reports in their own segment of the text.

It is important to remark that a targeted radionuclide therapy (TRT) protocol in the United States is regulated by the U.S. Food and Drug Administration (FDA). For therapy trials at this time, the FDA typically applies a limit on the activity per square meter of body surface area or activity per whole-body (WB) mass. These restrictions are by analogy with chemotherapy protocols. By improving the methodology for absorbed dose estimates, one would hope that the eventual restriction applied to a clinical radiotherapy trial would be a set of numerical absorbed dose limits for the patient's various major organs. By using the methodology outlined in Chapters 5 through 7, these specific limits would lead to a maximum permissible activity for a given patient. This analysis is generated via an imaging study done prior to therapy and would result in a treatment plan for the specific patient. In such a fashion, TRT approaches an ideal that has been developed over time in external beam therapy. We would hope that a similar strategy would be developed for chemotherapy in the future.

Some members of the medical oncology community are not pleased by the challenging prospect of radiation therapy treatment planning for TRT. Because of the complexity of dose estimation (Chapter 4), it is believed that few facilities would have the capability to actually perform an estimate in reasonable time. We point out that in a European context such plans are now required before treatment may be given. This sort of control is expected in a context where the therapy may possibly be injurious to normal tissues as well as to hospital personnel. Stabin (2008) estimated that the time required for such plans is relatively small given the data acquired via gamma camera, probe, and positron emission tomography (PET) scanner.

Chapters 5 and 7 mentioned that certain corrections need to be applied to the activity (and its time integral) or to the phantom **S** values to ensure that the estimates are done in an internally consistent fashion. Initially, data are usually supplied to the FDA from animal studies to obtain original approval of the clinical study. These activity integrals must be corrected using the method given in those two chapters. We term this modulation factor, which is an organ-by-organ multiplication, to be a Type I correction. Changing the **S** matrix diagonal elements, on the other hand, is required in generating a patient-specific therapy absorbed dose estimate during the actual clinical trial. This second correction (Type II) is more unusual and may be problematic due to lack of anatomic data on the individual to be treated. For the latter reason, many clinical computations in the historical dose estimation literature are of a "mixed" type that does not fall into the Type II format due to lack of organ size data. Examples given in this chapter are predominantly of this mixed format.

In some of the studies in the literature, these corrections are clearly included in the report. As noted already, Type I corrections are standard and are expected by the FDA reviewers. Type II corrections, while ideal, may not be possible due to a lack of anatomic information. Historically, this sort of special information was often impossible to obtain, and many of the early Medical Internal Radiation Dose (MIRD) reports do not even mention an attempt to correct for patient organ size. Even today, there may not be enough data to allow the Type II correction; thus, the values presented in the report may have limitations. Because of the advent of hybrid scanners (Chapter 2), this gap in patient information is closing, and better-absorbed dose estimates should be forthcoming. As mentioned in Chapter 7, Type II corrections may be factors of two or more, so the uncorrected estimates may indeed be far from the actual results. These corrections are due to random anatomic variation as well as disease-induced changes in organ size.

Recall that the red marrow absorbed dose calculation is a particularly difficult example of absorbed dose estimation. The red marrow (RM) dose includes both Type I and Type II corrections. In the former case, the activity in the blood is generally used as a surrogate for that in the RM so that $\tilde{A}$ must be corrected. Yet the mass of an individual's RM will probably not coincide with that of any phantom. In effect, RM mass is usually unknown due to multiple combinatory factors including age, sex, body size, and, probably most importantly, prior therapies known to be toxic to marrow stroma. These various issues make red marrow absorbed dose calculations one of the most difficult to make. It is ironic that the RM is also the tissue of greatest sensitivity to ionizing radiation, so toxic absorbed doses may be in the range of only 1.5 to 2.0 grey (Emami, Lyman, et al. 1991).

8.2 ABSORBED DOSE ESTIMATES FOR ANIMAL MODELS

Possibly the most accurate absorbed dose estimates are those done for small animals such as mice and rats. These two laboratory animals have been involved in internal-emitter radiation assays since the earliest days of nuclear medicine. As mentioned in Chapters 4 through 7, the MIRD philosophy has been the most common technique for calculating animal dose on a per organ basis. As seen in Chapter 7, this strategy requires that mouse and rat **S** matrices be determined by Monte Carlo or other methods. In the following, we describe one of the earliest results for these dose functions involving the use of ^{90}Y (Hui, Fisher, et al. 1994).

To perform MC calculations, a spatial model of the mouse was required as the first step. At City of Hope, we set up a collaboration with the Battelle Institute, which performed the measurement of organ sizes and separations for the major mammalian organs in the nude mouse. Both Berger's point kernel and the EGS4 electron transport codes were then employed to find the absorbed fraction (φ) for each source–target organ pair. Notice that this computation involves a beta spectrum and not the usual set of photon emissions. A total of 14 tissues were involved, including liver, spleen, lungs, and tumor spheres. A more complete description of the computational method is included in Chapter 7.

Given the φ values, absorbed doses were estimated for nude mice receiving the murine anti-carcinoembryonic antigen (anti-CEA) antibody mT84.66 labeled with ^{90}Y. Biodistribution data were taken from sacrificed animals over a period out to 7 days after antibody injection. For these studies, the ^{111}In surrogate label was used instead of the ^{90}Y label. Table 8.1 gives the dose estimates in units of Gy/MBq, the mouse-sized relative unit. It was found that tumor doses were the largest at 5.1 Gy/MBq, with the spleen being the next highest at 4.4 Gy/MBq (Beatty, Kuhn, et al. 1994). In the latter case, a large fraction (55%) of the splenic dose came from adjacent organs such as the liver. Considering red marrow, nearly 2/3 of the dose came from adjacent source organs. The reader should keep the magnitude of these dose values in mind and refer to them as we now turn to the discussion of human results using TRT.

TABLE 8.1 Absorbed Dose Estimates for a Murine Model and ^{90}Y-ZCE-025 Antibody

ORGAN	SELF-DOSE (Gy/MBq)	CROSS-DOSE (Gy/MBq)	TOTAL DOSE (Gy/MBq)
Liver	3.2	0.2	3.4
Spleen	2.0	2.4	4.4
Kidneys	1.2	0.6	1.8
Red marrow	0.50	1.1	1.6
WiDr tumors	4.6	0.5	5.1

Source: Beatty, B. G. et al., *Cancer*, 73, 1994. Reprinted with permission of John Wiley and Sons.

8.3 ABSORBED DOSE ESTIMATES FOR ¹³¹I-MIBG THERAPY

A standard treatment for pheochromocytoma and other neuroendocrine tumors is the use of ¹³¹I-metaiodobenzylguanidine (¹³¹I-MIBG). Such therapies are conceptually simpler than those typically found in TRT. Generally, ¹³¹I-MIBG is given as a sole treatment to the patient without any external beam or chemotherapy adjuncts. The group at the University of Michigan (Koral, Wang, et al. 1989) reported on using the quantitative single-photon emission computed tomography (QSPECT) technique described in Chapter 5 to predict specific patient doses for this type of therapy. As mentioned in that chapter, a prospective sequence of geometric mean (GM) images (with ¹³¹I as the MIBG label) was taken as well as a single quantitative SPECT study (with ¹²³I as the MIBG label) at one time point near maximum tumor uptake of the agent. This last measurement allowed absolute calibration of the geometric mean image set. Three patients were studied for dose estimation with a total of seven lesions. Table 8.2 gives absorbed dose results for these tumor sites as well as their individual masses as determined by computed tomography (CT) or magnetic resonance imaging (MRI) scanning.

The results in the table indicate that estimated absorbed tumor doses varied by one order of magnitude from a minimum of 18 Gy to a maximum of 180 Gy. By dividing through by the total injected activity, some lesions were shown to have results exceeding 20 Gy/GBq—a value larger than typical results obtained with TRT, as the following studies will show. Thus, treatment of pheochromocytoma offers one of the most promising venues for TRT. The worldwide interest in ¹³¹I-MIBG therapy reflects these high levels of absorbed tumor dose. Other, more common malignancies have been treated with TRT. The next section considers the therapy of non-Hodgkin's lymphoma (NHL) with this strategy.

TABLE 8.2   Absorbed Dose Estimates for ¹³¹I-MIBG

PATIENT	TUMOR	VOLUME (cm³)	ABSORBED DOSE (Gy/GBq)
1	A	5.8	4.1
	B	5.6	2.9
2	A	7.0	20.4
	B	2.6	21.6
	C	17	9.1
3	A	26	2.3
	B	8.5	3.4

Source: Koral, K. F. et al., *Int. Jour. Rad. Phys.*, 17, 1989.
With permission of Elsevier.

8.4 LYMPHOMA THERAPY ABSORBED DOSE ESTIMATES

8.4.1 Treatment of Lymphoma Using Lym-1 Antibody

As our first example in TRT applied to lymphoma, we consider the clinical trial of the Lym-1 antibody against B cells (DeNardo, Kukis, et al. 1999). These researchers at the University of California–Davis (UC–Davis) labeled their agent with either ^{131}I or ^{67}Cu. The latter radionuclide, although a radiometal, is comparable to ^{131}I in both its particle and photon emissions with betas of 577 keV (20%), 484 keV (35%), and 395 keV (45%). The most common gamma ray is at 184 keV (49%). Recall that ^{131}I has a dominant beta at 606 keV (86%) and a principal gamma at 364 keV (81%). Production of ^{67}Cu is not a simple task. The reaction generally requires an accelerator bombarding a purified ^{67}Zn target. Copper-67 is not readily available for this reason, and the UC–Davis investigators had to pursue their radiolabel via various U.S. government laboratories including Los Alamos and Brookhaven National Lab. In the case of radioactive copper, the BAT chelator was used to hold the metal ion while radioactive iodine was attached to the intact antibody with the chloromine-T method.

Separate NHL patient trials were conducted with the two labeled compounds. For ^{67}Cu, there were a total of 14 patients, whereas 46 were accrued for Lym-1 with the ^{131}I label. Patient activity was quantitated via the geometric mean method as outlined in Chapter 5. The standard MIRD adult phantom was used for all tissues with one exception. Because the spleen is often greatly enlarged in NHL, the authors directly measured its size with anatomic imaging. As summarized in Table 8.3, the radioactive copper label led to greater estimated dose values for both normal tissues as well as the

TABLE 8.3   2IT-BAT-Lym-1 Dose Estimates with Two Labels

TISSUE	^{67}Cu-2IT-BAT-Lym-1 DOSE ESTIMATES (Gy/GBq)	RATIO OF TUMOR/ORGAN DOSE	^{131}I-I-Lym-1 DOSE ESTIMATES (Gy/GBq)	RATIO OF TUMOR/ORGAN DOSE
Liver	1.57 ± 0.47	1.62	0.31 ± 0.09	3.35
Spleen[a]	0.97 ± 0.41	2.62	0.58 ± 0.25	1.79
Lungs	0.52 ± 0.08	4.88	0.34 ± 0.12	3.06
Kidneys	0.62 ± 0.20	4.10	0.30 ± 0.09	3.47
Red marrow	0.09 ± 0.03	28.2	0.11 ± 0.04	9.45
Tumor	2.54 ± 1.46 (n = 38)	1.0	1.04 ± 0.75 (n = 98)	1.0

Source: DeNardo, G. L. et al., *Clin. Cancer Res.*, 5, 3, 1999. With permission.
[a] Note that splenic mass was corrected using anatomic imaging.

tumors that could be assayed for uptake. In the copper case, the 38 lesions demonstrated an absorbed dose/activity ratio of 2.54 ± 1.46 Gy/GBq. For radioiodine, the corresponding result ($n = 98$) was only 1.04 ± 0.75 Gy/GBq. This difference is more than a factor of two. One cannot account for this difference using a strict comparison of the average beta energies of [67]Cu and [131]I. In fact, the copper average energy is slightly less than that of iodine. Instead, the difference is probably best attributed to the loss of iodine from the tumor site due to dehalogenation.

Increased tumor absorbed dose values with the [67]Cu label do not necessarily mean that it is the superior of the two radioactive labels. Notice that the liver, in particular, showed increased absorbed dose with the radiometal label. This was the only tissue that demonstrated decreased tumor/organ absorbed dose ratio when compared with iodinated Lym-1. The decrease of iodine accumulation in the various tissues, including tumor, may be attributed to dehalogenation of the label due to local enzymes. To establish the mechanism of reduced uptake with iodine, the authors did a separate study with four patients whereby each patient was given both agents sequentially within a relatively short period of time. In these studies, it was found that uptake (%ID/g) in essentially every tumor imaged was less for the radioactive iodine-labeled compared with [67]Cu-labeled Lym-1. Iodine-labeled Lym-1 also showed a monotonic decrease of uptake with time in the lesions indicative of a clearing mechanism. After several days, in fact, the uptake of the iodinated Lym-1 was less than that of the metal-labeled-Lym-1 by an order of magnitude in many of the tumors being imaged. Use of iodination labeling, while simple and efficient in practice, can lead to greatly reduced absorbed dose values in tumors.

In an almost simultaneous protocol, the DeNardos (DeNardo et al. 1999; DeNardo, O'Donnell, et al. 2000) also evaluated ^{90}Y as a third possible therapeutic label for Lym-1. Here, the authors used the alternative chelator BAD to hold the radiometal ^{111}In to the antibody. A 13-patient study of NHL was evaluated for dose estimation using ^{111}In-BAD-Lym-1 as a surrogate for ^{90}Y-BAD-Lym-1, a pure beta emitter. A summary of the dose estimates is included in Table 8.4. As in the case of [67]Cu-BAT-Lym-1, the

TABLE 8.4   Lym-1 Absorbed Dose Estimates ([90]Y Label)

TISSUE	[90]Y-2IT-BAT-Lym-1 DOSE ESTIMATES (Gy/GBq)	RATIO OF TUMOR/ORGAN DOSES
Liver	6.10 ± 1.43	1.08
Spleen[a]	5.89 ± 2.65	1.12
Lungs	1.37 ± 0.19	4.80
Kidneys	2.70 ± 0.76	2.26
Red marrow; method 1	0.10 ± 0.09	61.0
Red marrow; method 2	1.50 ± 0.89	4.07
Tumor	6.57 ± 3.18	1.00

Source: DeNardo, G. L. et al., *J. Nucl. Med.*, 41, 5, 2000. Reprinted with permission of the Society of Nuclear Medicine.

[a] Splenic mass was corrected using anatomic imaging.

tumor doses were the largest of all tissues. Liver and spleen absorbed doses, however, were comparable in magnitude. One feature of this work was that two methods were used to evaluate the red marrow dose. Method 1 was the traditional one involving the use of the blood curve as the surrogate for the RM (Chapter 7). The second method used the lumbar vertebrae as the sources of RM activity. Red marrow dose was estimated to be considerably larger in the case of vertebral imaging. It may be, of course, that the vertebral accumulation may not be entirely due to the marrow within it.

By comparing the two radiometal-labeled Lym-1 antibodies, it is seen that absorbed doses approached 7 Gy/GBq with ^{90}Y. This enhanced value reflects the larger average beta energy of the pure beta emitter compared with either ^{131}I or ^{67}Cu as alternative labels for Lym-1. As is conventional in such computations, the authors assumed that all beta emissions were stopped in the organ or tissue of origin.

It is important to compare Lym-1 results with those obtained with other anti-B-cell therapies of NHL. As of this writing, two commercial protocols are now available that involve treating NHL with anti-CD20 antibodies. Note that CD-20 is a 35 kDa antigen that is on the membrane of both diseased and normal B cells. A number of antibodies to this antigen have been taken into clinical trials, and the two commercial applications—in turn—are discussed starting with the ^{90}Y-labeled agent. Before getting to these two examples, the various nonradiological methods used to treat NHL should be considered.

8.4.2 Zevalin Absorbed Dose Estimates for Lymphoma Patients

Traditionally, NHL patients have been treated with chemical agents that are toxic to B cells. Because of biological engineering efforts, there is now the possibility of aiding lymphoma victims using purely antibody methods that reduce the number of B cells in the blood and at other sites in the body. One of the most significant of such purely immune therapeutics is rituximab. This agent is a chimeric anti-CD20 antibody that has had wide application in NHL therapy. In this context, rituximab is provided to the patient as an unlabeled agent that reduces the number of B cells by stimulation of the patient's remaining immune system. The unlabeled antibody may reduce cell numbers by a variety of mechanisms. These processes are postulated to include antibody-dependent cellular cytotoxicity (ADCC), human idiotype two formation, and complement-dependent cytotoxicity (CDC) (Postema, Boerman, et al. 2001). As is the case for essentially all cancer therapies, it is found that some individuals are resistant (or develop resistance) to this treatment and require other—or additional—methods. One of these combinations of drugs uses both rituximab and another anti-CD20 antibody that carries the beta emitter ^{90}Y.

Wiseman, Kornmehl, et al. (2003) reported absorbed dose estimates on a total of 179 patients who have been studied using the anti-CD20 antibody ibritumomab tiuxitin. In the four Phase I to Phase III groups tabulated, this murine antibody—called Zevalin by the developer—was labeled with ^{90}Y and used to augment rituximab therapy. The chelator was MX-DOTA—called tiuxitin by the developer. Zevalin is one of two radiopharmaceuticals (RPs) commercially available for TRT of non-Hodgkin's lymphoma. As in other applications of ^{90}Y-labeled agents, it was necessary to determine the clinical biodistribution using an ^{111}In-labeled surrogate. By surrogate, we mean the same antibody labeled with ^{111}In instead of ^{90}Y to permit gamma camera planar and SPECT imaging. On Day 1, the patients received 250 mg/m^2 of rituximab with ^{111}In-Zevalin imaging following immediately thereafter (within 4 hours). Images and blood samples were obtained at 4 to 6 days after the injection of the surrogate. The GM method outlined in Chapter 5 was used to determine activity in the various major organ systems over this interval. On Day 7, the therapy aspect of the study was begun; as in the imaging case, 250 mg/m^2 of rituximab was given just prior to beginning the TRT to assure that the biodistribution would not be changed by going from the imaging to the therapy aspect of the study. According to the specific protocol, each NHL patient was given 7.4, 11, or 15 MBq/kg of ^{90}Y-Zevalin with a maximum activity of 1.2 GBq. A sample imaging study is shown in Figure 8.1.

In the dose estimation process, the MIRDOSE3 (Stabin 1996) program was used given the $\tilde{A}$ values for the various source organs. The authors comment that two target

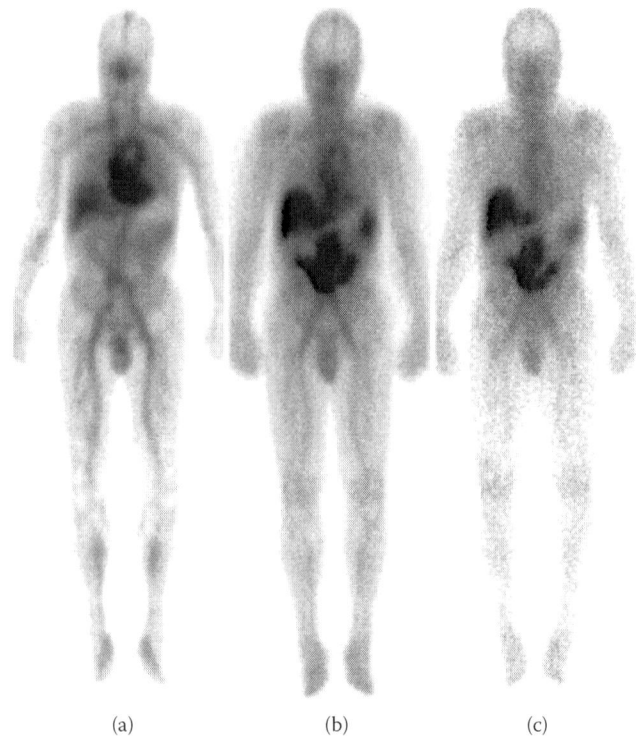

(a)   (b)   (c)

FIGURE 8.1 Whole-body camera images obtained using ^{111}In-ibritumomab tiuxetan. (a) 24 h postinjection. (b) 72 h postinjection. (c) 120 h postinjection. Note blood pool at early time but increasing accumulation in the abdominal NHL mass at 72 and 120 h. Liver uptake is seen throughout due to the ^{111}In label. (From Wiseman, G. A. et al., *J. Nucl. Med.*, 44, 3, 2003. Reprinted with permission of the Society of Nuclear Medicine.)

TABLE 8.5 Zevalin Absorbed Dose Estimates ([90]Y Label)

ORGAN[a]	ABSORBED DOSE (Gy/GBq)	RANGE OF ABSORBED DOSE (Gy/GBq)
Spleen	7.4	0.37–30.0
Liver	4.3	0.85–17.6
Lungs	2.1	0.59–4.9
Bladder wall	0.89	0.38–2.3
Red marrow	0.59	0.1–1.8
Total body	0.54	0.27–0.78
Kidneys	0.22	0–0.95

Source: Wiseman, G. A. et al., *J. Nucl. Med.*, 44, 3, 2003. Reprinted with permission of the Society of Nuclear Medicine.

[a] Hepatic and splenic masses were corrected for actual patient organ sizes using anatomic imaging.

organs had their mass values corrected using CT scan data: the liver and spleen. Other organ sizes were as taken from MIRDOSE3 tables for the male or female adult phantom. Red marrow dose estimates were done by either using the blood as a surrogate or imaging lumbar vertebrae. Clinical results for Zevalin are included in Chapter 10; here we only describe the absorbed dose estimates for the major organ systems.

Table 8.5 gives a listing of the estimated absorbed doses for the important organs. It was found that the greatest dose (on a per-activity basis) was to the spleen with a resultant ratio of 7.4 Gy/GBq. This elevated value is not surprising on two accounts: the movement of the RP into the spleen due to its involvement with NHL and the subsequent retention of activity due to the radiometal label. Keep in mind that the true organ mass was employed in the computations thanks to having a CT image of the patient's spleen available to the dose estimators. Liver was next in magnitude, with a ratio of 4.3 Gy/GBq. Targeting to the liver is expected for antibodies; the use of a radiometal as the effector radionuclide, as was the case with the spleen, implies a greater time at that site than would have been seen with radioiodine. Hence, the absorbed dose is relatively high. In order of descending magnitudes, lungs were the third organ on the list, with a ratio of 2.0 Gy/GBq. While perhaps indicative of disease in the pulmonary system, this elevated value probably resulted from a characteristically long antibody residence time in the lungs. No absorbed dose estimates for tumors were published in this summary over the various four clinical protocols. Hence, we cannot directly compare this Zevalin result with that of the group at UC–Davis cited earlier in this chapter.

The authors attempted to find a relationship between hematological toxicity and two absorbed dose estimates: those of the red marrow or the total body. They also attempted correlating the blood area under the curve (AUC) and the blood effective half-life with this toxicity. In no case were they able to find a result at or better than the classical $p < .05$ value.

We should add, in this context of anti-CD20 treatments, that subsequent reports on other clinical trials of Zevalin have highlighted the variability of absorbed dose estimates for a given TRT protocol. For example, a Milan study (Cremonesi, Ferrari, et al. 2007) described the dose estimates for 22 NHL patients treated with ^{90}Y-ibritumomab tiuxetan. Three methods were used to quantitate the amount of absorbed dose in the major source organs. Method A used the actual target organ masses as seen via CT imaging of every patient. In the other two methods (B and C), various simplifications were assumed to expedite computation. With Method B, standard organ masses from OLINDA (Stabin, Sparks, et al. 2005) software were substituted into the analysis. A similar simplification was made in Method C with several added features. Among these was the assumption that the blood curve could be used to extrapolate the number of decays in the red marrow—that is, organ activity values occurring after the last recorded data point. Also, all curves were modeled using single exponential functions instead of the biexponentials used in Methods A and B. As noted in Chapter 6, using a single exponential to represent a time–activity curve can be only an approximation when the actual function must initially rise and then fall as the RP moves through the tissue of interest.

The Milan authors found that their results varied among the three methods as shown in Table 8.6. Most notable among these variations are the changes in splenic and renal absorbed dose estimates. These organ absorbed dose values varied by factors of two-fold to threefold in going from the most patient-specific method (A) to the least specific (Methods B and C). Most of this variation is due to the use of standard patient masses in the B- and C-type calculations. As noted in Chapter 4, the denominator of the canonical MIRD equality used in these beta-ray computations is the mass of the target (= source) organ. If this size is incorrect due to assuming phantom-based tissue sizes, the computation of absorbed dose will also be in error. Notice, as mentioned in Chapter 4, that Method A is essentially that of a Type II computation (i.e., intentionally patient specific).

More recently, the MIRD Committee of the Society of Nuclear Medicine has also filed a report on absorbed dose estimates with Zevalin (Fisher, Shen, et al. 2009). In MIRD Report 20, Fisher and coworkers gave estimated absorbed dose values for both ^{111}In and ^{90}Y labels on iibritumomab tiuxetan for a set of 10 patients who were carefully followed with imaging at three different experimental venues. Because organ sizes were measured via CT imaging, we again have a Type II computation that is patient specific for these 10 individuals. Unlike the Milan group (Cremonesi et al. 2007), however, the

TABLE 8.6 Several Methods of Estimating Zevalin ([90]Y Label) Absorbed Doses

ORGAN	METHOD A (Gy/GBq)	METHOD B (Gy/GBq)	METHOD C (Gy/GBq)
Liver	2.8	3.0	2.5
Spleen	1.9	3.9	6.1
Kidneys	1.7	2.3	4.0
Lungs	1.7	1.7	3.2
Red marrow	0.8	0.8	0.8
Whole body	0.5	0.6	0.6

Source: Cremonesi, M. et al., *J. Nucl. Med.*, 48, 11, 2007. Reprinted with permission of the Society of Nuclear Medicine.

Note: Median values are shown.

TABLE 8.7 Comparison of Three Reports on Zevalin (^{90}Y Label) Absorbed Dose Estimates (Gy/GBq)

ORGAN	METHOD A[a]	WISEMAN ET AL.[b]	MIRD NO. 20[c]
Kidneys	1.7 (0.6–3.8)	0.22 (0–0.95)	2.4 (1.4–3.9)
Liver	2.8 (1.8–10.6)	4.32 (0.9–18)	3.1 (2.3–6.6)
Spleen	1.9 (0.8–5.0)	7.4 (0.4–30)	4.3 (1.0–9.0)
Red marrow	0.8 (0.4–1.0)	0.59 (0.1–1.8)	2.4 (1.7–4.5)
Whole body	0.5 (0.4–0.8)	0.54 (0.3–0.8)	0.6 (0.4–0.8)

Source: Fisher, D. R. et al., *J. Nucl. Med.*, 50, 4, 2009. Reprinted with permission of the Society of Nuclear Medicine.
Note: Data range in parentheses.
[a] Cremonesi, Ferrari, et al., 2007.
[b] Wiseman, Kornmehl, et al. 2003.
[c] Fisher, Shen, et al., 2009.

MIRD Committee used uptake in the L3 to L5 vertebrae to establish a red marrow time–activity curve. A comparison of these results using Method A from Milan and the earlier synopsis from Mayo Clinic (Wiseman et al. 2003) is given in Table 8.7.

It is found that the report from the Mayo Clinic (Wiseman et al. 2003) is generally in fairly good agreement with the pair of later Type II organ absorbed dose calculations. There are two manifest exceptions to this conclusion: dose estimates to the kidney and spleen. In the former case, the two patient-specific results are approximately an order of magnitude larger than that of the Wiseman et al. value (0.2 Gy/GBq). The MIRD Report 20 (Fisher et al. 2009) gave an estimate of 2.4 Gy/GBq, whereas the Milan result is 1.7 Gy/GBq. The magnitude of this difference cannot be explained by the variation of patient renal masses compared with the standard phantom value. Instead, there is probably a methodological issue of kidney visibility in the gamma camera images as taken by the three groups. In the earlier Wiseman work, the kidneys may not have been highly visible and hence had to be grouped into a category of organs that were not used as sources. This apparently was not the case with the MIRD review of 10 patients or the Milan study involving 22 patients at the various Italian institutions.

Splenic variation among the three reporting groups is another difference seen among the three estimates of absorbed dose. Estimates for the patient-specific calculations were 1.9 and 4.3 Gy/GBq for the Milan study (Cremonesi et al. 2007) and MIRD Report 20 (Fisher et al. 2009), respectively. Yet the Wiseman et al. (2003) review has the larger value of 9.4 Gy/GBq. In this case, all three groups used a correction for splenic mass so that organ size variation cannot be invoked as an explanation of the differences. This disparity is not as large as that seen with the renal analysis. Notice also that the range of splenic dose estimates of the Mayo Clinic result does overlap numerically with the other two values.

We wish to present the case that dose variations seen in the three reports are indicative of what may be found when absorbed doses are estimated using different assumptions. The authors of the MIRD Report 20 (Fisher et al. 2009), in fact, used the word estimates in its title. It would be appropriate if everyone in our dose estimation enterprise were to use the same descriptor in their verbal identification of work and results. Dosimetry, as noted in Chapter 4, may logically be used only in the context of dose measurement—not in the estimation process. To do otherwise is to mislead the reader into believing that something has been measured when, in fact, it has not. To paraphrase Gertrude Stein, an estimate is an estimate is an estimate—and not a measurement.

8.4.3 Bexxar Absorbed Dose Estimates for Lymphoma Patients

A second TRT (Bexxar) has become commercially available for NHL patients. This agent is also an anti-CD20 intact antibody (Tositumomab) that uses ^{131}I as its radio-nuclide. Bexxar is a radiolabeled form of the anti-B1 anti-CD20 antibody developed earlier. Since ^{131}I has gamma and beta emissions, the user does not need a surrogate imaging label—and associated radionuclide cost—as in the Zevalin example. In addi-tion, the intrinsic price of ^{131}I is considerably less than that of comparable activities of ^{90}Y necessary for Zevalin therapies. Such economic reasons have been at the heart of the Bexxar development since its early stages.

A multicenter Phase II trial of Bexxar was reported out of the University of Nebraska (Vose, Wahl, et al. 2000). By definition of the protocol, the initially estimated whole body (WB) absorbed dose was used to calculate the amount of activity to be given individual patients. A pretherapy diagnostic study with the same agent (5 mCi) was used to limit WB absorbed dose estimates to 75 cGy for each patient with sufficiently high pretherapy platelet counts (> 150 k cells/ml). If, however, the counts were between 101 k cells/ml and 149 k cells/ml, the WB dose estimate had to be reduced to 65 cGy. Using the WB dose estimate to give total Bexxar activity is in sharp distinction with the strategy of treatment of the Zevalin investigators who used activity per body weight (MBq/kg) as the deter-mining factor for calculating total injected activity. As in the case of Zevalin, a pre-TRT therapy of unlabeled anti-B1 antibody (450 mg) was given to each patient to decrease the targeting of the labeled material to normal tissues such as the liver and spleen.

Absorbed doses were estimated by the study group for 47 patients at the eight par-ticipating institutions. Activities were documented by using single probes and gamma cameras for the whole body. Geometric mean counts, obtained with the gamma camera, were used to determine activity versus time in several specific organs including the kid-neys, spleen, liver, lungs, and bladder. Results for these tissues are given in Table 8.8. It

TABLE 8.8   Bexxar Absorbed Dose Estimates

ORGAN	ABSORBED DOSE/75 cGy WHOLE-BODY DOSE (Gy)	RATIO OF TUMOR/ORGAN DOSE
Kidneys	5.0	1.6
Spleen	3.8	2.1
Liver	2.2	3.6
Lungs	1.8	4.4
Bladder	2.1	3.8
Tumors	8.0	1.0

Source: Based on Vose, J. M. et al. *J. Clin. Oncol.* 8, 6, 2000.

was seen that the maximum absorbed dose estimate occurred for the kidneys at 5.0 Gy with the WB dose being fixed at 75 cGy. Clinical results of the Bexxar Phase II trial are given in Chapter 10 as an example of combination therapy.

8.5 INTERVENTIONAL THERAPY OF HEPATIC MALIGNANCIES USING MICROSPHERES

Treatment of lymphoma with systemic injection is a relatively recent application of TRT. Two other protocols, also currently approved by the FDA, have a much longer history in cancer therapy using targeted radioactivity. This older strategy, originally developed shortly after the Second World War, involves permanently lodging man-made, radioactive spheres into the capillary bed of the liver in an effort to treat local malignant disease. Hepatic carcinoma would be a possible target for primary therapy, whereas colon cancer metastatic to the liver is probably the most common form of secondary disease. Other applications are possible due to the occurrence of secondary lesions in the liver from a number of primary sites.

Targeting is done mechanically by catheter intervention inside the patient. Two objectives of this form of therapy are desired. First is the blockage of blood flow into the downstream tumor sites to deprive the lesions of oxygen and nutrients. The second objective is to treat with ionizing radiation any malignancies at the sites with the attached beta emitter. Because of pulmonary and other normal organ risks, no intravenous (IV) injections may be used as in the case involving NHL therapy. Instead, the radiation oncologist teams with the radiology division to pass a catheter into the hepatic artery prior to the injection of the labeled spheres. Once that catheter is correctly lodged, as seen on the fluoroscope, the radioactive spheres are injected in a controlled fashion. In recent times, the traditional radioactive label has been ^{90}Y due to its high energy (cf. Chapter 1) that enables radiation dose well beyond the points of lodgment. For ^{90}Y, that maximum distance is 11 mm from a source.

The anatomic hypothesis underlying this form of targeted therapy is that tumor lesions in the liver receive their blood flow predominantly from the arterial supply and not from the portal vein. In practice, the interventional radiologist (IR) can see that contrast media is targeting into the tumor sites during confirmation of catheter placement. We should note that it is difficult to document quantitatively how localized the eventual therapy injection will be by using x-ray contrast media alone. Therefore, prior to the therapy, the IR will also inject ^{99m}Tc microaggregated albumin (MAA) surrogate into the system and observe if there are any "leaks" into the gut vasculature or, more importantly, into the lungs. In both commercial protocols, the fractional amount of injected activity going into the lungs must be strictly less than 20% of the total. The GM method is generally used to make this last assessment. If the lung fraction is at or exceeds the 20% figure, the study shall not be done on the patient.

Modern revival of this relatively old technology is due to the two commercial suppliers producing ^{90}Y-microspheres. The older technology is based on resin microspheres that are, on average, 32 μm in diameter (range 20 to 60 μm) (Lau, Ho, et al. 1998). More recently, glass spheres of approximately 25 (±10) μm size have become available (Lewandowski, Thurston, et al. 2005). In these two products, the spheres are biocompatible and do not break down in vivo such that there is essentially no leakage of the radionuclide from the encapsulating media. Such losses had been on of the limitations in the earlier manifestations of this technology that date to the 1960s (Kennedy, Coldwell, et al. 2006). Recall that yttrium is a bone-seeking element such that released ^{90}Y will deliver an absorbed dose to RM. MAA particle size is comparable to the sphere values previously cited (i.e., approximately 35 μm average diameter). Thus, users of the MAA surrogate anticipate that the albumin agent will distribute similarly to the yttrium-labeled spheres.

In both commercial protocols, a total catheter-delivered activity is calculated to generate a required dose to the liver lesions. Often, only a single microsphere treatment can be performed due to the severity of the disease. It is generally the case that the primary disease site in these patients (e.g., in the colon) has been surgically removed. Thus, it is the liver lesions—and more particularly their size—that actually threaten the life of the individual. Surgery, in other words, is not possible for this patient cadre at the time of enrollment in the clinical procedure. We should add that external beam therapy has not been used in treatment of hepatic patients due to the manifest sensitivity of the liver to ionizing radiation (Emami et al. 1991).

Absorbed dose estimation for implanted microspheres is based on what is termed the partition method (Ho, Lau, et al. 1996). Here, the advocates calculate the distribution of radioactivity among tumors, normal liver, and lung using the geometric mean method and the injection of ^{99m}Tc MAA as a surrogate. Absorbed dose is then given by a constant times the ratio of activity in the tumor divided by the mass of the lesion. The numerical value of the constant is 49.7×10^3 if activity is in GBq and mass is in grams (Lau et al. 1998). An elaboration of the partition method using the MIRD approach has also been described (Gulec, Mesoloras, et al. 2006). With the canonical formalism, the authors use a gamma camera to determine the fractions of injected ^{99m}Tc-MAA activity going to each of three important tissues: tumor, liver, and lung. The very important lung portion, called the shunting fraction (SF), is given by

$$SF = \frac{A(\text{lung})}{A(\text{lung}) + A(\text{liver})} \qquad (8.1)$$

where A represents activity measured using the GM method. Again, one assumes that the distribution of ^{99m}Tc MAA is identical to that of the subsequent ^{90}Y-labeled spheres. Using $A(mCi)$ and mass (g), the estimated absorbed doses (cGy) are then

$$\text{Dose(Lung)} = A_0 * SF * 184 \times 10^3 / m(\text{lung})$$

$$\text{Dose(Liver)} = A_0 * f_L(\text{Liver}) * 184 \times 10^3 / m(\text{liver}) \qquad (8.2)$$

$$\text{Dose(Tumor)} = A_0 * f_T(\text{tumor}) * 184 \times 10^3 / m(\text{tumor})$$

TABLE 8.9 Absorbed Dose Estimates with ^{90}Y Microsphere Therapy of Hepatic Disease

NUMBER OF TREATMENTS	NUMBER OF PATIENTS	TUMOR (Gy)	LIVER (Gy)	LUNGS (Gy)
1	56	225 (83–748)	52 (25–136)	6 (0–24)
2	10	482 (279–801)	119 (64–143)	13 (5–19)
3	4	578 (409–788)	116 (85–116)	13 (6–25)

Source: Lau, W. Y. et al., *Int. Jour. Rad. Phys.*, 40, 1998. With permission of Elsevier.

where f_L and f_T refer to the fractional accumulation in liver and tumor, respectively, and A_0 is the total activity injected via catheter. These last two fractions are also determined using the MAA injection.

The group in Hong Kong described the absorbed dose estimates using these shunting calculations (Lau et al. 1998). Table 8.9 contains their results for a total of 70 patients having from one to three separate treatments with resin microspheres. All the patients were initially documented to have primary hepatocellular carcinoma (HCC). It was found that median tumor dose estimates were on the order of 200 Gy per application with corresponding liver values of approximately 50 Gy. Lung median absorbed doses were 5 Gy per treatment. The amount of injected activity was determined for each patient using a proposed tumor dose above 120 Gy while keeping the lung shunt at less than 15% and the gastric shunt below 5%.

There was no direct correlation attempted by the Hong Kong group between their absorbed dose estimates and clinical outcome. Median survival was only 9.4 months during the 3-year study. The severity of the HCC was such that surgery was precluded—at least at the beginning of the protocol. After one or more treatments, four of the patients had sufficient reduction in the size or extent of disease so that a surgical procedure became possible. Two of these appeared to be long-term survivors; that is, there was no pathological evidence of disease at the time of subsequent resection. Only fibrotic tissue was found as a residue of the radiation and occlusive damage.

Most of the HCC patients that had elevated alpha fetal protein (AFP) prior to the sphere-based treatments showed abrupt reduction of these protein levels within 5–20 days of therapy. In some cases, the resultant level was so low as to be undetectable. Using AFP as their tumor marker, the authors felt that there would be extended intervals, sometimes lasting for months, over which levels would slowly climb, implying the necessity of additional interventional treatments. Notice that no control group receiving only standard chemotherapy was present in the Hong Kong protocol.

An Australian clinical group from Perth (Gray, Van Hazel, et al. 2001) reported a two-armed study whereby 70 patients with liver metastatic sites were treated either by resin microspheres and chemotherapy or chemotherapy alone. Following standard clinical practice, the chemotherapy was floxuridine administered as a 12-day infusion via the same hepatic artery input. All patients in this Phase III trial had primary colon cancer and were followed with several criteria: tumor response, time to progression, survival, quality of life, and toxicity. Table 8.10 gives the results of the trial. Using

TABLE 8.10   Microsphere Therapy of Colon Cancer Metastatic Sites in the Liver

CRITERION	FLOXURIDINE + MICRO SPHERES	FLOXURIDINE THERAPY	p-VALUE
PR + CR by area	44%	18%	0.01
By CEA Level	72	47	0.004
Time to progress by area	15.9 months	9.7 months	0.001
Two-year survival	39%	29%	0.06
Three-year survival	17%	6.5%	0.06

Source: Gray, B. et al., Ann. Oncol., 12, 12, 2001. With permission of Oxford University Press.

measured cross-sectional areas of the hepatic lesions, it was found that the overall response rate (ORR = partial response [PR] + complete response [CR]) was 44% for the spheres plus chemotherapy but only 18% for conventional chemotherapy alone ($p = .01$). Survival was also improved, but with a probability of difference of only 0.06, so that this sphere-based TRT therapy by definition was not significant at the traditional 5% level of confidence.

A more recent Phase I study has been reported on resin microspheres and simultaneous multiagent chemotherapy (Sharma, Van Hazel, et al. 2007). In this protocol, a single injection of the radioactive agent was applied along with the current standard combination chemotherapy treatment for metastatic colorectal cancer FOLFOX (oxaliplatin, fluorouracil, and leucovorin). Of the 20 patients in the study, PR was seen in 18. The authors comment that a Phase II or III trial is warranted based on the high probability of partial response.

8.6 COLORECTAL CANCER THERAPY USING TRT

This chapter's two previous sections described clinically successful procedures involving NHL and malignant liver lesions, respectively. Yet both of these examples are limited due to the relative rarity of both diseases. One could, in particular, argue that the intervention via the hepatic artery is a technique with obviously restricted application outside the liver. We now consider the results obtained with intravenous TRT of primary solid tumors and their various metastatic sites.

As an example, we consider the estimated absorbed doses to patients involved in Phase I trial at the City of Hope (Wong, Chu, et al. 2006) involving the intact cT84.66 Mab. Because of the variation of scan results form one patient to another, we used only blood, liver, and residual body as compartments that were measured during the ^{111}In-cT84.66 imaging phase of the study. Other organs that were occasionally seen were represented by trapezoidal functions to allow their activity to be integrated to long times.

TABLE 8.11   Absorbed Dose Estimates
for ^{90}Y-cT84.66 and ^{131}I-MN-14 in Patients
with CEA-Positive Lesions

ORGAN	^{90}Y-cT84.66 (Gy/GBq)	^{131}I-MN-14
Liver	9.7	0.77
Kidneys	1.5	0.77
Lungs	0.30	0.77
Red marrow	0.59	0.82
Whole body	0.57	0.21
Tumors	26.7 ($n = 18$)	2.6 ($n = 33$)

Sources: Wong, J. Y. C. et al., *Cancer Biother.
Radiopharm.*, 21, 2, 2006; and Behr, T. M.
et al., *J. Nucl. Med.*, 38,3, 1997, published
by Mary Ann Liebert, Inc. With permission.

For modeled organs, the ADAPT II software package (D'Argenio and Schumitzky 1979) was used to fit individual patient data sets to allow rate constants to be estimated. Predicted absorbed doses for major organ systems were then calculated using the MIRDOSE3 software package (Stabin 1996) for the ^{111}In-cT84.66 agent and predicted for ^{90}Y-cT84.66. A total of 13 patients were available for study of CEA-positive tumors that had metastasized. No correction was made for patients' actual organ masses, and the phantom values from MIRDOSE3 were used directly. Thus, this was a mixed-type calculation that was not of either Type I or Type II.

Table 8.11 contains a summary of the ^{90}Y-cT84.66 clinical study. The liver was the normal organ found to have the highest estimated absorbed dose on a per GBq basis. Included in Table 8.11 is a set of estimates from a similar agent, MN-14, which is likewise an intact antibody to CEA. A total of 59 patients were evaluated in this report out of the Garden State Cancer Center (GSCC) (Behr, Sharkey, et al. 1997). Two differences between the two agents were that MN-14 is a murine antibody and that it was labeled with ^{131}I. As in the City of Hope study, absorbed doses were estimated using a mixed method with activities determined with the patient, but organ geometry and separation were obtained from a standard MIRD phantom. Likewise, tumor sizes were determined using either CT or, if not available, nuclear images.

It is seen that the two anti-CEA antibodies led to similar tumor to normal organ dose ratios but that larger absolute absorbed doses were obtained using the ^{90}Y-labeled agent. There are, as mentioned with the UC–Davis NHL results, two obvious reasons for this advantage: higher beta energy with yttrium and reduced loss of this radionuclide from the tumor. Tumor doses were essentially one order of magnitude larger with the yttrium-90 label: 26.7 Gy/GBq versus 2.62 Gy/GBq. Most of this difference was due to the mean energy ratio (approximately threefold) between yttrium-90 and iodine-131. We would assume that the remainder, approximately another factor of three, could be attributed to the average amount of dehalogenation occurring over the course of the 7 days of imaging required in both of these colorectal patient trials.

8.7 SUMMARY

We began the chapter with a discussion of use of a Monte Carlo-derived **S** matrix to estimate absorbed doses for the nude mouse. In this small-scale geometry, the ^{90}Y pure beta emitter contributed significant cross-organ radiation doses. It was seen in the spleen, for example, that more than one half of the total organ absorbed dose was due to other tissues such as the liver. For the marrow, approximately 2/3 of its dose was due to outside (nonmarrow) tissues. Absolute magnitudes of these estimates are also of significance. Typical values were in the range of 1 to 5 Gy/MBq—essentially three orders of magnitude larger than the corresponding values estimated later for the patient studies. As we anticipate from dilution arguments, this gross variation is due to the much smaller mass of the test animal compared with a patient. In other words, the patient presents approximately a 3,000-fold larger volume into which the RP is diluted upon IV injection. Therefore, the ratio of absorbed dose per amount of activity must be reduced by this same factor in the human subject in a cancer therapy trial.

The majority of such trials do not attempt to demonstrate clinical correlations with absorbed dose estimates. For some protocols, the absorbed doses are simply not calculated, so the outcome, while of clinical interest, cannot be directly related to the concept of ionizing energy divided by target–tissue mass. In the few cases where doses are assessed, there are usually no positive correlations with disease outcome or normal organ toxicity. This chapter, which selects from this latter literature, finds that there may be several reasons for lack of correlation. Some of these factors are methodological; that is, the predicted absorbed dose can be varied by factors of two or more depending on exactly what algorithm is used in the calculation.

To demonstrate this variation, let us turn our attention to the Zevalin NHL trials. As the Milan group (Cremonesi et al. 2007) showed, the various methods used in the estimation process lead to a range of final results. Most importantly, the mass of the target tissues must be obtained—usually from CT or MRI scans—and then put into the algorithm as the denominator. If, instead, one uses phantom values, the results may differ by factors of twofold or more. A reader from outside the pharmacology field may wonder why a dose predictor would fall back on phantom values. An obvious reason is simplification of the estimation process. Organ masses obtained via CT (or MRI) require computed tomographic imaging and slice-by-slice integration to find total volume. If one discusses this issue with medical oncologists, they may assert that these complications of absorbed dose estimation lead to a lack of clinical interest in TRT methods. In their world, unless the estimation process is logistically as simple as chemotherapy, it is not worth doing. Recall that typical chemical therapy is prescribed in units of millimoles (mM) or mg of agent per total body mass or body surface area. No specific "doses" in terms of mM or mg per unit mass of the tumor or normal tissue are available since there is no radiolabel or other method to externally track pharmaceutical movement within the patient. In fact, in chemotherapy, there is no demonstrable proof of the targeting of the agent to disease sites.

Tumor-specific localization is clearly of vital importance in all therapies of malignancy. This chapter described using man-made spheres to occlude vessels in the cancer-infested liver. Here, the targeting is via catheter so that the normal hepatic tissues and other organs, such as stomach and lungs, can be largely protected. In these studies, there is, again, no specific correlation being attempted between the absorbed dose estimates and the outcome of the patients. One would hope that such efforts will be forthcoming. At present, there is a task group (TG 144) of the American Association of Physicists in Medicine (AAPM) that is assigned to this project. As noted by the Australian investigators (Gray et al. 2001), there is a clear clinical advantage to this form of interventional TRT when comparing it with standard chemotherapy given alone.

We have seen that NHL can be treated by TRT using IV injection due to the sensitivity of B cells to relatively low doses of ionizing radiation. Solid tumors are not as tractable even though similar absorbed radiation doses have been obtained in therapy trials on diseases such as colorectal cancer. In the case of liver primary or metastatic disease, both types of tumors may be treated with catheter placement of the ^{90}Y-beta sources to eliminate bone marrow dose due to circulation of the therapeutic agent. This mechanical trick cannot be easily generalized, and solid tumors will usually need to be treated via IV—or perhaps intraperitoneal (IP)—injection. There is, in addition, the possible threat of multiple sites of the disease for some solid tumors. Thus, the colorectal patient may have lesions in the liver, lung, and various nodes. A similar result is possible for other solid malignancies such as breast and lung cancer. Some of these sites may be imaged, and some may not. Hence, as outlined in Chapter 1, there will always be a need for a molecule-specific TRT introduced via the general circulation.

REFERENCES

Beatty, B. G., J. A. Kuhn, et al. 1994. Application of the cross-organ beta dose method for tissue dosimetry in tumor-bearing mice treated with a 90Y-labeled immunoconjugate. *Cancer* **73**(3 Suppl): 958–65.

Behr, T. M., R. M. Sharkey, et al. 1997. Variables influencing tumor dosimetry in radioimmunotherapy of CEA-expressing cancers with anti-CEA and antimucin monoclonal antibodies. *J Nucl Med* **38**(3): 409–18.

Cremonesi, M., M. Ferrari, et al. 2007. High-dose radioimmunotherapy with 90Y-ibritumomab tiuxetan: comparative dosimetric study for tailored treatment. *J Nucl Med* **48**(11): 1871–9.

D'Argenio, D. Z. and A. Schumitzky. 1979. A program package for simulation and parameter estimation in pharmacokinetic systems. *Comput Programs Biomed* **9**(2): 115–34.

DeNardo, G. L., D. L. Kukis, et al. 1999. 67Cu-versus 131I-labeled Lym-1 antibody: comparative pharmacokinetics and dosimetry in patients with non-Hodgkin's lymphoma. *Clin Cancer Res* **5**(3): 533–41.

DeNardo, G. L., R. T. O'Donnell, et al. 2000. Radiation dosimetry for 90Y-2IT-BAD-Lym-1 extrapolated from pharmacokinetics using 111In-2IT-BAD-Lym-1 in patients with non-Hodgkin's lymphoma. *J Nucl Med* **41**(5): 952–8.

Emami, B., J. Lyman, et al. 1991. Tolerance of normal tissue to therapeutic irradiation. *Int J Radiat Oncol Biol Phys* **21**(1): 109–22.

Fisher, D. R., S. Shen, et al. 2009. MIRD dose estimate report No. 20: radiation absorbed-dose estimates for 111In- and 90Y-ibritumomab tiuxetan. *J Nucl Med* **50**(4): 644–52.

Gray, B., G. Van Hazel, et al. 2001. Randomised trial of SIR-Spheres plus chemotherapy vs. chemotherapy alone for treating patients with liver metastases from primary large bowel cancer. *Ann Oncol* **12**(12): 1711–20.

Gulec, S. A., G. Mesoloras, et al. 2006. Dosimetric techniques in 90Y-microsphere therapy of liver cancer: The MIRD equations for dose calculations. *J Nucl Med* **47**(7): 1209–11.

Ho, S., W. Y. Lau, et al. 1996. Partition model for estimating radiation doses from yttrium-90 microspheres in treating hepatic tumours. *Eur J Nucl Med* **23**(8): 947–52.

Hui, T. E., D. R. Fisher, et al. 1994. A mouse model for calculating cross-organ beta doses from yttrium-90-labeled immunoconjugates. *Cancer* **73**(3 Suppl): 951–7.

Kennedy, A. S., D. Coldwell, et al. 2006. Resin 90Y-microsphere brachytherapy for unresectable colorectal liver metastases: modern USA experience. *Int J Radiat Oncol Biol Phys* **65**(2): 412–25.

Koral, K. F., X. H. Wang, et al. 1989. Calculating radiation absorbed dose for pheochromocytoma tumors in 131-I MIBG therapy. *Int J Radiat Oncol Biol Phys* **17**(1): 211–8.

Lau, W. Y., S. Ho, et al. 1998. Selective internal radiation therapy for nonresectable hepatocellular carcinoma with intraarterial infusion of 90yttrium microspheres. *Int J Radiat Oncol Biol Phys* **40**(3): 583–92.

Lewandowski, R. J., K. G. Thurston, et al. 2005. 90Y microsphere (TheraSphere) treatment for unresectable colorectal cancer metastases of the liver: response to treatment at targeted doses of 135–150 Gy as measured by [18F]fluorodeoxyglucose positron emission tomography and computed tomographic imaging. *J Vasc Interv Radiol* **16**(12): 1641–51.

Postema, E. J., O. C. Boerman, et al. 2001. Radioimmunotherapy of B-cell non-Hodgkin's lymphoma. *Eur J Nucl Med* **28**(11): 1725–35.

Sharma, R. A., G. A. Van Hazel, et al. 2007. Radioembolization of liver metastases from colorectal cancer using yttrium-90 microspheres with concomitant systemic oxaliplatin, fluorouracil, and leucovorin chemotherapy. *J Clin Oncol* **25**(9): 1099–106.

Stabin, M. G. 1996. MIRDOSE: personal computer software for internal dose assessment in nuclear medicine. *J Nucl Med* **37**(3): 538–46.

Stabin, M. G. 2008. Update: the case for patient-specific dosimetry in radionuclide therapy. *Cancer Biother Radiopharm* **23**(3): 273–84.

Stabin, M. G., R. B. Sparks, et al. 2005. OLINDA/EXM: the second-generation personal computer software for internal dose assessment in nuclear medicine. *J Nucl Med* **46**(6): 1023–7.

Vose, J. M., R. L. Wahl, et al. 2000. Multicenter phase II study of iodine-131 tositumomab for chemotherapy-relapsed/refractory low-grade and transformed low-grade B-cell non-Hodgkin's lymphomas. *J Clin Oncol* **18**(6): 1316–23.

Wiseman, G. A., E. Kornmehl, et al. 2003. Radiation dosimetry results and safety correlations from 90Y-ibritumomab tiuxetan radioimmunotherapy for relapsed or refractory non-Hodgkin's lymphoma: combined data from 4 clinical trials. *J Nucl Med* **44**(3): 465–74.

Wong, J. Y., D. Z. Chu, et al. 2006. A phase I trial of (90)Y-DOTA-anti-CEA chimeric T84.66 (cT84.66) radioimmunotherapy in patients with metastatic CEA-producing malignancies. *Cancer Biother Radiopharm* **21**(2): 88–100.

Dose Estimates and Correlations with Laboratory and Clinical Results

9

9.1 INTRODUCTION

This chapter discusses the relatively rare situations reported in the literature in which estimated absorbed doses correspond to biological outcomes. These results include tumor regression and toxic effects in normal tissues as, for example, the red marrow (RM) or kidneys. As noted in the previous chapter, most dose estimates are not compared with clinical outcomes. If such comparisons are made, there is often a lack of correlation with any measured effects. The probable lowest-order reason for this last discrepancy is that the calculations have inherent uncertainties, often unknown to the estimator, that render significant correlations unlikely.

A seeming lack of correspondence between dose estimates and outcomes has been probably the most difficult issue in the history of targeted radionuclide therapy (TRT). As described in Chapters 4, 5, and 6, estimates of absorbed dose in TRT require relatively complicated experimental data acquisitions as well as difficult mathematical operations on often unknown geometries. Again, the unfortunate term *dosimetry* has been linked with these processes, thus giving rise to a negative view on the entire assessment of absorbed dose. Using the correct term *dose estimation* allows the analyst greater credibility in any subsequent discussion of why these numerical values may differ from those actually found in nature. We must also add that most of the published estimates do not carry with them any associated error or absorbed dose uncertainty. This feature is a further confounding item in why there may be little belief in the absorbed dose calculation.

Until relatively recently, there has also been relatively sparse clinical outcome data with which to correlate the various dose values. Historically, going back to the 1960s, most analysts' estimates have involved U.S. Food and Drug Administration (FDA) applications that refer to diagnostic radiopharmaceutical (RP) agents. With resultant maximum organ doses on the order of 5 to 10 cGy, it is very unlikely that the individual patient will demonstrate any untoward effects with such exposures either immediately or later due to stochastic developments. There remains, of course, the chance that a large enough cadre of individuals will show statistically significant effects. While currently of interest to epidemiologists studying the computed tomography (CT) imaging situation, long-term studies have not been initiated in nuclear medicine.

This chapter discusses one murine and several clinical trials in which absorbed dose or biologically effective dose is shown to be correlated with cellular, tumor, or normal tissue outcome. In most of these assays, the results are direct; that is, they are seen shortly after the radiation exposure. In the renal case using targeting small proteins, the effects are seen much later, corresponding to what is called a stochastic result. The possibility that untoward results may occur much later than the exposure to ionizing radiation is another reason for the difficulty of obtaining correlations between dose estimates and clinical outcomes. With older and often seriously ill patient populations, there simply may not be sufficient time available to observe some of the possible outcomes.

We first describe animal studies in which estimates of absorbed dose values are shown to correlate with tumor regression. Such correspondences are expected and occur with regularity in both single-cell (in vitro) and tumor (in vivo) experiments with mice and other rodents. Given these concordances, it is frustrating that no similar results generally hold for clinical situations. It is the lack of such correlations that has been so difficult to deal with in the radiological physics and radiation oncology contexts. Admittedly, calculating a human dose result, as seen in Chapters 4, 5, and 6, is no trivial task. Thus, we can rationalize our failures. First, though, let us demonstrate that, at least in simple enough geometries, there is good correlation of dose estimate and cellular and animal outcomes.

9.2 ANIMAL RESULTS CORRELATING ABSORBED DOSE AND EFFECTS

A direct method to demonstrate that absorbed dose is a relevant computation is to contrast tumor-growth results between specific and nonspecific antibodies to a given tumor marker. Such experiments would be unethical in humans but have been carried out in animals. In this context, the investigator would give the same amount of radioactivity by intravenous (IV) injection into two groups of tumor-bearing animals. One group of nude mice would receive the activity attached to a specific antibody to the tumor, whereas the other group would receive a nonspecific antibody. Buras, Beatty, et al. (1990) used ^{90}Y-ZCE025 and ^{90}Y-ZME018 as the specific and nonspecific agents to target LS174T colon xenografts (carcinoembryonic antigen+ [CEA+]) in nude mice. The former RP is based on an anti-CEA antibody, whereas the latter is an antimelanoma protein that will not find its antigen in colon cancer cells. Absorbed doses were estimated using methods of Chapter

TABLE 9.1   Absorbed Dose Estimates for LS174T
Xenografts in Nude Mice

ORGAN	NONSPECIFIC THERAPY (Gy/MBq)	SPECIFIC THERAPY (Gy/MBq)
Tumor	3.2	7.7
Liver	4.1	8.3
Spleen	6.1	6.6
Kidneys	3.6	3.9
Lung	3.4	1.9
Muscle	1.6	2.2
Blood	5.0	3.6

Source: Buras, R. R. et al., *Archives of Surgery*, 125, 1990.
Copyright 1990 American Medical Association. With
permission. All rights reserved.

4 along with biodistribution data obtained with the relevant agent. No cross-organ doses were considered in this simple trial. Table 9.1 gives the dose estimates for both therapeutic radiopharmaceuticals. Aside from the liver, it was seen that the estimated doses were comparable for most unaffected (normal) organs but were approximately a factor of 2.5-fold higher in the CEA-positive tumors for ^{90}Y-ZCE025. This enhancement reflected higher uptake in the lesions for this specific agent. At uptake maximum, ZCE025 demonstrated a u_T result of 17 ± 9.0 %ID/g, and ZME018 had a corresponding value of 5.8 ± 0.45 %ID/g. Taking the ratio of these two maximal tumor uptake values yields a quotient of 2.9, which is similar to the ratio of estimated absorbed doses.

Increase of liver dose estimates in the case of the ^{90}Y-ZCE025 antibody was presumably due to secreted CEA leaving the tumor site and moving, via the blood, to hepatic tissue. This trafficking does not occur with most antigens used in TRT but does present as an issue in anti-CEA imaging and therapy. In fact, presence of CEA in a patient's bloodstream can be used to diagnose the recurrence (or occurrence) of colorectal and other cancers. This phenomenon will be described more extensively in Chapter 12 in the examination of modeling the biodistribution of multiple LS174T tumors in the nude mouse.

Figure 9.1 shows the corresponding tumor growth curves for the average of the seven mice treated in each of three groups: control and TRT with the specific agent and with the nonspecific therapy. In the Buras et al. (1990) results, tumor growth rate was determined using the concept of lag time. By definition, this is the time required to achieve a mass of 1 gram post LS174T tumor cell implantation in the nude mouse. Measured lag time for ^{90}Y-ZCE025 was 42 ± 4 days but only 25 ± 3 days for nonspecific therapy. The control group (no radiation) had a corresponding value of 18 ± 1 days.

In the case of animal studies such as those previously described, the amount of activity, activity per body surface area, or activity per body total mass would not correlate with tumor regression. This is because these three traditional chemical indices of therapy would be identical in the two treatment arms of the experiment. Both the specific and nonspecific agents were given at the same total activity (120 μCi) to animals having the same surface area and body mass. Thus, using chemotherapy-type

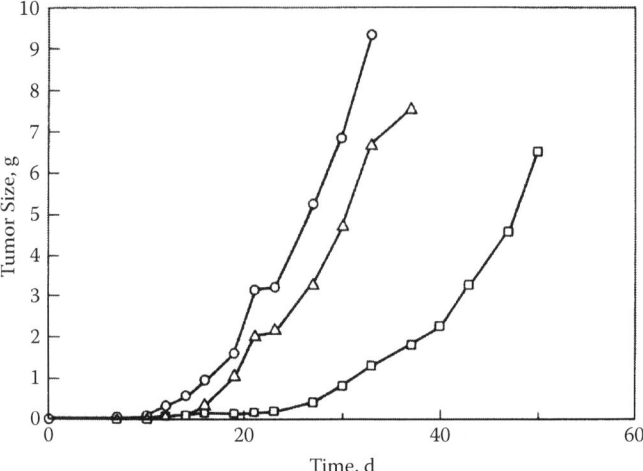

FIGURE 9.1   Tumor growth delays in nude mice bearing LS174T implanted colorectal tumors. Squares = the specific antibody treatment group. Triangles = the nonspecific protein treatment group. Circles = the control animals. (From Buras, R. R. et al., *Archives of Surgery*, 125, 1990. Copyright 1990 American Medical Association. With permission. All rights reserved.)

parameters, there would be no obvious way to discriminate between the treatment arms, and there would be no correlation with tumor lag time. It is only when we get to the specifics of targeting the lesion that we can see the difference between agents and that the absorbed dose actually did correlate with increased lag time. Implicit in this argument is the assumption that neither protein has any intrinsic effect on tumor growth. By giving these antibodies unlabeled, there is no evidence that either one of them can impact tumor cell replication.

It is of interest to ask why use of the nonspecific agent (^{90}Y-ZME018) would lead to measurable tumor growth retardation. By analyzing the biodistribution for this protein, it was found that the maximum tumor uptake was on the order of the value cited in Chapter 2 for a uniformly distributed agent in a 20 g mouse: 5 %ID/g. Upon performing integration over the tumor uptake curve for this antimelanoma antibody, absorbed doses as given in Table 9.1 were obtained. Thus, there was a finite absorbed dose due to the use of a nospecific agent and, correspondingly, a positive effect on tumor growth delay. Use of the specific anti-CEA protein ^{90}Y-ZCE025 led to a higher uptake, higher absorbed dose estimate in the tumor, and, correspondingly, an increased growth delay. In other words, absorbed dose estimates did correlate with a manifest effect on tumor size. Because of modulation of the tumor marker, there may be instances where the concentration of the target antigen decreases posttherapy. Thus, subsequent nonspecific therapy of a tumor may be useful in a situation involving variegated antigen expression.

It is important to consider the consequences when a single radiation therapy treatment does not suffice for a given subject. In our opening discussion of the human xenografts of colon cancer in the nude mouse, we commented on the increase in tumor lag time due to specific therapy. As demonstrated in the figures the initial TRT did not cure the tumors in the animal; it only delayed their growth. Thus, some retreatment would have been necessary to rid the animals of the disease. Generally, if TRT has been originally successful in reducing a lesion's size or the amount of a measurable marker, it would be tempting to try the same strategy repeatedly on any evidence of tumor recurrence. It is often assumed that the presence of the marker is unaltered by the specific nature of the therapy. Yet applying the species-selection concept developed by Charles Darwin and Alfred Wallace in the mid 1850s, this assumption may not be valid due to the possibly preferential therapy of cells containing relatively larger amounts of the target molecule in an initially heterogeneous tumor.

To test the continuing persistence of a molecular marker, an animal study was performed at City of Hope (Esteban, Kuhn, et al. 1991) involving nude mice treated with TRT for implanted human LS174T colon cancer tumors. These lesions produced CEA (105 µg/g) and were, correspondingly, treated initially with ZCE025, an anti-CEA murine antibody labeled with ^{90}Y at 120 µCi as described already. After 6 weeks of a single therapy, residual tumors were removed and passed, by implantation on the flank, to other nude mice. At the same time, samples of these residual lesions were examined pathologically and homogenized for specific CEA content. It was found that the concentration of carcinoembryonic antigen decreased by a factor of approximately fourfold to a value of 25 µg/g after the initial therapy. Over repeated 6-week intervals, aliquots of the treated lesions were sequentially passed from mouse to mouse out to 4 months. Over the time interval, concentration of CEA slowly increased but did not return to the value found in the naïve tumors not treated with an anti-CEA antibody. After three passages, levels of CEA were still only one half their initial value (45 µg/g). Because of the relative increase in the CEA-bearing clones, the authors postulated that the presence of CEA induced more rapid proliferation in those cells that are strongly CEA positive.

This xenograft evaluation demonstrated that selective killing of the malignant cells was specific to those that carried the highest concentration of the CEA target molecule. Cells not expressing this antigen at relatively high levels were spared selectively and allowed to grow. In a clinical context, the corresponding result is difficult to demonstrate but may prove to be similar—at least for this and other secreted antigens. Therefore, to justify retreatment with the same TRT anti-CEA agent, the oncologist must demonstrate that the residual lesion is still expressing comparable concentrations of the antigen of interest. A diagnostic scan using the imaging form of the agent (^{111}In-ZCE-025) would be appropriate for this testing. Notice that timing of the subsequent treatment may also be important. The City of Hope researchers could not find densities of CEA comparable to the initial values even 4 months after therapy in this human tumor xenograft in the nude mouse model.

As was pointed out in the previous study, analogous results were not seen by the group at the National Institutes of Health (NIH; Schlom, Molinolo, et al. 1990) with the TAG-72 glycoprotein as their tumor target. Using the ^{131}I-B72.3 antibody, no difference was seen between the initial tumor density of antigen and the resultant density measured after the first pass of therapy. Thus, the NIH team felt that a dose-fractionation scheme was justified in treating lesions expressing TAG-72—a nonsecreted protein. One may conclude that the antigen itself must be considered prior to implementing subsequent therapies. Again, an imaging study may be the optimal way to determine the continued presence of the particular antigen after the initial treatment.

Having described some animal studies, we now turn to clinical outcomes that exhibit positive correlation with absorbed dose estimates. We will cover blood markers, red marrow dose, lymphoma regression, and renal toxicities. In the last example, that of long-term renal toxicity, the result is shown to more strongly correlate with dose estimates if the biologically effective dose (BED) is invoked as described in Chapter 4.

9.3 LYMPHOCYTE CHROMOSOME DEFECTS OBSERVED FOLLOWING TRT

One of the first clinical trials to show a direct linear correlation between absorbed dose estimates and a demonstrable clinical effect was reported by City of Hope in 2000 (Wong, Wang, et al. 2000). This study was a Phase I protocol using an anti-CEA antibody. The authors used a chromosomal assay involving white cells taken from the exposed patient's blood. A baseline blood sample had been obtained prior to beginning the TRT treatment. It is important to point out that every patient had undergone prior solid-tumor therapies of various types, including exposure to external beam irradiation. Thus, each person had his or her own intrinsic amount of chromosome translocation prior to beginning radionuclide treatment. It is relatively rare for any patient to enter into targeted radionuclide therapy without this prior exposure to therapeutics. One would assume that this missing or often poorly documented history is another of the fundamental reasons for the lack of dose estimate correlation with clinical effects.

Chromosome (Xsome) translocation evaluation has, for a long time, been applied in assaying human blood effects following exposure to a variety of types of ionizing radiation. In these cases, no baseline data are usually available since the subject is generally a member of an exposed group within a general population without workplace or clinical history. There are, of course, control group measurements possible on individuals who are ethnically similar and were not exposed in these ionizing radiation incidents. Among other applications, chromosomal assays have been used in evaluating absorbed doses resulting from nuclear attacks (Stram, Sposto, et al. 1993), from radiation accidents such as at Chernobyl (Salassidis, Georgiadou-Schumacher, et al. 1995) as well as from external beam therapy as in the treatment of Hodgkin's disease (Smith, Evans, et al. 1992).

In the Wong et al. (2000) assay, the 18 patients in a Phase I trial were treated using TRT for CEA-positive disease—primarily colorectal cancer. Up to three treatments were possible before the onset of immune response (human antichimeric antibodies [HACA]) to the chimeric cT84.66 intact antibody. Three patients had two cycles of TRT therapy, and two patients received three cycles. In repeat treatments, individuals would receive the same dosage of activity as in their initial therapy. Because of FDA experience, it was necessary to treat patients using units of activity per meter squared. This value ranged between 5 and 22 mCi/m^2 with ^{90}Y-DTPA-cT84.66. As is necessary in such studies, an ^{111}In-DTPA-cT84.66 surrogate injection with 5 mCi of ^{111}In was used to determine the biodistribution. These indium data were then used to predict the absorbed dose per organ for the ^{90}Y-labeled antibody. The American Association of Physicists in Medicine (AAPM) algorithm (Chapter 5) was used to calculate the RM absorbed dose.

Two blood samples were necessary: one prior to the treatments and one at 5 to 6 weeks after the last therapy. After 72 h of phytoemaglutinin-induced cell expansion, approximately 1,000 cell spreads were read for each patient at each of the two time points. The probes were then used to document that a given segment of DNA had relocated from Chromosome 3 to another chromosome or from Chromosome 4 to another chromosome in the patient's lymphocytes. Each of the 18 patients had the number of such translocations per the total number of cell spreads measured, for each chromosome, at the end of therapy. This frequency value was then corrected by subtracting the initial frequency of translocations found in that Xsome during the baseline sample. The net number of translocations was plotted against the estimated RM dose in cGy for the patient. A linear correlation was attempted in the resultant graph. Results are shown in Figure 9.2 for Chromosomes 3 and 4.

The correlation coefficients (ρ) seen in the two Xsome studies were similar: 0.79 for Xsome 3 and 0.89 for Xsome 4. Both are significant at the .05 level of confidence. It is important to realize that the subtraction of the baseline level of translocation has been done in each of these examples. Thus, the varied radiation history of each patient was

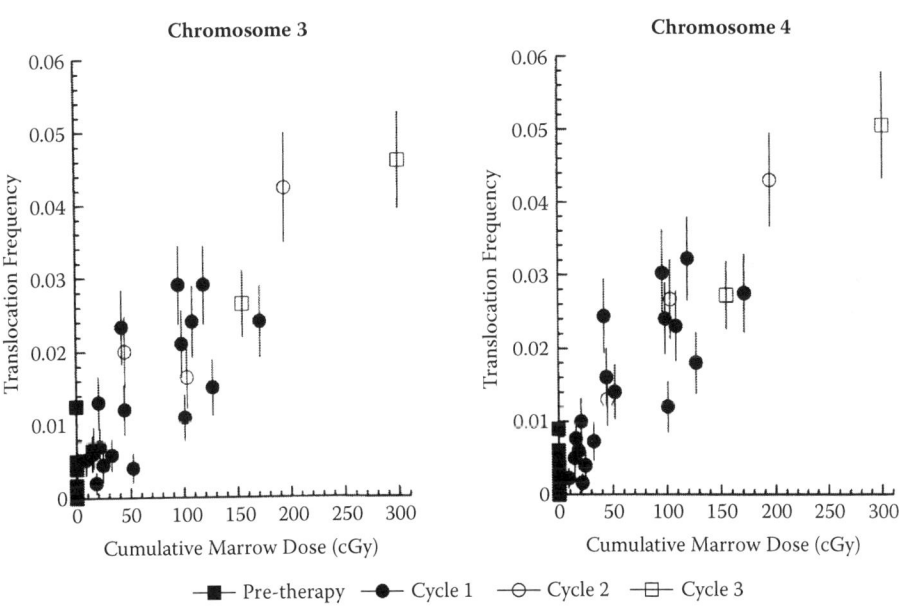

FIGURE 9.2 Increase of Xsome 3 and Xsome 4 translocations as a function of estimated RM absorbed dose using the AAPM algorithm. Cycles are indicated by the data point representation. Linear correlation coefficients of 0.79 and 0.89 were found for these two chromosomes, respectively. (From Wong, J. Y. et al., *Int. J. Radiat. Oncol. Biol. Phys.*, 46, 3, 2000. With permission.)

TABLE 9.2   Correlation of Chromosomal Translocation Frequency and Treatment Parameters

EXTERNAL VARIABLE	CHROMOSOME 4 FREQUENCY CORRELATIONS (ρ)	CHROMOSOME 3 FREQUENCY CORRELATIONS (ρ)
RM estimated absorbed dose	0.89	0.79
Activity/m^2	0.94	0.88
Activity/kg	0.92	0.84
Whole-body absorbed dose	0.92	0.82
Total activity	0.92	0.86

Source: Wong, J. Y. et al., *Int. J. Radiat. Oncol. Biol. Phys.*, 46, 3, 2000.

taken into account. As the authors showed, elevated levels of baseline translocations were seen in patients who had been exposed to external beam radiation therapy (EBRT) prior to entering the TRT trial.

Table 9.2 gives the various correlation coefficients for translocation frequency and other external variables beside the RM absorbed dose. Here, whole-body absorbed dose, total activity, total activity/m^2, and total activity per kg were also examined for significant correlations. In this case it was observed that multiple parameters did correlate with the incidence of translocations of the chromosome markers. This result presumably reflects the fact that the blood curve is the driving term in the marrow dose estimation process. Thus, other overall parameters such as whole-body dose and activity per unit area and per whole-body mass would also give good correlation with the number of chromosomal translocations.

Because of the tradition of looking for platelet decreases after TRT, the group also tabulated the correlation coefficients for the various clinical parameters and the fractional decrease in the platelet numbers. These results are in Table 9.3.

The authors concluded that Xsome translocation number in both Chromosomes 3 and 4 correlated with the estimated RM dose. This was a linear result in that the higher the estimated absorbed dose the larger was the number of translocations seen in either Xsome. Absorbed dose levels were not so large, however, that saturation of the system was evident in the chromosomal analysis. It was also reassuring that patients who had been exposed to EBRT exhibited increased baseline values for the number

TABLE 9.3   Correlation (ρ) of Fractional Decrease in Platelet Counts with Treatment Parameters

EXTERNAL VARIABLE	CORRELATION (ρ)
RM estimated absorbed dose	0.53
Activity/m^2	0.62
Activity/kg	0.65
Whole-body absorbed dose	0.59
Total activity	0.66

Source: Wong, J. Y. et al., *Int. J. Radiat. Oncol. Biol. Phys.*, 46, 3, 2000.

of translocations. Thus, a traditional in vivo "dosimeter," readable in each patient, was applied to evaluate the normal marrow toxicity induced by the TRT procedure.

9.4 LYMPHOMA TUMOR DOSE ESTIMATES AND DISEASE REGRESSION

The group at the University of Michigan has obtained some of the most interesting results for non-Hodgkin's lymphoma (NHL) TRT outcome and its correlation with tumor absorbed dose estimates (Koral, Dewaraja, et al. 2000). In this work, patients with advanced-stage malignant follicular lymphoma were treated with ^{131}I-tositumomab; this agent, an anti-CD20 intact antibody, is described more completely in Chapter 8. The target CD-20 protein is found on the surface of both malignant and normal B cells.

As noted in Chapter 8, each patient was treated with a specific amount of ^{131}I-labeled tositumomab (Bexxar) activity sufficient to give a whole-body estimated radiation dose of 75 cGy to that individual. To calculate the amount of required activity, a diagnostic study was initially performed using a pair of opposed gamma cameras over a 6- to 8-day interval to determine whole-body clearance of this labeled antibody to CD20. Generally, 5 mCi of ^{131}I-tositumomab was used for this aspect of the absorbed dose estimation. Therapy followed upon completion of the diagnostic study. Tumor size was measured with CT scanning before and after treatment. A typical patient would have between 1 and 11 lesions evaluable for absorbed dose correlation with clinical outcome.

A hybrid assay system, more completely described in Chapter 5, was developed to better estimate the amount of activity at each tumor location. A quantitative single-photon emission computed tomography (QSPECT) assessment of tumor activity at one time point during the therapy phase of the study was used to improve the accuracy of the activity values determine by routinely following the patient using the standard geometric mean (GM) strategy. The QSPECT image was generally taken at the maximum tumor uptake segment of this Phase II tositumomab TRT protocol. Typically, this time would be several days into the therapy phase of the study. The authors pointed out that the number of counts obtained at the comparable point in the diagnostic phase of the study would not be sufficient to allow adequate quantitation of activity at depth in the various tumor sites. Thus, the diagnostic study could not be used for predicting which patients were most likely to have a good therapy result. Since the radiolabel for this protocol was ^{131}I, there was the possibility of imaging during therapy; this is not possible using a ^{90}Y-labeled agent unless an ^{111}In-labeled surrogate is included during the therapy phase of the trial.

Perhaps the most important clinical aspect of the Michigan Phase II trial was that these NHL patients had not had any prior therapy for their disease. Such cohorts are rare in the development of RP materials. As mentioned already, most patients entering such trials have had one or more prior NHL therapies and have come to TRT as a method of last resort—having obtained no satisfaction with the most recent standard therapy method. At the time of their first publication on this cohort (Koral et al. 2000), only 20 of 55 patients had been available for follow up on their treatment outcome. For these 20 individuals, 5 were determined to have partial response (PR), and 15 were assigned to the complete response (CR) category. Corresponding to these two subpopulations of NHL patients, absorbed dose estimates for the tumors were 369 ± 54 cGy and 720 ± 80 cGy, respectively. In the case of PR subjects, there were 30 individual lesions, whereas the CR group had 56 tumors. Tumor volumes were determined using CT data that were fused to the SPECT imaging. An analysis of variance (ANOVA) of these PR and CR results gave a result of $p = .04$ if the dose were the only independent variable. If one expanded the independent parameters to include both absorbed dose and initial lesion size, the p-value increased to .06. The group at Michigan concluded that their analyses implied a trend toward significance. We would note that a choice of $p = .05$ as the threshold for statistical significance is arbitrary—but very traditional in the scientific literature.

In a subsequent article some two years later, Koral, Francis, et al. (2002) described use of a probit function to represent the percentage decrease in lesion size as plotted against the absorbed dose. Based on a Gaussian curve, the probit definition is given in Equation 9.1 and represents one of many forms possible (cf. Chapter 6) for generating a sigmoidal outcome

$$\text{Effect} = \frac{1}{\sqrt{2\pi}} \int_{-\infty}^{L} \exp(-z^2/2)dz \tag{9.1a}$$

$$L = C_1 + C_2 \bullet \text{Dose} \tag{9.1b}$$

Notice that the upper limit of the integral (L) is defined as a linear function of absorbed dose. Here, C_1 and C_2 are the fitted constants in the least-squares analysis of the patient data. Dose was estimated with the hybrid imaging systems as indicated above for a total of 43 NHL lesions. The "effect" of the TRT was defined as the percentage decrease in the tumor size. The authors took the initial tumor size, subtracted the volume at 12 weeks, and divided that difference by the original volume. All of the 43 lesions were analyzed initially using the probit analysis with the results shown in Figure 9.3. One interesting feature of this fit was that there was a cutoff of the probit function near the value of 55% volume reduction. The authors attributed this result to the antitumor effect of the intrinsic (unlabeled) tositumomab antibody. To put this explanation another way, had there been no ^{131}I attached to the protein, there would have been a 55% reduction in lesion volume. Notice the very rapid rise of the graph as one approaches 500 cGy. At 1,000 cGy, the response curve has become essentially flat so that all tumors treated at this dose or higher showed 100% decrease. Absorbed doses above this value did not contribute to enhanced tumor shrinkage.

Because of the general interest in treating smaller lesions (cf. Chapter 3), the Michigan group did a separate analysis of those 15 tumors having an initial mass less than 10 g. Figure 9.4 shows these results. Here, the probit function cutoff was much reduced, being only 11%, which the authors again attributed to the effect of the antibody itself. Also, as in the 43-lesion results of Figure 9.3, an absorbed dose of 1,000 cGy was

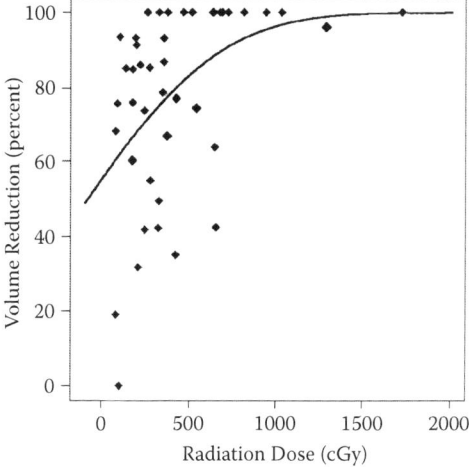

FIGURE 9.3   Probit results for the 43-lesion Michigan study of NHL. Note the rapid rise of the curve near 500 cGy. (From Koral, K. F. et al., *Cancer*, 94, 2000. Reprinted with permission from John Wiley and Sons.)

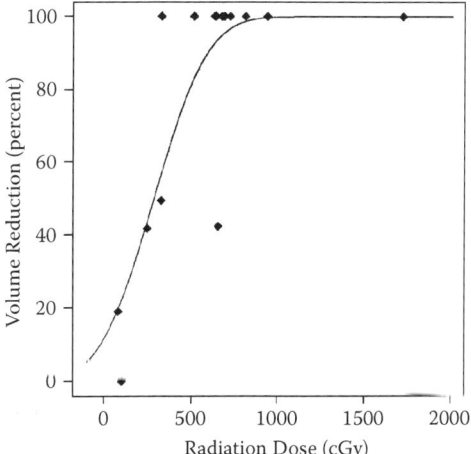

FIGURE 9.4   Michigan probit function analysis of the smaller (< 10 g) NHL lesions. (From Koral, K. F. et al., *Cancer*, 94, 2000. Reprinted with permission from John Wiley and Sons.)

sufficient to reduce all tumors to zero residual size. When this analysis was performed on the 28 lesions having a mass > 10 g, the probit results yielded a cutoff of 72%; that is, the cold antibody was responsible for nearly three quarters of the tumor size reduction over the 12-week time frame of this study.

The authors presented arguments to explain the vivid antitumor effect of the unlabeled antibody relative to the radiation effect. In larger lesions, because of limited penetration into the tumor mass (Chapter 3), the radiation dose was thought to be largely given to exterior cells at or near the tumor surface. Also there would have been radio-active decay and dehalogenation of the ^{131}I label, so the radiation dose rate becomes quite low after the first few days of treatment. Thus, the only long-term therapy was thought to be due to that produced by the cold antibody, which may remain *in situ* for extended periods. Because the unlabeled monoclonal antibody (Mab) has no emissions, the observers could not measure this persistence. Presumably, the antibody stays at the tumor site longer than the iodine label, which is cleaved by enzymes *in situ*.

Because of its relatively unique nature, the Michigan report of Koral et al. (2000) generated great interest—particularly among those treating NHL using TRT. Sgouros, Squeri, et al. (2003) at Memorial Sloan-Kettering Cancer Center (MSKCC) reported subsequently on their manifest lack of correlation between lymphoma lesion estimated absorbed dose and clinical outcome for patients undergoing tositumomab therapy.

To understand the differences between these two results, one explanation was the disparate nature of the two patient populations. While the MSKCC group had described a typical clinical population most of whom had one or more preexisting therapies, the Michigan report had concerned only previously untreated individuals (Koral, Kaminski, et al. 2003). When called on to write an editorial (Williams 2000) on the relatively unusual results from the initial Michigan study, I used the previously untreated nature of the Michigan patient group to explain its unique outcome. In other words, the individuals at MSKCC who had been given prior therapy may have had changes made to their tumor populations to cause a variable response to a given dose of ionizing radiation.

9.5 IMPROVING HEMATOLOGICAL TOXICITY CORRELATIONS WITH RED MARROW ABSORBED DOSE ESTIMATES

While tumor therapy is the fundamental objective, the activity-limiting toxicity seen in most TRT studies is that of the RM. This result is not surprising due to the sensitivity of the RM to ionizing radiation. Significant effects are known at external beam absorbed dose levels as low as 2.0 Gy (Emami, Lyman, et al. 1991). As seen already, NHL lesions may be completely removed at absorbed doses on the order of 5 to 10 Gy—values not much larger than those toxic to the RM. Yet a typical solid-tumor may require a 50 to 100 Gy absorbed dose treatment to reduce lesion size. Such an enormous disparity between RM toxic effects and solid-lesion treatment is a difficult issue to contend with in treating the cancer patient using internal emitters that move throughout the body. The therapist must make estimates of the RM absorbed dose and try to restrict the activity injected to keep this number below the toxicity level previously cited.

As noted in the discussion of the Xsome translocation evaluations, there has been a general lack of correlation between estimated red marrow dose and any subsequent decreases in the numbers of platelets or white cells. Many reasons are given for this disconnect; some are listed in the introduction to this chapter. Yet one wonders if there is any way to enable an agreement between RM absorbed dose estimates and the subsequent toxicities for a given set of patients. One such method, involving use of a patient-specific cytokine, has been reported by Siegel, Yeldell, et al. (2003).

This work was based on earlier investigations by Blumenthal and her colleagues at the Garden State Cancer Center (GSCC) (Blumenthal, Lew, et al. 2000) on five cytokines initially thought to be possibly correlated with hematological toxicity recovery following TRT. Two stimulatory proteins—FLT3-L (FMS-related tyrosine kinase 3 ligand) and stem cell factor—were included in their investigations. Three inhibitory cytokines—tumor necrosis factor, tumor growth factor beta, and macrophage inflammatory protein 1 alpha—were also studied. Of the five, only FLT3-L demonstrated a strong correlation with hematological toxicity. It was then assumed by the researchers that the presence of FLT3-L might be an indicator of the potential sensitivity of an individual patient to TRT. In their hypothesis, high levels of this particular cytokine may indicate a relatively large fraction of the bone marrow stromal cells and hence a relatively high sensitivity to ionizing radiation. In such cases, the absorbed dose may be expected to be more effective in reducing platelets and lymphocytes than in individuals having normal or subnormal levels of this particular cytokine.

A total of 30 patients were included in the study. All were treated using one of two anti-CEA antibodies in either intact or $F(ab')_2$ form. Most of the patients had either colorectal cancer or medullary thyroid cancer and were treated with TRT with ^{131}I-antibody activities in the interval 2.1 to 8.9 GBq. No direct marrow uptake and no bone metastatic sites were observed in the population. White blood cell (WBC) and platelet counts were measured posttherapy and compared with their pretreatment values. Comparisons of absorbed dose estimates and several blood cell variables were attempted. The latter included the traditional markers platelet toxicity grade (PTG), percent platelet decrease (PPD), and platelet nadir (PN). In addition, to establish positive correlations, the GSCC group also evaluated the relationship between RM dose and 1/PN. In this case, as the nadir of platelets decreased, 1/PN would increase (i.e., possibly be positively correlated with the estimated absorbed dose to red marrow).

In this study, photon contributions to the absorbed dose had to be included since ^{131}I was the radionuclide. The authors therefore posited the estimated RM dose as due to two source tissues: RM and the remainder of the body (RB). As outlined in Chapter 4, the canonical Medical Internal Radiation Dose (MIRD) method was used to predict RM dose via

$$D(RM) = \tilde{A}(RM) * S(RM \rightarrow RM) + \tilde{A}(RB) * S(RB \rightarrow RM) \tag{9.2}$$

In the equality, $\tilde{A}(RB)$ was computed by subtracting $\tilde{A}(RM)$ from the observed $\tilde{A}$(total body). Two techniques were used to estimate the $\tilde{A}(RM)$, the first one given in Chapter 4 where the humanoid phantom marrow mass is used to correct the measured blood curve $\tilde{A}$(blood) to give an estimate of the marrow total number of decays. In the second method, a one-compartment model (single exponential) was used to fit the blood clearance curve, and the same correction was applied. Either method required a correction factor (CF) that multiplied the results. Again, two methods for determining the CF were applied. Either a fixed value of 0.30 (cf. Chapter 4) or the Sgouros (1993) value was chosen based on the hematocrit of the patient

$$CF = 0.30 \tag{9.3a}$$

$$CF = \frac{0.19}{1 - \text{hematocrit}} \tag{9.3b}$$

Correlation coefficients were then determined for the two CF values using standard anthropomorphic models from either MIRD11 (Snyder 1975) or the MIRDOSE3 software (Stabin 1996). Table 9.4 gives the correlation coefficients for all the patients and the two CF values given in Equation 9.3. It was found that the modulation of absorbed dose by the FLT3-L concentration in the patient's blood substantially improved correlations of all four clinical parameters commonly used to document bone marrow toxicity. By comparing the standard (unmodulated) dose estimates with those determined using cytokine levels, one notices that the correlations were not significant in the former situation. Finally, it was seen that the CF value has little impact on these conclusions so that either Equation 9.3a or Equation 9.3b would suffice for the conversion of blood curve to marrow curve for the total number of marrow decays. Modulation of the dose was performed only in patients having elevated values of FLT3-L; those having normal or reduced cytokine levels did not have their absorbed dose values altered.

The authors also explored the (possible) improvement of the correlation if the same FLT3-L correction factor were applied to other standard clinical indicators of potential marrow toxicity. In the tradition of chemotherapy, these included the total body absorbed dose (cGy), the total activity given the patient (GBq), and the total activity divided by the patient's body weight (GBq/kg). Results for both corrected and uncorrected indicators are shown in Table 9.5. It was seen that the cytokine correction factor greatly enhanced the correlations for all three of these traditional marrow toxicity indicators. For example, if we select total body absorbed dose and 1/PN as our two variables of interest, it was seen that the ρ value went from 0.17 for the usual absorbed dose calculation up to 0.86 when the absorbed dose was corrected for cytokine FLT3-L concentration. Total activity as a toxicity indicator gave some interesting results as well. Here, the correlation coefficients were essentially nil for the activity per se but increased to values on the order of 0.50 if one corrected for cytokine. Figure 9.5 shows the scatter diagrams for the adjusted

TABLE 9.4 Correlation Coefficients for FLT3-L Adjusted and Simple Absorbed Red Marrow Dose Estimates (Unadjusted) and Clinical Toxicity

CLINICAL BLOOD PARAMETER	CF = 0.30 AND DOSE ESTIMATE ADJUSTED USING FLT3-L ρ	CF = 0.30 DOSE NOT ADJUSTED ρ	CF GIVEN BY EQ. 9.3b AND DOSE ADJUSTED USING FLT3-L LEVELS ρ	CF GIVEN BY EQUATION (9.3b) DOSE NOT ADJUSTED ρ
1/PN	0.86	0.20	0.85	0.18
PTG	0.68	0.31	0.68	0.27
PPD	0.46	0.15	0.45	0.11
PN	0.74	0.27	0.74	0.24

Source: Siegel, J. A. et al., *J. Nucl. Med.*, 44, 1, 2003. Reprinted with permission of the Society of Nuclear Medicine.

TABLE 9.5 Correlation Coefficients for FLT3-L Adjusted and Simple Toxicity Predictors and Clinical Toxicity

CLINICAL BLOOD PARAMETER	TOTAL BODY ADJUSTED DOSE ρ	TOTAL BODY DOSE ρ	TOTAL ACTIVITY ADJUSTED ρ	TOTAL ACTIVITY ρ	TOTAL ACTIVITY/kg ADJUSTED ρ	TOTAL ACTIVITY ρ
1/PN	0.86	0.17	0.53	0.21	0.79	0.04
PTG	0.68	0.23	0.61	0.05	0.72	0.28
PPD	0.46	0.06	0.51	0.15	0.59	0.33
PN	0.75	0.16	0.60	0.03	0.75	0.18

Source: Siegel, J. A. et al., J. Nucl. Med., 44, 1, 2003. Reprinted with permission of the Society of Nuclear Medicine.

parameters versus the 1/PN clinical indicator. Notice that, for the total of 30 patients, there are relatively few points (8 to 10) at very elevated dose levels. These examples drive the analysis and lead to the strong correlations. This result implies that observers restricted to situations occurring at relatively low absorbed dose values might have greater difficulty finding any correlation between hematopoetic toxicity and clinical parameters.

One can conclude, using the results of GSCC group, that patient-specific information can make substantial differences in the absorbed dose estimate to the red marrow. Rather than attempting to use the marrow mass (which is essentially unknown for a given person) as their adjustment variable, the authors applied the level of a hematopoetic cytokine in the patient's blood to determine individual susceptibility to the radiation dose. The calculated absorbed dose was then multiplied by the level of the cytokine to achieve a new parameter—essentially an effective dose for that individual.

9.6 RENAL TOXICITY FOLLOWING PEPTIDE RADIONUCLIDE THERAPY

Beside red marrow, the kidneys have been identified as another organ system possibly at risk in TRT. A basic reason for this tendency is clearance of many radioactive agents, or their metabolic products, via the renal system. As we have noted in Chapter 2, an RP that has relatively low molecular weight is a particular example of an agent that is likely to be removed from the body via the urinary excretion route. In a multistep process of labeling a lesion, for example, the final step is generally one whereby a low molecular weight (MW) ligand carrying the radiolabel is injected into the patient. This material is designed to clear rapidly via the kidneys to reduce the blood curve and thereby decrease RM absorbed dose. There are, in addition, single-step targeting processes that use radiolabeled small molecules. One of these techniques has been found to exhibit a good correlation between renal absorbed dose and clinical effects if kidney size and absorbed dose rate are taken into account.

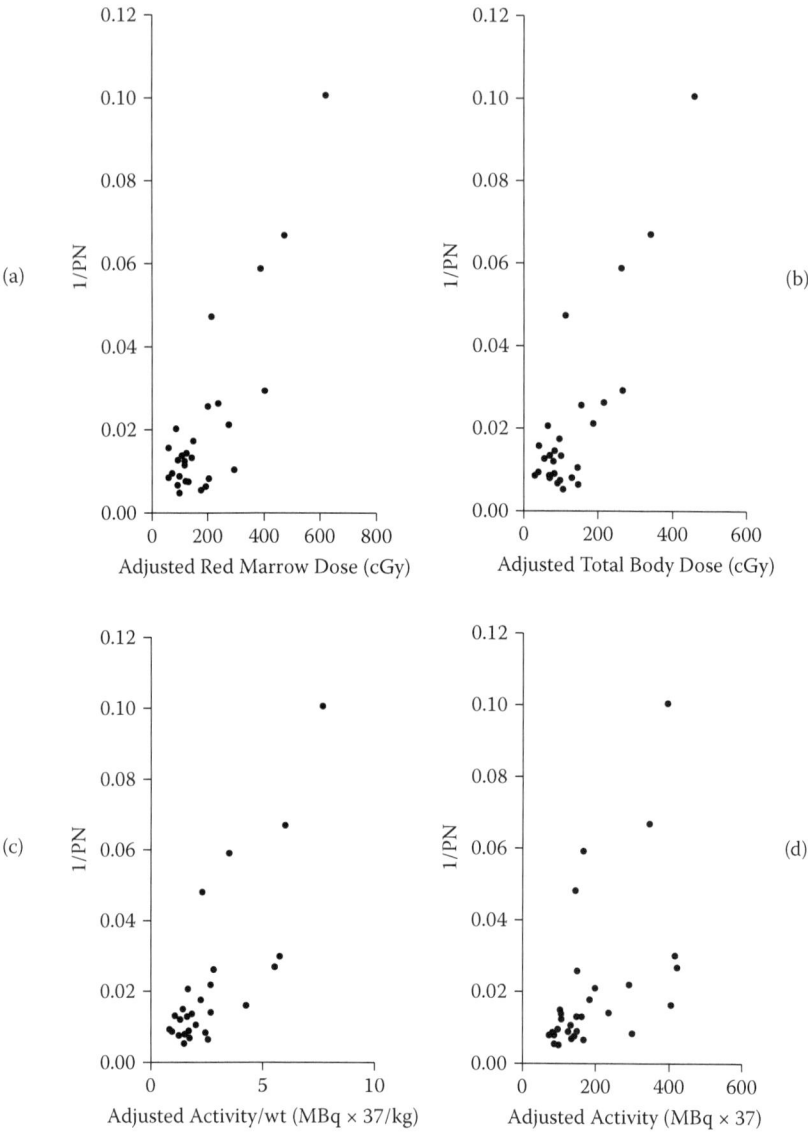

FIGURE 9.5 Scatterplot of 1/PN (on the ordinate) versus four possible indicator of RM toxicity. (a) the adjusted RM absorbed dose is shown on the abscissa. (b) the adjusted whole-body dose. (c) adjusted activity per body weight on the abscissa. (d) the adjusted activity on the abscissa. Notice the linearity of the relationship and the lack of saturation at these dose and activity levels for this 30-patient group. (From Siegel, J. A. et al., *J. Nucl. Med.*, 44, 1, 2003. Reprinted with permission of the Society of Nuclear Medicine.)

One of the most interesting clinical results of TRT has occurred in the treatment of somatostatin receptor (SR)-positive tumors. These are usually neuroendocrine lesions with one example being pheochromocytoma as mentioned in Chapter 8. In that earlier discussion, we indicated the historical use of ^{131}I-MIBG as a possible therapeutic agent. In the more general case of SR-positive tumors, one RP of widespread use has been DOTA-D-Phe[1]-Tyr[3]-octreotide (DOTATOC)—a variant of octreotide with the DOTA chelator holding ^{90}Y as the therapy label. There are only eight amino acids in octreotide, so the MW is on the order of 1 kDa with a correspondingly rapid loss from blood into the kidneys.

An international group (Barone, Borson-Chazot, et al. 2005) reported on the renal absorbed dose estimates for ^{90}Y-DOTATOC and the possible correlation with loss of kidney function. In the associated activity measurements, the authors utilized [86]Y-DOTATOC as a surrogate. The use of [86]Y as the radiolabel permitted positron emission tomography (PET) imaging over the initial phase of the targeting. In determining the toxicity, creatinine levels and creatinine clearance rate were evaluated over an extended period of time posttherapy in this treated population of 18 patients. Values of the two creatinine-based parameters were taken at baseline and at 6-month intervals posttreatment. Dose estimates to the kidneys were performed in three separate ways. In a first approximation, the authors used the standard phantom masses found in MIRDOSE3 (Stabin 1996). Next, the analysts added specific information on the patient's own kidney sizes using CT data and a spatially segmented model of the kidneys (Bouchet, Bolch, et al. 2003). Finally, the authors combined the latter results with BED estimates (Chapter 4) using the α/β value of 2.6 Gy and a renal cell repair time of 2.8 h. In this last step, the radioactivity effective half-life was set at 30 h. Explicitly, the BED was computed from an earlier suggestion (Konijnenberg 2003):

$$\mathrm{BED} = \sum_{i=1}^{N} D_i + \beta/\alpha \frac{T_{1/2}^{\mathrm{repair}}}{T_{1/2}^{\mathrm{repair}} + T_{1/2}^{\mathrm{effective}}} \sum_{i=1}^{N} D_i^2 \tag{9.4}$$

where D_i is a given (single treatment) absorbed dose estimate and $T_{1/2}^{\mathrm{repair}}$ and $T_{1/2}^{\mathrm{effective}}$ are the repair and effective half-times, respectively. In the latter case, the group assumed a single half-time to describe clearance of DOTATOC from the kidneys. As noted in Chapter 6, clearance is generally more complicated than a single exponential so that the latter assumption is an approximation.

Table 9.6 gives the three absorbed dose estimates. It was seen that there was considerable difference among predictions made with the various methodologies. Positive, significant correlations ($\rho = 0.54$; $p < .02$) became possible for the method that used absorbed dose but with specific renal masses being taken into account. By far the best correlation ($\rho = 0.93$; $p < .0001$) was found with the BED being used as the independent variable in the analysis. In these three correlation analyses, the dependent variable was the percentage loss of creatinine clearance per year.

To represent the sigmoidal response of renal tissue damage to absorbed dose or BED, the authors proposed the function

$$\mathrm{Effect}(D') = (1 - \exp(-C_3 D'))^b \tag{9.5}$$

TABLE 9.6   Three Estimates of Renal Absorbed Dose or BED for ^{90}Y-DOTATOC

PATIENT	KIDNEY TOTAL VOLUME (c.c.)	MIRD PHANTOM DOSE (Gy)	DOSE USING PATIENT RENAL MASSES (Gy)	BED WITH PATIENT RENAL MASSES (Gy)
1	278	26.6	31.9	38.1
2	231	27.5	39.6	54.8
3	283	27	31.7	40.5
4	337	27.5	27.2	33.8
5	503	27	19.4	27.7
6	411	26.2	21.2	30.2
7	365	26.8	26.6	36.3
8	299	27.1	32.8	56
9	278	27.1	35.2	56.1
10	343	27.2	28.6	50.8
11	408	30	26.6	31
12	288	25.6	29.6	35.5
13	427	26.1	22.1	31.5
14	433	38.6	32.2	59.3
15	404	28	23	28.1
16	299	29.2	35.4	41
17	321	26.7	27.6	33.1
18	302	27	29.7	42.7

Source: Barone, R. et al., *J. Nucl. Med.*, 46(Suppl 1), 2005. Reprinted with permission of the Society of Nuclear Medicine.

where C_3 and b are the adjustable parameters, and D' is either dose or BED. While no explicit values for the two quantities were listed in the publication, the authors did comment that b is most obvious at low values of D, whereas C_3 becomes dominant at the rapidly rising portion of the S-shaped response curve. Notice that as D' becomes large, the effect becomes saturated (i.e., equal to unity or 100%). In the case of creatinine clearance loss, a value of 100% implies no remaining renal function.

Following this initial report, the MIRD Committee (Wessels, Konijnenberg, et al. 2008) published an analysis of these 18 patients as well as 25 others contributed by Bodei (Bodei, Cremonesi, et al. 2004). Using the same analytic approach, a sigmoidal graph of end-stage renal disease versus BED was generated for the total of 43 patients receiving ^{90}Y-DOTA Octreotide. This result is shown in Figure 9.6. Two parameters defining this function are TD$_{50}$ of 35 Gy and $m = 0.18$. Included in the figure are data obtained using external beam exposures where both kidneys were in the radiation field. By using the BED concept, these two modalities (TRT and external beam) can be represented by very similar curves that essentially overlie each other. Thus, by invoking BED instead of absorbed dose, the authors have placed the two treatments (TRT and EBRT) into close correspondence for the stochastic results of renal toxicity. This last clinical outcome was documented using the decrease in creatinine clearance rate.

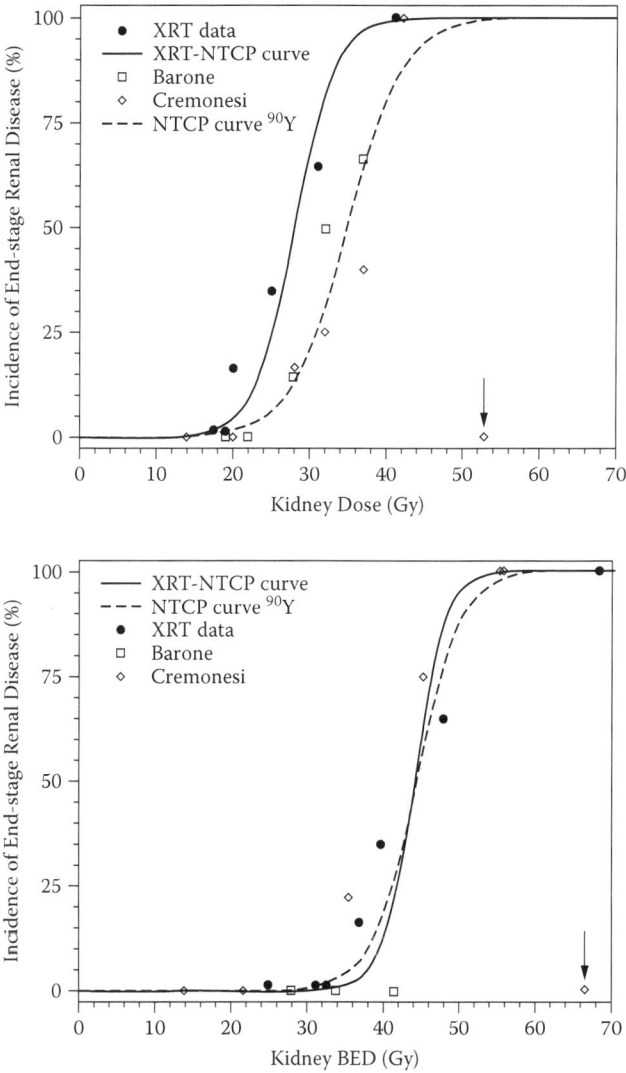

FIGURES 9.6 Sigmoidal response curves of creatinine loss per year versus the dose (Gy) or BED in Gy. Larger dots represent repeated treatments with the range being between one and five DOTATOC therapies. Notice the onset of renal damage near the threshold value of 40 Gy in the BED curve. (From Wessels, B. W. et al., *J. Nucl. Med.*, 49, 11, 2008. Reprinted with permission of the Society of Nuclear Medicine.)

9.7 SUMMARY

Correlations between absorbed dose estimates and outcome in animals and patients are relatively rare in the literature. In the former case, the appropriateness of absorbed dose can be established by following the effects of a specific antibody on tumor growth. By comparing that specific therapy with a comparable nonspecific agent, it is seen that the retardation of tumor size depends on absorbed dose. No direct correlation can be demonstrated with traditional chemotherapy parameters such as total activity, activity per unit area, or activity per unit volume of the animal. In each of these last three descriptors, the two agents are identical, yet the effect of tumor size reduction is given only by the absorbed dose.

Compared with animal experiments, significant correlations between clinical outcome and absorbed dose estimates are more difficult to find. This chapter briefly described several of these results. One of the more interesting is the analysis of chromosomal translocations done by Wong et al. (2000) at City of Hope. Frequency of both Xsome 3 and Xsome 4 translocations correlated strongly with a number of parameters—including the red marrow absorbed dose estimate. In this assay, it was also found that patients who had earlier been exposed to EBRT had elevated numbers of translocations compared with individuals without such dose increments. This result points out probably the most significant fact in trying to obtain a correlation between clinical outcome and treatment parameters: individual details of the patient must, if possible, be entered into the study from the beginning.

The group at the University of Michigan (Koral et al. 2000) demonstrated that absorbed dose estimates for NHL tumors did correlate with tumor size regression in previously untreated patients. Here, explicit cognizance was taken of the size of each lesion. Quantitative SPECT was employed to normalize GM planar camera images so that absolute activity in each site could be better obtained. A probit analysis of the decrease of tumor size with tumor estimated dose revealed a positive correlation and evidence of saturation at or above 10 Gy lesion absorbed dose. One interesting fact shown in this study was the effect on tumor regression of the unlabeled tositumomab antibody. For larger lesions, this protein itself contributed a large majority of NHL tumor regression. A similar analysis should be considered in general cases of combination therapies as are described in Chapter 10.

A theme of attention to patient-specific information also appears in the work of Siegel et al. (2003) in which the GSCC group was able to substantially improve the correlation between hematological toxicity and RM absorbed dose in CEA-positive patients following anti-CEA radioimmunotherapy. In this clinical example, the patient's cytokine FLT3-L was used to modulate the radiation dose estimates to the red marrow. A standard FLT3-L value (80 pg/ml), obtained from normal individuals, was required to be measured prior to making this correction. Moreover, the authors did not reduce radiation dose estimates if the amount of the cytokine were below the standard value. Thus, all dose estimates were modulated only upward.

Finally, in the DOTATOC analyses of renal toxicity observed clinically, it was seen that two patient–organ-specific steps were necessary to estimate realistic absorbed doses. First, the actual size of both kidneys was used as taken from anatomic (CT) data. Secondly, a BED was determined assuming certain parameters for RP clearance, cell repair rate, and the α/β ratio for the renal system. Notice that the last three parameters cannot be made specific to the patient but are presumed to be generic for the normal human. Among the very satisfying outcomes of this analytic approach was again the appearance of a sigmoidal dose–response function. In other words, as the BED increased, the reduction in creatinine clearance approached a logical maximum of 100%. This last result is very important because the range over which the TRT-derived absorbed dose is effective has now been established for the kidneys. One would assume that similar results could be derived for other normal tissues.

One can safely conclude that absorbed dose, like politics, is a local phenomenon. The analyst must take into account details of each patient. Primary among these are the sizes of the tissues at risk. Beside that fundamental measurement, one would need to include blood cytokine levels for RM estimates. Finally, we would need to determine other details at the organ level such as the rate of tissue repair and dose rate in the target organ structures. Overall, the final result appears to necessarily be a BED value, not simply an absorbed dose. Thus, one could not transpose the effectiveness of a cGy from the liver to a cGy to the kidney. Instead, the organ's α/β ratio, the effective half-time of the loss of activity and the cellular turnover rate must all be included in the estimation process.

REFERENCES

Barone, R., F. Borson-Chazot, et al. 2005. Patient-specific dosimetry in predicting renal toxicity with (90)Y-DOTATOC: relevance of kidney volume and dose rate in finding a dose-effect relationship. *J Nucl Med* **46 Suppl 1**: 99S–106S.

Blumenthal, R. D., W. Lew, et al. 2000. Plasma FLT3-L levels predict bone marrow recovery from myelosuppressive therapy. *Cancer* **88**(2): 333–43.

Bodei, L., M. Cremonesi, et al. 2004. Receptor radionuclide therapy with 90Y-[DOTA]0-Tyr3-octreotide (90Y DOTATOC) in neuroendocrine tumours. *Eur J Nucl Med Mol Imaging* **31**(7): 1038–46.

Bouchet, L. G., W. E. Bolch, et al. 2003. MIRD Pamphlet No 19: absorbed fractions and radionuclide S values for six age-dependent multiregion models of the kidney. *J Nucl Med* **44**(7): 1113–47.

Buras, R. R., B. G. Beatty, et al. 1990. Radioimmunotherapy of human colon cancer in nude mice. *Arch Surg* **125**(5): 660–4.

Emami, B., J. Lyman, et al. 1991. Tolerance of normal tissue to therapeutic irradiation. *Int J Radiat Oncol Biol Phys* **21**(1): 109–22.

Esteban, J. M., J. A. Kuhn, et al. 1991. Carcinoembryonic antigen expression of resurgent human colon carcinoma after treatment with therapeutic doses of 90Y-alpha-carcinoembryonic antigen monoclonal antibody. *Cancer Res* **51**(14): 3802–6.

Konijnenberg, M. W. 2003. Is the renal dosimetry for [90Y-DOTA0,Tyr3]octreotide accurate enough to predict thresholds for individual patients? *Cancer Biother Radiopharm* **18**(4): 619–25.

Koral, K. F., Y. Dewaraja, et al. 2000. Tumor-absorbed-dose estimates versus response in tositumomab therapy of previously untreated patients with follicular non-Hodgkin's lymphoma: preliminary report. *Cancer Biother Radiopharm* **15**(4): 347–55.

Koral, K. F., I. R. Francis, et al. 2002. Volume reduction versus radiation dose for tumors in previously untreated lymphoma patients who received iodine-131 tositumomab therapy. Conjugate views compared with a hybrid method. *Cancer* **94**(4 Suppl): 1258–63.

Koral, K. F., M. S. Kaminski, et al. 2003. Correlation of tumor radiation-absorbed dose with response is easier to find in previously untreated patients. *J Nucl Med* **44**(9): 1541–3; author reply 1543.

Salassidis, K., V. Georgiadou-Schumacher, et al. 1995. Chromosome painting in highly irradiated Chernobyl victims: a follow-up study to evaluate the stability of symmetrical translocations and the influence of clonal aberrations for retrospective dose estimation. *Int J Radiat Biol* **68**(3): 257–62.

Schlom, J., A. Molinolo, et al. 1990. Advantage of dose fractionation in monoclonal antibody-targeted radioimmunotherapy. *J Natl Cancer Inst* **82**(9): 763–71.

Sgouros, G. 1993. Bone marrow dosimetry for radioimmunotherapy: theoretical considerations. *J Nucl Med* **34**(4): 689–94.

Sgouros, G., S. Squeri, et al. 2003. Patient-specific, 3-dimensional dosimetry in non-Hodgkin's lymphoma patients treated with 131I-anti-B1 antibody: assessment of tumor dose-response. *J Nucl Med* **44**(2): 260–8.

Siegel, J. A., D. Yeldell, et al. 2003. Red marrow radiation dose adjustment using plasma FLT3-L cytokine levels: improved correlations between hematologic toxicity and bone marrow dose for radioimmunotherapy patients. *J Nucl Med* **44**(1): 67–76.

Smith, L. M., J. W. Evans, et al. 1992. The frequency of translocations after treatment for Hodgkin's disease. *Int J Radiat Oncol Biol Phys* **24**(4): 737–42.

Snyder, W. S. 1975. *S, absorbed dose per unit cumulated activity for selected radionuclides and organs.* New York, Society of Nuclear Medicine.

Stabin, M. G. 1996. MIRDOSE: personal computer software for internal dose assessment in nuclear medicine. *J Nucl Med* **37**(3): 538–46.

Stram, D. O., R. Sposto, et al. 1993. Stable chromosome aberrations among A-bomb survivors: an update. *Radiat Res* **136**(1): 29–36.

Wessels, B. W., M. W. Konijnenberg, et al. 2008. MIRD pamphlet No. 20: the effect of model assumptions on kidney dosimetry and response—implications for radionuclide therapy. *J Nucl Med* **49**(11): 1884–99.

Williams, L. E. 2000. Clinical results and the necessity of estimating patient-specific radiation absorbed dose in radioimmunotherapy. *Cancer Biother Radiopharm* **15**(4): 301–3.

Wong, J. Y., J. Wang, et al. 2000. Evaluating changes in stable chromosomal translocation frequency in patients receiving radioimmunotherapy. *Int J Radiat Oncol Biol Phys* **46**(3): 599–607.

Multiple-Modality Therapy of Tumors

10

10.1 INTRODUCTION

Until this chapter, we have limited the discussions to an emphasis on a single type of therapy: absorbed dose delivery via a targeting agent. This strategy may be called targeted radionuclide therapy (TRT) (Williams, DeNardo, et al. 2008). In some instances, such as non-Hodgkin's lymphoma (NHL), we have mentioned in passing the associated use of unlabeled antibodies to the CD20 target molecule on the human B-cell surface. As seen in Chapters 8 and 9, absorbed doses achievable with radioimmunotherapy (RIT) are effective in treating lymphoma in NHL patients but generally not so in the case of solid tumors. Estimated dose values are, in fact, comparable to the two types of disease as seen in Chapter 8. The elevated sensitivity of B cells to ionizing radiation is the reason for the wide clinical application of TRT to lymphoma as part of standard hematology practice. Aside from using catheter placement of activity, there is, at this time, no analogous application of RIT or other therapeutic radionuclide strategy to solid tumors.

The frustration of the radiation oncologist with the lack of internal emitter therapy for solid tumors has led to a number of radiation-augmentation and radiation-enhancement ploys. As yet, none of these strategies have proven successful clinically, but the imagination of the oncology practitioners has been considerable. The present chapter discusses these ideas and how they may be justified based on cellular studies, animal experiments, and some early clinical applications. Two approaches to this augmentation may be considered. One method involves the use of increased radiopharmaceutical (RP) uptake at tumors sites; the other method enhances the given radiation dose by linear if not synergistic effects due to combinations of agents. While a combination of two therapies is the emphasis in the following, we must admit that still more complicated strategies may be attempted involving three or more modalities. These higher-level multiples are relatively uncommon but may be expected to increase in the future.

Initially, an appealing concept is manipulation of tumor environment so that greater access or greater accumulation of the radioactive agent at the lesion can be obtained. Probably the oldest form of this type of enhancement is using preliminary surgery to reduce the number and size of the lesions in the patient. As Chapter 12 will show, having fewer and smaller sites increases uptake (%ID/g) at the remaining targets. In addition, there is the mass law as demonstrated in Chapter 3. If we assume that the uptake is inversely proportional to the tumor mass, smaller lesions should receive proportionally larger absorbed radiation doses. A second application of this type is use of another modality or an unlabeled agent to allow more of the labeled material to have access to the interstitial tumor space. One of the earliest of such access-increasing concepts was the use of external beam radiation treatment (EBRT) prior to RIT. In addition, in the case of an antigen target, it may even be possible to increase the production of that antigen at the disease sites. By up-regulation of expression of the tumor-associated target molecule, the therapist may be able to locate more of the RP within the tumor space.

We should reiterate the general rule that a tumor is not a collection of clones of the original malignant cell. In principle, there may be relatively few tumor cells expressing a given antigen or marker at any time point during the therapy. Thus, the enhancement of molecular expression may be a useful concept in general. Some observers will point out that the tumor may be a so-called autocrine system whereby the target antigen acts as a self-stimulator on other similar cells in the primary as well as in metastatic lesions. In this situation, up-regulation may prove to be a decidedly mixed blessing.

Radiation enhancement methods are a second type of augmentation of a given radiation dose. Traditionally, one employs a chemotherapeutic that can act alone (as a tumor-cell killer) or as a radiation sensitizer at the tumor sites. This material may be given before or even during the internal emitter treatment and provides an effective increase in the lethality of a given cGy absorbed dose. A large number of such enhancers are now known and will be discussed herein. Hyperthermia is another example of an absorbed dose enhancer. By elevating tumor temperatures to approximately 42°C, a given absorbed dose may be more effective than if delivered at the normal body temperature of 37°C. A more modern form of enhancer is stimulation of the inherent immune system of the patient. Here, lysed tumor cells, produced by the TRT agent, may be thought of as releasing novel molecules into the interstitial fluids. These tumor cell residues are then brought under the purview of the patient's immune system such that a T-cell-driven response is now possible. Prior to the use of the ionizing radiation from the TRT source, these novel molecules would have been locked within the tumor cell or otherwise not available to the patient's inherent immune system.

Methods outlined so far are generally not mutually exclusive. One could imagine the use of chemotherapeutics as well as hyperthermia as enhancers of RIT or other internal emitter radiation therapy. Clearly, many questions arise in such combination therapies. These uncertainties include the amount (e.g., in mM/kg or mg/kg) and the timing of any enhancer. In addition, there is the possibility of untoward cross-effects when one enhancer may not interact favorably with a second agent to reduce the amount of either therapy tolerated by the patient.

In the following, we consider five types of enhancement of internal emitter radiation therapy. The order followed will be, in a rough sense, the historical one seen in single modality non-TRT cancer therapy: surgery, hyperthermia, external beam treatment, chemotherapy, and immune manipulation. We will devote one segment of this chapter to each of the five, and, if available, clinical applications are described for each.

10.2 SURGERY AND TARGETED RADIONUCLIDE THERAPY

Perhaps the simplest use of combined modalities is to follow a surgical procedure with targeted radionuclide therapy that is dedicated to clearing out residual disease at the former tumor sites. It should be obvious that cellular and animal studies are not necessary for justification of this strategy. Use of TRT following surgery has had several applications in the clinic. In the following, we consider therapy of thyroid, breast, brain, and liver cancers in patients. Some of these reports have included radiation dose estimates as well as clinical outcomes.

10.2.1 Treatment of Residual Thyroid Tissue

An obvious prototype for TRT following surgery is treatment of residual thyroid tissue in the thyroid cancer patient. Here, we restrict ourselves to describing therapy of follicular (80% of cases) and papillary (15% of cases) thyroid carcinoma. The remaining patients are primarily victims of medullary cancer, of neural cell origin, which does not take up iodine. We should mention in passing that the medullary cancer produces carcinoembryonic antigen (CEA) so that anti-CEA therapy is possible for these patients. An example is given in Figure 1.1. While treatment with $Na^{131}I$ is a worldwide standard for follicular and papillary disease, there are relatively few analyses of the results—particularly with respect to the absorbed dose to residual thyroid tissues. We should note that this entire organ is the target although some of the remaining thyroid cells are almost certainly normal. Endocrinologists have the liberty to destroy the remaining gland completely since the patient may be given artificial hormone orally as a replacement; that is, the thyroid is expendable.

The group at the University of Cincinnati (Maxon, Englaro, et al. 1992) reported on a sequence of 70 patients who received a treatment for adenocarcinoma of the thyroid over an approximate 12-year interval (1978–1990). Imaging of the neck area using rectilinear scanners (cf. Chapter 2) and 2 mCi of ^{131}I was used to determine the geometric area of residual thyroid while surgical notes sufficed for thickness. Using the resultant mass values and the standard MIRD method (cf. Chapter 4), absorbed doses to both residual thyroid and lymph nodes were computed. The geometric mean method was employed to determine $A(t)$ for these two tissues. Integration of activity over time was done with an exponential approach (Chapter 6) assuming only radiodecay after the last time point (usually 72 h). Activities (A_0) given to the patients were then determined by the prescription that the residual thyroid would receive 300 Gy, whereas any nodes would receive 80 Gy. It should be emphasized that, since the Cincinnati group actually prescribed a patient-specific amount of ^{131}I, its approach was distinct from that of the usual nuclear medicine practice in which a fixed activity, typically 100–150 mCi, would be given to every patient regardless of total body mass, body area, or any other specific parameter.

Results of the Cincinnati group (Maxon et al. 1992) were encouraging. In the 70-patient study, 86% of those with only residual thyroid foci were treated successfully in a single administration of radioiodine (122/141 sites). By doing multivariate analyses, it was seen that the two most important variables in achieving success were the residual tissue mass and the effective half-time of iodine in the mass. In particular, it was found that if the mass were less than 2 grams, the likelihood of a single-treatment success was significantly higher ($p < .005$) than if the mass were greater than 2 grams. In the case of nodal disease, there were no clear correlations between successful therapy and radiation dose or lesion size.

10.2.2 Breast Cancer Treatment Postsurgery

Paganelli and coworkers in Milan (Paganelli, De Cicco, et al. 2009, 2010) described a two-step procedure to distribute radiation dose into the residual surgical cavity after breast cancer resection. These investigators did not use an antibody approach but rather based the targeting on the high affinity of the avidin–biotin complex ($k = 10^{-15}$ M). The protocol is being followed even though there is evidence that many individuals have antiavidin antibodies due to the ingestion of chicken eggs. In the first step of the procedure, unlabeled egg avidin with a MW of 66 kD was injected into the cavity and allowed to move throughout the volume. After approximately 24 h, 3.7 GBq of ^{90}Y-DOTA-biotin was injected intravenously. This activity level is felt to be below that at which renal toxicity would occur; biotin due to its very low molecular weight (MW) (224 Da) is primarily excreted via the kidneys. Absorbed dose estimates (D) were made using a coinjection of 185 MBq of ^{111}In-DOTA-biotin. Gamma camera images were obtained at 1.5, 5, 16, and 24 h post the radioactive biotin injection. These investigators used a computed tomography (CT) scan of the patient to determine tumor site volumes as well as those of normal tissues of interest. Because of a lack of hybrid imaging at the clinical sites, image fusion was required in the reported study. Breast volumes undergoing treatment were calculated using image sizes from a fused single-photon emission computed tomography (SPECT)–CT image set.

The authors made their dose estimates using the linear-quadratic model of Chapter 4; that is, a biologically effective dose (BED) value is generated assuming α, β, $T_{1/2rep}$, and $T_{1/2cl}$ for each tissue. Table 10.1 gives the assumed values for several of these parameters. Table 10.2 lists both the estimated absorbed dose as well as the BED for breast lesions in one group of 15 women. Each of the patients received 100 mg of avidin. Other groups received a lower amount (50 mg; $N = 10$) and a higher amount (150 mg; $N = 10$) with similar results in terms of the dose and BED. We should note that there was a slight tendency for the estimated absorbed dose to be less for the lowest level of

TABLE 10.1 BED Values Assumed for Breast Irradiation Using ^{90}Y-DOTA-Biotin

Tissue	α/β (Gy)	$T_{1/2rep}$ (h)
Breast	10	1.5
Kidney	2.6	2.8

Source: Paganelli, G. et al., *Eur. J. Nucl. Med. Mol. Imaging*, 37, 2, 2010.

TABLE 10.2   Absorbed Dose and BED Values for Breast Cancer Patients Treated with
^{90}Y-DOTA-Biotin Following Avidin Administration

AMOUNT OF AVIDIN	PATIENT REGION	ABSORBED DOSE (Gy)	DOSE/ ACTIVITY (Gy/GBq)	BED (Gy)	BED/ ACTIVITY (Gy/GBq)	MASS OF TREATED LESION (g)
100 mg	Breast; Higher uptake	19.5 ± 4	5.3 ± 1.1	21 ± 4	5.7 ± 1.2	250 ± 100
	Breast; Lower uptake	13.3 ± 7.6	3.6 ± 2.1	14.3 ± 9.0	3.9 ± 2.4	350 ± 120
	Kidneys	3.8 ± 1.1	1.0 ± 0.3	4.3 ± 1.7	1.2 ± 0.5	Obtained at CT

Source: Paganelli, G. et al., *Eur. J. Nucl. Med. Mol. Imaging*, 37, 2, 2010. With kind permission form
         Springer Science and Business Media.

avidin. However, increasing the level to 150 mg did not significantly improve the breast
dose values.

As Table 10.2 demonstrates, the two-step avidin–biotin process leads to residual
breast dose estimates of approximate 20 Gy or 5 Gy/GBq. These estimates are fivefold
higher than the corresponding renal dose value. Notice also that, in this application, the
dose and BED are relatively similar numerically with the latter only about 10% higher
than the former. Not shown in the table is the bone marrow estimate, which is 0.2 Gy or
0.054 Gy/MBq. This value is well below that thought to be toxic to the marrow.

The investigators concluded that such ^{90}Y-based TRT could have impact in the clinical
therapy of breast cancer survivors—particularly those who cannot obtain external beam
treatment due to the difficulty of repeated travel to clinically overburdened accelerator
sites. The EBRT technique (in Italy) consists of five daily fractions for a total of 5 to 7
weeks. Because of the logistical complications of such multiple visits to the limited number
of radiation therapy venues, approximately 30% of present-day patients do not receive the
recommended course of EBRT. In addition, the start of such therapy may be delayed due to
issues of wound healing and other complications of the initial surgery. We should mention
that the Milan investigators (Paganelli et al. 2010) included a course of EBRT for 32 of their
study group of 35 breast patients. A total external beam dose of 40 Gy was delivered over a
4-week interval. This value, for each patient, was converted to a BED and added to the TRT
absorbed dose for a total BED value of between 60 and 70 Gy for each breast site.

Two advantages of TRT of breast disease merit separate comments. First, the lymphatic system of the patient is seen by gamma camera imaging in a majority of the
avidin–biotin injections. This implies possible ^{90}Y-irradiation along the lymphatics
draining the tumor: a result that may reduce the likelihood of metastatic disease subsequent to the initial treatment. Second, there is one aspect of their work that the Milan
group did not discuss, namely, the lack of untoward cardiac effects with the potential
use of pure TRT in breast cancer. A well-known complication of EBRT is the danger
to the heart muscle in patients having left breast disease. Incidental external irradiation
of the heart (tangential fields) may lead to damage that requires some months to years
to clear up. Since many of the patients are older women, this time of cardiac recovery
is not necessarily available, and the quality of life suffers. It is also possible that the
patient's heart simply does not recover from the external beam radiation insult.

10.2.3 Brain Tumor Therapy Postsurgery

Analogous to the Milan results, the group at Duke University (Reardon, Akabani, et al. 2002) developed a method to locally irradiate the walls of the cavity left after initial brain tumor resection. Early work concentrated on the ^{131}I-anti Tenacsin antibody 81C6. Tenacsin is overexpressed in gliomas but not in normal brain, which makes this an almost ideal situation for specific-molecule therapy. Tenacsin is, however, produced in many nonneurological and critical tissues such as the liver and kidneys. Thus, the clinician must be careful to keep the anti-Tenacsin sequestered at the former brain tumor site. We should add that these patients, because of the nature of their disease, also receive EBRT and chemotherapy as components of their treatments.

To localize only within the tumor space, antibody is introduced into the surgically created resection cavity (SCRC) via a reservoir that communicates directly from outside the patient's head. Gamma camera measurements confirm a lack of trafficking between the SCRC and the blood supply as very little activity is seen in normal organs. A dose estimation model has been developed to permit calculations of the absorbed dose to various contiguous layers of brain tissue lying on the surface of the residual surgical cavity.

Reardon et al. (2002) reported a 33-patient Phase II study using 120 mCi of [131]I-81C6 injected into the reservoir. These individuals were newly diagnosed and thus had no prior therapies. Subsequent to the TRT trial, EBRT and chemotherapy were given over the following 12 months. Survival was an endpoint of the study. Patients ($N = 27$) with glioblastoma multiforme (GBM) had a median survival of 79.4 weeks, whereas the median for all brain cancer patients was slightly longer at 86.7 weeks. The former value exceeds the 56-week survival observed for newly diagnosed GBM patients receiving carmustine polymers (a chemotherapeutic) that were also placed directly into the SCRC.

Because of the extended range of the beta radiation coming from ^{131}I, the Duke group decided to attempt labeling (Zalutsky, Reardon, et al. 2008) with [211]At, an alpha particle emitter with a 7.2 h half-life. Astatine is a halogen and shares a column of the periodic table with iodine and bromine. Here, the range of the ionizing alpha is on the order of 50 micrometers (i.e., approximately 1 to 2 cell diameters). It was felt that this shortened range would produce less general toxicity in the normal brain cells lying outside the surgical volume. An 18-patient study has been reported with a majority ($N = 14$) of the subjects being GBM victims. Unlike the previous [131]I-81C6 group, all of these patients had at least one prior therapy and were therefore considered recurrent cases. In the [211]Astatine study, a median survival time of 52 weeks was obtained for the GBM group—a value significantly better than the survivals obtained for recurrent disease and either no treatment (23 weeks) or using the chemical therapy agent carmustine (31 weeks).

One may conclude that the Duke research shows promise for treatment of GBM and other brain cancer patients. Because of the aversion to taking too much normal tissue during surgery, the oncologist may now use the anti-Tenacsin targeting to irradiate tumor cells lying at the rim of the major lesion. Survivals are extended, and there seems little toxicity to normal tissues since the radioactivity is confined to the tumor bed because of the method of injection. Notice how the question of blood-borne targeting that is at the center of most TRT applications is avoided by having the activity placed directly into the space of interest.

10.2.4 Hepatic Tumor Therapy to Expedite Subsequent Surgery

Logically, we can also consider the initial use of TRT to expedite subsequent surgery. One example is to locally irradiate liver lesions prior to subsequent laparotomy in the treatment of primary liver cancer. In this application, the sequence is reversed due to the initially unresectable nature of the patient's disease. To direct the anti-CEA antibody to the primary site, the patient is catheterized to receive a direct injection into the hepatic artery. Thus, the method bears some resemblance to the Duke brain tumor therapy procedure previously outlined and to the use of targeting radioactive spheres described in Chapter 8. The group at Shanghai Medical University (Zeng, Tang, et al. 1998) reported on a paired study involving 32 initially unresectable hepatocellular carcinoma (HCC) patients who received TRT using ^{131}I-Hepama-1 antibody. Antibody therapy was delivered, in a Phase I and II trial, via arterial infusion into the affected lobe of the liver. A control group of 33 HCC patients received conventional chemotherapy using the same injection method.

Five-year survival of the experimental group (28.1%) was significantly higher than that of the control (9.1%) with $p < .05$. The authors pointed out that this outcome was a result of the fact that the TRT cadre often had opportunities for resection following the tumor shrinkage induced by the targeted therapy. For example, in the TRT patients, 17 (53%) were resected after the initial treatment, whereas only 3 (9%) in the control group could be operated on. Another interesting outcome of the trial was that the induction of human anti-mouse antibody (HAMA) was positively correlated with survival at the 1-year interval. For the 32 TRT patients, 11 developed HAMA within 4 weeks of the initial TRT treatment. For these individuals, 8 of 11 survived at 12 months, whereas only 6 of 21 HAMA-negative patients were alive at that time point ($p < .05$).

These results imply that there may be a role for targeted radiation therapy as a tumor-reduction agent prior to possible later surgery. Again, it is important to notice that the Shanghai group used a direct injection into the diseased lobe of the liver and did not attempt to use the general circulation to obtain tumor access. Thus, metastatic sites were not accessible to the TRT. Chapter 8 described other methods of hepatic-artery directed radiation therapy of hepatic disease.

10.3 HYPERTHERMIA AND TRT

Hyperthermia has a very long and somewhat checkered history in medical oncology. Heating of the tumor volume was probably used prehistorically as a treatment modality for superficial sarcomas and various skin lesions such as melanomas. In principle, these relatively exposed tumors would be amenable to having hot sources, such as heated stones, placed against them to induce accelerated cell death. Just as in those very early examples, the modern advocate has two fundamental issues to contend with in any clinical application of heating.

First is the difficulty of placing thermal energy only at the lesion sites so that normal tissue damage is minimal. As in the case of ionizing particles, the clinician must be careful that normal tissues surrounding the tumor do not achieve elevated temperatures during the course of a therapy. Multiple modern methods are available including heated solutions (lavage), focused ultrasound, radiofrequency, and microwave radiation. Generally, the tumor is thought of as being as sensitive to hyperthermia as is normal, adjacent tissue. We should add that the *hyperthermia dose* used in the medical literature is a combination of two parameters: the temperature and the length of time the tissue is held at that value. For example, a treatment plan may call for 43°C for a prescribed time of 30 minutes.

The second issue that confounds the application of locally applied heating is measurement of the actual temperature rise—or, equivalently, the local temperature value. Such thermometry is, in a very direct sense, analogous to the measurement of ionizing radiation dose at the tumor or other relevant sites in the body. Chapter 4 described microthermometry as one of the fundamental techniques used in ionizing radiation dosimetry. For hyperthermia, temperature measurement is best done if one or more sensors (e.g., thermistors) may be placed into and around the treatment volume. Yet these measurements imply invasive procedures and may lead to complications in the patient's treatment. As noted with ionizing radiation dosimetry, placing the detector (thermistor or dosimeter) inside the lesion volume may cause damage to the circulation and may release tumor cells into the blood and lymphatic systems. No validated indirect method has yet been developed for thermometry. As in the case of ionizing radiation, the simplest application of hyperthermia dose evaluation is generally done in cell culture whereby temperatures are more readily measured and controlled.

Sakurai and associates (Sakurai, Mitsuhasi, et al. 1996) at Gunma University performed experiments involving cultured cells in a combination of simultaneous hyperthermia and low-dose rate ionizing radiation from a ^{137}Cs source. This photon emitter was used as a surrogate for a radionuclide label on an RP. In one experimental setup, the authors exposed Yoshida sarcoma cells to simultaneous heating and ionizing radiation at dose rates of 50 cGy/h over an 8-hour treatment time. Highly synergistic effects are seen at temperatures 40, 41, and 41.5°C but not at $T \geq 42$°C. Some indication of thermal tolerance is seen in such experiments if the cells are exposed only to hyperthermia. Here, the investigator is presumably selecting those clones that are thermo-tolerant over those that are sensitive to elevated temperatures. One can recall an analogous situation that can occur with TRT directed against a specific molecule in the lesion as described in Chapter 9.

In a similar experiment (Sakurai, Mitsuhashi, et al. 1998) using human lung carcinoma cells in culture, thermo-tolerance was also observed when no concurrent ionizing radiation was present. If the two modalities were combined, there is an increasing sensitivity to temperature as indicated by the D_0 value. Recall that this is the amount of ionizing radiation dose required to reduce the cell survival by a factor of e (i.e., to 37% of the original value). Table 10.3 gives these D_0 values for a constant dose rate of ^{137}Cs irradiation (50 cGy/h) and various fixed temperatures. Because of the cellular exposure situation, the temperature could be kept constant throughout the culture volume so that local variations were not significant. Experiments were carried out over

TABLE 10.3 Human Lung Adenocarcinoma Cell
Survival with Concurrent Hyperthermia and [137]Cs
Ionizing Radiation at 50 cGy/h

TEMPERATURE OF CELL CULTURE (°C)	D_0 VALUE MEASURED (Gy)
37	6.55
38	5.25
39	4.24
40	3.99
41	3.46
41.5	1.83
42	0.70

Source: Sakurai, H. et al., *Anticancer Research*, 18, 4A, 1998.
Reprinted with permission of the International Institute
of Anticancer Research.

times out to 48 h or 24 Gy total absorbed radiation dose. It was seen that the presence of elevated tumor cell temperatures greatly enhanced the effect of the ionizing radiation. Going from 37°C (normal mammalian cell temperature) to 42°C reduces the D_0 value by approximately one order of magnitude. This result implies that a given ionizing radiation dose is essentially 10 times more effective at 42°C than at normal body temperature. This is an example of the power of the method and why a number of investigators have attempted to combine these two modalities.

Some difficulties may arise in the expression of proteins on the surface of cells due to elevated temperatures. Such changes could adversely affect the ability of an RP to target a given tumor-associated antigen. In one example, Norsk Hydro Institute researchers documented (Davies Cde, Rofstad, et al. 1985) that three separate human melanoma-associated proteins in cell culture were strikingly reduced following a course of hyperthermia. Techniques were variable: 42°C for 1, 2, or 3 hours exposure; 43°C for 1 and 2 hours; and 43.5°C for only 20 minutes. Generally, the effects were more marked as the temperature and time were increased. Antigen expression went through an initial reduction (sometimes to as low as 20% of the initial cell values) out to times on the order of 3 days postheating. Then a rebound phenomenon occurred whereby the expression increased above its original value. Such overshooting is characteristic of a feedback system trying to achieve equilibrium. The amount of overshoot was as large as a factor of threefold by 6 days postheating. Thus, there arises a question of the optimal timing of hyperthermia relative to the injection of a targeting (via an antibody) radiopharmaceutical.

The Norsk Hydro Institute results (Davies Cde et al. 1985) showed that all three antigens on the cell surface are primarily being denatured by heating. To some extent, there is likely to be a simultaneous disruption of the melanoma cell membrane so that there is less antigen accessible to the targeting antibody. These melanoma results indicate that the simultaneous use of hyperthermia and that low-dose ionizing radiation may not be effective if the latter is being supplied via a targeting RP. Instead, for the melanoma example described here, the optimal time to introduce the antibody would be several days after heating at which time the antigen expression would probably be elevated compared with the level found in untreated melanoma cells. Clearly, the investigator must test human tumor cells of a given type in culture before beginning to use hyperthermia as an adjunct to ionizing radiation. By mistiming the heating, one could inadvertently reduce the therapeutic effect of a given amount of activity prescribed for the patient.

Workers at Northwestern University performed both animal (Mittal, Zimmer, et al. 1992) and patient (Mittal, Zimmer, et al. 1996) studies involving a combination of hyperthermia and radioactive antibody targeting. Their interest concerned therapy of colorectal cancer and involved ^{131}I-labeled anti-CEA antibodies. As mentioned in Chapter 8, this common malignancy is difficult to treat using TRT alone due to the relatively small tumor absorbed doses produced by the radioactive antibodies.

In nude mouse studies at Northwestern University (Mittal et al. 1992), the investigators found that the combination therapy led to an extended tumor regrowth delay (TRD) compared with either type of therapy used alone. TRD refers to the time required to achieve a given tumor size posttreatment; this parameter is measured from the time that the untreated (control) tumor would reach that size. For example, if the LS174T human colon xenografts used in the nude mouse study required 6 days to reach 4 gram mass and the therapy delays the treated tumors such that they achieve this mass in 16 days, the TRD is 10 days for the treated group. It is found that hyperthermia alone or TRT alone significantly increased the TRD compared with the untreated group of mice. Maximum TRD was achieved using the combination therapy whereby the animals had their leg tumors heated to 42°C for 45 minutes after having received 186 µCi of ^{131}I-anti-CEA antibody 48 hours previously. The doubling time for these xenografts is given in Table 10.4. It is seen that the combined therapy led to the maximum TRD of 26 ± 8 days.

The Northwestern researchers followed up their animal studies several years later with a clinical trial involving nine colorectal cancer patients (Mittal et al. 1996). Of these, only six showed significant uptake at the tumor sites (i.e., uptake above that of adjacent soft tissue or blood). These six individuals were treated with combination therapy involving hyperthermia at 40.3°C. In terms of tumor size, only one of the six exhibited a reduction in lesion dimensions at 1 month after the completion of the therapy. Another patient did, however, develop a hematoma due to the insertion of a thermal-

TABLE 10.4　Volume Doubling Times for LS174T Xenografts in Nude Mice Receiving Various Therapies

THERAPY	VOLUME DOUBLING TIME (VDT)
Control	1.9 ± 1.0 d
Hyperthermia	3.7 ± 1.8
131I-anti-CEA antibody	17.0 ± 5.7
Hyperthermia + antibody	26.0 ± 8.2

sensing catheter into the lesion. This study proved the feasibility of performing a combination of hyperthermia and TRT with a labeled antibody—but also showed one possible untoward result of catheter-induced damage at the tumor site.

10.4 EXTERNAL BEAM AND TRT

EBRT has been a standard method of tumor treatment since the first part of the twentieth century. As in the case of the other augmentation methods described earlier in this chapter, there is the traditional pair of questions: how much ionizing radiation dose to give to the tumor volume, and when that dose should be delivered relative to the injection of the RP? In more recent times, there is the additional issue of the dose rate of the ionizing radiation. External beam rates will, in general, be much higher than those seen using radioactive molecules at the tumor sites.

Order's group at Johns Hopkins University (JHU) was one of the first to explore the possibility of change in tumor uptake of an antibody following variable absorbed doses of EBRT. In these early experiments, doses of 2, 6, or 10 Gy were given, via a ^{60}Co teletherapy unit, to the livers of rats bearing hepatomas. The group (Msrikale, Klein, et al. 1987) reported on four rat hepatic cancer cell lines (H4IIE, 3924A, 7800, and 7777) that were implanted into the livers of normal ACI rats. Injection of the ^{131}I-anti-ferritin antibody was performed 24 h after the end of the external beam irradiation. Treatments were done at 80 cm source-skin distance (SSD) with a beam rate of 1.02 Gy/m. Because of the early date of this work, the antibody-to-rat ferritin was derived by challenging New Zealand white rabbits with rat ferritin extracts. Biodistributions were then obtained at 24 h after the injection of the RP. Comparisons of u_T (%ID/g) were made between the control animals (no EBRT) and the three EBRT dose levels.

The JHU group found that, in all four tumor types, increasing the external beam treatment dose beyond 2 Gy enhanced uptake in the hepatomas ($p < .05$). The most dramatic effect occurred with the H4IIE cell line, where the uptake increased from 1.67 %ID/g to 2.18 %ID/g in going from 0 Gy to 10 Gy delivered by external beam. The result was approximately a 30% enhancement. Keep in mind that this is a rat experiment, so tumor uptakes will be reduced compared with a typical mouse model by approximately one order of magnitude (cf. Chapter 2 for typical mouse uptake values). A control study, done with a nonspecific rabbit IgG, also revealed increased tumor uptakes in the cases of the H4IIE and 7800 tumor lines. Except for the 3924A cell line, no significant changes were seen in the normal liver, which was also in the radiation field (4×5 cm). The authors attributed their increased tumor uptakes to enhanced vascular permeability at the lesion site.

Approximately 2 years after the JHU results, Wong, Williams, et al. (1989) reported on similar experiments done with the nude mouse as the test species. Tumor inoculation (10^7 cells) was done either 6 or 11 days prior to the external beam therapy. Here, the LS174T human colorectal tumor was studied with a ^{60}Co irradiation prior to injection of

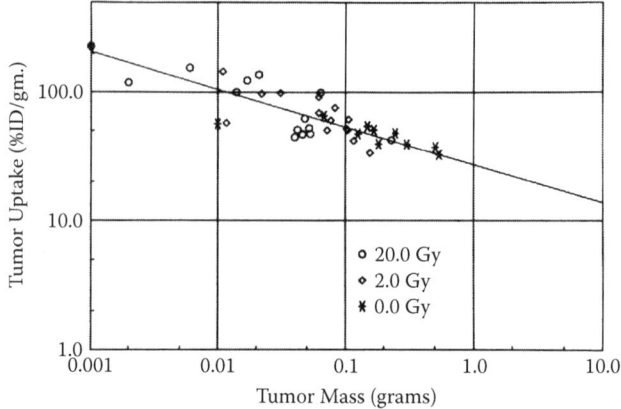

FIGURE 10.1 Tumor (LS174T) uptake as a function of mass for several EBRT dose levels. Mice were irradiated with ^{60}Co at 80 cm SSD at 6 days posttumor inoculation. A power-law relationship is shown with slope = −0.29 and intercept 18.1. Notice that the greater external beam dose correlates with the smaller tumor sizes. (From Wong, J. Y. et al., *Int. J. Radiat. Oncol. Biol. Phys.*, 16, 3, 1989.)

the anti-CEA antibody ^{111}In-T84.66. Absorbed doses were 0 (control), 2, or 20 Gy given as a single fraction at 0.937 Gy/m with an 80 cm SSD. Animals were injected with the RP between 0 and 2 hours posttermination of the EBRT. Biodistributions were taken at 48 hours after the radioactive drug was administered.

Consistent with the JHU results, irradiation prior to antibody injection did significantly increase LS174T tumor uptake at the 48 h time point for the 6-day-old tumors. It was found, however, that the effect could be explained completely by the regression in tumor size induced by the EBRT. In other words, tumor shrinkage due to the initial ^{60}Co irradiation explained the enhanced uptake in the lesions. Thus, the hypothesis of increased vascular permeability at the tumor site postirradiation was not needed. It was also determined that the enhancement with radiation dose was not significant for the 11-day-old tumors. In this case, the uptakes were independent of the EBRT dose; that is, the 2 Gy and 20 Gy cohorts had the same tumor u_T values (approximately 21 %ID/g). It was felt that the 11-day lesions were not susceptible to the external beam therapy as there was no correlation between tumor mass and external beam radiation dose. Graphs of the two situations—6-day and 11-day lesions—are included as Figures 10.1 and 10.2.

We should point out that the Johns Hopkins group did not control for tumor size in its work Thus, it may be that both of these experiments could have the same explanation. EBRT, over a limited time interval, may therefore induce a regression in tumor size, which allows a consequently greater uptake measured in %ID/g. As was mentioned in Chapter 3, it is found again that tumor size must be included in any general analysis of tumor uptake. Without this control, understanding of any outside intervention may be difficult to interpret. Unless there is explicit evidence to the contrary, one must assume that the tumor uptake will be a function of tumor mass.

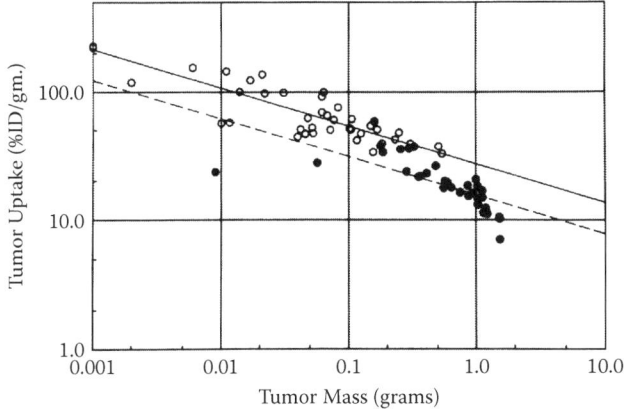

FIGURE 10.2   Tumor (LS174T) uptake as a function of mass for 6-day-old (open symbols) and 11-day-old tumors (closed symbols). Notice that the older lesions have greater mass and correspondingly reduced uptake. Two power-law correlations are shown for the respective lesion ages. (From Wong, J. Y. et al., *Int. J. Radiat. Oncol. Biol. Phys.*, 16, 3, 1989.)

10.5 CHEMOTHERAPY AND TRT

Chemical agents have a tradition in medical oncology that is probably even longer than that of ionizing radiation. One can imagine situations that would go back into prehistory where the medical personnel would attempt to treat a malignant disease by means of an agent taken topically, by mouth or, in the nineteenth century and later, by injection. By introducing chemical agents into the patient either before or during TRT, the investigator can do one of the two manipulations already described: either the antigen expression is increased, or the given absorbed radiation dose is made more effective. Chemical agents that lead to the latter situation are called radiation sensitizers and have a long history in oncology. Sensitizers come in two flavors: chemical agents that only enhance effects of ionizing radiation but are not inherently toxic by themselves (e.g., taxanes); and those that are toxic in their own right (5-FU). Table 10.5 contains a partial list of both types of the sensitizers that have proven popular in oncology. A brief survey of some combinatory treatments is given next.

10.5.1 TRT and Cisplatin

Kievit, Pinedo, et al. (1997) at the Vrije University in Amsterdam reported the use of the toxin cisplatin as a sensitizer in the case of human ovarian cancer in a nude mouse model. Three different human ovarian cancer cell lines were studied with one of three

TABLE 10.5   Partial List of Cytotoxic Radiation Sensitizers

AGENT	TOXICITY MECHANISM
Doxorubicin	Intercalating DNA
Hydroxyurea	Toxic in S-phase
5-Fluorouracil	Inhibits thymidylate synthetase; S-phase inhibitor
Mitomycin C	Inhibits DNA, RNA synthesis; cross-linking agent
Cisplatin	Cross-linking agent
Taxanes	Stabilizes microtubules in cells; G_2/M cell-cycle blocker
Gemcitibine	S-phase sensitivity; blocks DNA synthesis

treatment formats: cisplatin alone; TRT alone; or a combination of both modalities. The targeted therapy used ^{131}I-Antibody 323/A3 against ovarian cancer cells. As in the case of the Northwestern University studies, the growth delay was measured for each treatment type. By dividing by the control growth delay, a specific growth delay (SGD) was determined. The results showed that the two therapy modalities were additive, but not synergistic, in their effects. Thus, the SGD could be increased by adding the cisplatin to TRT in a linear way with the combined therapies, yielding a total SGD that was the sum of each modality taken separately. Such results, in general, are of fundamental interest in the implementation of combination therapies.

10.5.2 TRT and Taxanes

O'Donnell and the group at the University of California–Davis (O'Donnell, DeNardo, et al. 2002) performed a similar experiment with prostate cancer xenografts and taxanes in nude mice. Taxanes do not cause tumor volumes to decrease and can be considered a nontoxic sensitizer for ionizing radiation. The antibody was ^{90}Y-DOTA-ChL6 against the human PC3 prostate carcinoma. This antibody reacts with a cell-surface antigen on adenocarcinomas including prostate tumors. Either one of two taxanes were injected 24 h after the start of TRT: docetaxel at 300 µg per animal; or paclitaxel at either 300 or 600 µg. TRT activity level was fixed at 150 µCi of ^{90}Y-DOTA-ChL6 given intravenously.

Giving the taxanes at 24 h was considered to be an optimal strategy since, at that time, most of the radioactive agent had cleared from the blood to reduce any normal organ sensitization effects in the mice. Multiple groups ($N = 5$ to 10) of animals were followed, including those given only taxanes, those receiving TRT alone, and those receiving a combination of both agents. It was seen that both taxanes, as expected, produced no tumor response at 330 µg levels and that all animals treated in this fashion died by 84 d, the nominal end of the experiment. TRT alone led to 50% survival, whereas TRT plus 600 µg of paclitaxel achieved 100% survival at 84 d. If the paclitaxel dose were reduced to 300 µg and combined with the TRT, survival was reduced to 60% at the endpoint. The authors concluded that use of taxanes enhanced the number of surviving mice.

10.5.3 TRT and Gemcitabine

Treatment of pancreatic cancer has had a dismal outcome in most patients with limited treatment options available due to the generally late discovery of the disease. A group at Garden State Cancer Center (GSCC) investigated combination therapies of CaPan1 human pancreatic cancer xenografts in the nude mouse (Gold, Modrak, et al. 2004). Again, ^{90}Y was the radionuclide of choice. The label was attached, via DOTA, to the antiMUCI antibody cPAM4. A nominal activity level of 100 µCi was given to each animal on Day 0. It was known that the maximum tolerated dose (MTD) of cPAM4 was 250 µCi from earlier experiments. Thus, the investigators were attempting to treat the animals at activity levels well below those known to cause hematological toxicities.

In this research study, a set of gemcitabine (GEM) injections (2 mg each) was given on Days 0, 3, 6, 9, and 12 after the injection of the TRT. A second group of mice received, in addition to this regimen, a second round of TRT plus GEM 5 weeks later. A control antibody therapy was also performed using the nonspecific hLL2 (anti-B cell) antibody labeled with ^{90}Y. The primary endpoint was mouse survival, which was defined as having a lesion < 5.0 g in volume. Initial xenograft size was intentionally made large—approximately 1.0 g—to simulate the initial clinical situation with a pancreatic cancer patient.

Survival times were largest for the combination therapy with a value of 12 weeks; this extent was twice that of the control group ($p < .022$). Mice treated with GEM alone had survival times of only 4 weeks, whereas those treated only with TRT survived for 9 weeks. Following the group treated with hLL2 (nonspecific monoclonal antibody [Mab]) plus gemcitabine led to a time of 7 weeks. Differences in these last three alternative treatments were all highly significant ($p < .009$). Thus, the combination of TRT and GEM was the most effective method in treating these relatively large xenografts.

Repeating the therapy using a second treatment at 5 weeks later also demonstrated that the ^{90}Y-DOTA-cPAM4 combined with GEM was the optimal therapy. In this case, the only comparisons made were between pure TRT and the combination of both agents since the control and pure gemcitabine groups had not survived long enough. As in the single treatment example, the specific antibody and GEM combination had significant survival advantages ($p < .001$ in all comparisons).

The use of GEM as a sensitizer was by no means true for all forms of ionizing radiation and all cell types. In a study involving multiple myeloma cells in culture, a European group (Supiot, Thillays, et al. 2007) reported that sensitization was possible with low linear energy transfer (LET) radiation but did not occur with alpha rays coming from a RP based on an antimyeloma antibody. In these studies, three human myeloma cell lines—LP1, RPMI 8226, and U266—were irradiated 24 h post 10 nM amounts of GEM. Radiation was either EBRT using ^{60}Co gamma rays at 1 Gy/min or the ^{213}Bi labeled antibody CHX-DTPA-B-B4. As noted in Chapter 1, this radionuclide is an alpha emitter. Cell survival after exposure to radiation alone and after exposure to GEM plus radiation were compared. In the case of the LP1 and U266 lines, there was clear enhancement of cell death due to the combination of the two modalities using

[60]Co gamma rays. With this low LET radiation, no sensitization was seen for the RPMI 8226 cell line. Tumor type may thus also be a factor even for photon radiation. Effects induced by the alpha emitter [213]Bi were not enhanced in any of the three cell types; in fact, using gemcitabine reduced the sensitivity of both the LP1 and RPMI 8226 lines, whereas it had no effect with U266.

The authors concluded that, by comparing the area under the curves (AUCs) for the cell survival curves, sensitizations of 1.55 and 1.49 (radiation enhancement ratios) were observed for the LP1 and U266 examples, respectively, using ^{60}Co radiation. No such effects were seen with any of the myeloma lines with alpha particle radiation emitted by [213]Bi. It was felt that the lack of effect with the higher LET radiation may be due to the high value of dE/dx exhibited by the alphas, which did not permit repair of damaged DNA. In the case of gamma rays, the damage was less profound and the intervention of GEM inhibited possible cellular DNA repair.

An early clinical trial involving gemcitabine and TRT was reported in an American multicenter study (Pennington 2009) involving advanced pancreatic cancer patients. Patients were treated in 4-week cycles at 200 mg/m^2 GEM dose level. The hPAM4 antibody was used to deliver either [111]In (for imaging and dose estimation) or ^{90}Y for therapy. The latter material has been dose escalated and has now achieved 9.0 mCi/m^2 with only transient hematological toxicity. Of the eight patients, two have shown tumor responses with significant decreases in [18]FDG positron emission tomography (PET) imaging uptake at the lesion sites.

10.5.4 TRT and 5-Fluorouracil (5-FU)

Wong, Shibata, et al. (2003) at City of Hope performed a Phase I trial involving 5-FU and TRT on 21 patients with metastatic colorectal cancer. In the protocol, the TRT activity per body surface area was fixed at 16.6 mCi/m^2 due to earlier single-agent trials with the same cT84.66 antibody labeled with ^{90}Y using a DTPA chelator. The 5-FU was administered continuously using a venous catheter over a 5-day period beginning 4 hours prior to the TRT injection. Cohorts of patients were studied at 700, 800, 900, and 1000 mg/m^2/day. To be entered into the study, the patient usually had to demonstrate at least one lesion with an injection of 5 mg of the same antibody labeled with 5 mCi of [111]In. An exception occurred if individuals had only hepatic uptake, in which case they were admitted to the study even if the liver lesions did not show uptake above that of the normal surrounding tissue.

Maximum tolerated doses were found to be 1000 mg/m^2/day for the 5-FU and 16.6 mCi/m^2 for the ^{90}Y-cT84.66. Limiting toxicities were generally of the hematological types. Other effects traditionally associated with 5-FU such as mucositis, nausea, and fatigue were seen. Radiation dose estimates for tumors and normal tissues are given in Chapter 8 for this trial. To summarize, these values were found to be comparable to those seen in the earlier trial using just ^{90}Y-cT84.66. One interesting result was that the incidence of human anti-chimeric antibody (HACA; note this was a chimeric antibody) was significantly reduced compared with the therapy trial involving only TRT.

10.6 IMMUNE MANIPULATION AND TRT

10.6.1 Increasing the CEA Content of Colorectal Tumors

It is certainly possible that a given lesion shows great spatial (and temporal) variability in the expression of any possible target antigen. Thus, by selecting a particular molecule for the directed radionuclide therapy, radiation oncologists may become inherently self-limiting in their ability to treat the tumor. Cells expressing the antigen will be selectively killed whereas others deficient in that protein will be spared. In fact, one traditional argument for using high-energy beta particles in the therapy protocol is to have sufficient cross-cell radiation to damage cells that have relatively poor expression of the target molecule. By the same reasoning, using alpha-emitter therapy agents is not a good strategy for solid tumors. Here, we are assuming lesions having sizes on the order of 1 cm or more. For very small tumor sites, the argument is often reversed, and alpha particles are selected as optimal since damage to nearby normal structures is essentially nil.

Because of the probable variation in antigen density in a lesion, an attractive idea is to try to maximize the average amount of target present before beginning, as well as during, the TRT directed against that molecule. One of the first experiments was the use of γ-Interferon (γ-IFN) to up-regulate CEA at the colorectal cancer cell surface (Kuhn, Beatty, et al. 1991). Recombinant γ-IFN also raises the amount of HLA class I and II antigens in a number of normal and tumor cell lines. City of Hope researchers were able to increase the CEA content of inherently low-CEA (WiDr) xenografted human tumors from 809 ng/g up to 5600 ng/g by treating nude mice with 10^5 units of γ-IFN every 8 hours over a 4-day period. The treatment was given by intraperitoneal (IP) injection. When these tumors were excised, it was found that the lesion uptake of ^{90}Y-ZCE025 anti-CEA antibody had been enhanced from 10.3 %ID/g up to 28.5 %ID/g at the end of the treatment (ibid.).

A murine therapy trial was then begun with groups of nude mice bearing WiDr human xenografts being given the TRT by itself, γ-IFN treatment by itself, interferon with the TRT, or phosphate buffered saline (PBS) alone. The recombinant interferon (10^5 units) was given by IP injection 2 days before and 4 days during the TRT therapy. Treatments were given every 8 hours by a dedicated surgical fellow (Dr. Joe Kuhn). In addition, a control antibody (ZME018) against melanoma was labeled with ^{90}Y and used in the trial with and without γ-IFN. Thus, there were a total of six different therapy regimens. Table 10.6 gives the results of the regimens in terms of tumor doubling times in the nude mouse model.

It was found that the doubling time for small tumor (< 1.0 g) growth was essentially 5 to 7 days for the mice receiving the nonspecific TRT, PBS, or γ-IFN. However, those treated with TRT (120 μCi of ^{90}Y-ZCE025) and γ-IFN had a time of 11.3 days—more

TABLE 10.6 Tumor Doubling Times: Use of Gamma-Interferon to Enhance anti-CEA Targeted Radionuclide Therapy in WiDr Colorectal Tumors

REGIMEN	TUMOR SIZE < 1.0 g	TUMOR SIZE > 1.0 g
TRT with specific Mab ZCE025 and γ-INF	11.3 ± 0.43 d	14.0 ± 0.51 d
TRT with specific Mab ZCE025 and PBS	6.84 ± 0.30	13.82 ± 0.54
RT with nonspecific Mab ZCE018 and γ-INT	7.18 ± 0.23	14.43 ± 0.62
RT with nonspecific Mab and PBS	6.69 ± 0.32	14.42 ± 0.49
Gamma interferon	5.28 ± 0.25	12.8 ± 0.37
PBS	6.69 ± 0.32	14.42 ± 0.49

Source: Kuhn, J. A. et al., *Cancer Res.,* 51, 1991. With permission of the American Association for Cancer Research.

than twice that of the nonirradiated (control) groups. If the trial were repeated with larger lesions (> 1.0 g), the results were equivocal with all six regimens leading to tumor doubling times on the order of 13 to 14 days. The conclusion was that the CEA content had been up-regulated and that, if the tumors were sufficiently small, the tumor size doubling time was essential increased twofold with the combination therapy of anti-CEA TRT and γ-IFN.

10.6.2 Using Cold Anti-CD20 Antibody to Enhance TRT in Lymphoma Therapies

TRT's most notable success has been in the treatment of NHL using antibodies to CD20—a B-cell surface marker. While NHL is now only the sixth leading cause of cancer death in the United States, its incidence, for unknown reasons, is increasing at a relatively rapid rate. We should point out that CD20, the target antigen in most B-cell therapy studies, is present on both malignant and normal cells. This fact is probably at the core of the decision to use preinjections of cold anti-CD20 antibodies to increase the effectiveness of the radioactive therapy agents. Two RP agents have been approved by the U.S. Food and Drug Administration (FDA) for treatment of non-Hodgkin's lymphoma. Because of their manifest success in treating this disease of B cells, Zevalin, and Bexxar may now be considered among frontline of therapies for NHL.

10.6.2.1 Zevalin Therapy

One example of the use of unlabeled material is the ibritumomab tiuxetan (Zevalin) protocol for NHL. In this case, the preinjection is rituximab (Rituxan) at 250 mg/m² some 4 hours before the radiolabeled material. Rituxan is a well-known anti-B-cell antibody in its own right and has been used for some years as a therapy for non-Hodgkin's lymphoma. Its application in this situation is as both a therapeutic as well as an agent to improve the subsequent biodistribution of the labeled ibritumomab tiuxetan. A time delay of up to 4 hours is permitted between the two injections. Zevalin is given initially with a ^{111}In label at an activity level of 5 mCi. The therapy activity level is 0.4 mCi/kg using ^{90}Y as the radiolabel. If the patient has platelet levels between 1.0×10^{11}/L and 1.49×10^{11} /L, the therapy ratio is reduced to 0.4 mCi/kg. In any event, the total ^{90}Y activity is limited to 32 mCi per patient.

The initial study, done with ^{111}In ibritumomab tiuxetan, was used as a biodistribution confirmation whereby uptake in blood pool, liver, spleen, and kidneys is expected. Because of the preadministration of the rituximab, blockage of some of the normal organ uptake is anticipated to emphasize disease areas in the NHL patient. If lung uptake is observed, however, the biodistribution is taken as being suboptimal, and the therapy study is not performed. It is not expected that tumor sites will necessarily show up on any of the two or three images taken at 2–24 h, 48–72 h, and 90–120 h if desired. Generally, the images are done with a whole-body gamma camera to show the distribution of the ^{111}In-ibritumomab tiuxetan over time.

The group at Mayo Clinic (Witzig, Gordon, et al. 2002) has reported on a 143-patient Phase III study involving ^{90}Y-ibritumomab tiuxetan TRT. These patients had relapsed or refractory follicular or transformed NHL. Approximately half of the group ($n = 70$) was assigned to the control arm and received 375 mg/m^2 of rituximab per week for a total of 4 weeks. The experimental arm ($n = 73$) was given the preinjections of 250 mg/m^2 of rituximab before receiving 0.4 mCi/kg of Zevalin. As indicated already, a 5 mCi injection of ^{111}In-ibritumomab tiuxetan was given prior to the ^{90}Y-ibritumomab tiuxetan to assess biodistribution and to make absorbed dose estimates. An independent panel then determined the response rates in each of the two arms of the study. A result was termed a complete response (CR) if the patient demonstrated no residual disease after the completion of therapy. Partial response (PR) was achieved if at least one lesion showed more than a 50% reduction in both lateral dimensions. Overall response rate (ORR) is the sum of these two (CR + PR) parameters.

The Mayo group found that the control population had an ORR of 56% whereas the experimental cadre had an ORR of 80%; the difference was significant at $p = .002$. Considering CR rates, the rituximab-treated group had a rate of 16%, whereas the ^{90}Y-ibrititumomab tiuxetan patients had a CR of 30%. This difference was significant at the 4% level of confidence. Two other clinical parameters, the duration of response and time to progression, were reported as not significant at the 5% level of confidence.

10.6.2.2 Tositumomab (Bexxar) Therapy

Tositumomab therapy, also for NHL, likewise consists of two sequential administrations. First is 450 mg of the tositumomab antibody given as an unlabeled protein injection over approximately 1 hour. This is immediately followed with 35 mg of tositumomab labeled with [131]I. Effectiveness of the double application of antibody is demonstrated in Figure 10.3 where the same patient is shown without (left-hand image) and with cold antibody pretreatment (right-hand image) (Seldin 2002). Notice how the normal organ uptake, so vivid in the left-hand image, is reduced by using the cold Mab preinjection. Of particular interest was the reduction of uptake in the patient's spleen when cold anti-CD20 was applied.

In the case of tositumomab, the amount of activity in the second injection is determined by a prior imaging test to determine whole-body clearance of ^{131}I-tositumomab.

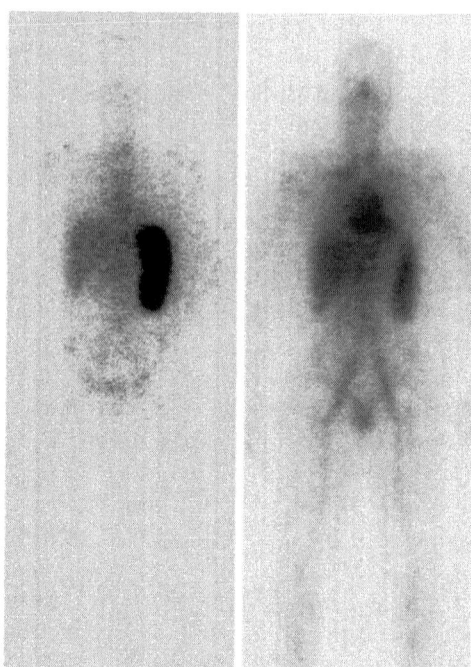

FIGURE 10.3 Bexxar protocol. Anterior images of the same individual after receiving purely radioactive tositumomab (left) and unlabeled as well as labeled tositumomab (right). Notice the reduction of splenic uptake in this NHL patient when the prior cold dose of this anti-CD20 antibody is used. Large splenic size is a frequent feature of NHL due to B-cell invasion of the spleen. (From Seldin, D. W., *J. Nucl. Med. Technol.*, 30, 3, 2002.)

Here, only 5 mCi of the radiolabel is used, but the amounts of antibodies (both cold and hot) are the same as in the subsequent therapy application. By measuring the specific patient's clearance, the radiation oncology staff may calculate the whole-body dose on a per mCi basis. The time–activity curve for the whole body is determined by imaging at Day 0, sometime in the Day 2 to Day 4 interval, and once in the Day 6 to Day 7 interval. Activity values are then fitted to a single exponential function, and the integral under the curve is determined analytically. The clinical protocol then requires that the whole-body dose be fixed at 75 cGy for the patient. This limit presumably reflects a bone marrow dose limitation of approximately the same magnitude if we assume that the whole-body and red marrow (RM) doses are comparable in magnitude (Chapter 8). Notice how this strategy differs from that of Zevalin, whereby the latter is given without regard to any *a priori* radiation dose estimates. Another unique feature of the Bexxar protocol is that the patient must be given KI over a 2-week period following the radioactive injection to block possible uptake of free radioiodine into the thyroid.

A summary of some 60 patients treated with Bexxar (Kaminski, Zelenetz, et al. 2001) was reported out of the University of Michigan. The population had essentially the same diagnosis as in the Mayo Clinic summary. Unlike the Mayo study, the Michigan review, summarizing results from nine institutions in both the United States and in

TABLE 10.7   Unlabeled and Labeled Antibodies for Treating NHL

PROTEIN	LABEL	PATIENT TYPE	RESPONSE (%)	COMPLETE RESPONSE (%)
Last chemotherapy	None	Pretreated	28	3
Bexxar	^{131}I	Pretreated	65	20
Rituximab	None	Pretreated	56	16
Zevalin	^{90}Y	Pretreated	80	30

Sources: Kaminski, M. S. et al., *J. Clin. Oncol.*, 19, 19, 2001; and Witzig, T. E. et al., *J. Clin. Oncol.*, 20, 10, 2002.

Europe, did not have a prospective control arm. Instead, these investigators used the most recent chemotherapy for each patient as that individual's internal control. Comparisons were then made between the Bexxar result and the last chemical treatment. The ORR for the TRT was 65% (i.e., 39 of 60 patients). Yet the ORR for the most recent chemotherapy was only 28% (17/60). This difference of response led to a probability of the null hypothesis of $p < .001$. Median duration of response was also significantly better with Bexxar: 6.5 months versus 3.4 months for the latest chemotherapy. This difference was significant at the .001 level of confidence.

Table 10.7 gives the clinical results for the two commercial anti-CD20 protocols. It is seen that the response rate can approach 80%, whereas the complete response rate is on the order of 30%. One of the most encouraging aspects of these large-trial outcomes is that the NHL patient population now has a novel form of treatment that can be introduced either as a follow up to prior chemotherapy or as an initial method of reducing B-cell numbers. Keep in mind that both commercial protocols require simultaneous use of cold antibody (to the same antigen) to expedite targeting of the subsequent radiolabeled material. We will return to this strategy in the final chapter.

10.6.3 Vaccination and TRT in Colorectal Cancer Therapy in Mice

One of the oldest strategies in the therapy of cancer has been the concept of a vaccine against some specific marker or epitope on the tumor cell surface. Some recent solid-tumor results have been encouraging in the use of anti-CEA vaccine in combination with antibody therapy using ^{90}Y as the radiolabel. In this work (Chakraborty, Gelbard, et al. 2008), a mouse strain was developed that carried the human CEA gene so that the animals were a more exact match to the clinical situation. Unlike the nude mouse so favored by many investigators, these animals had a fully developed immune system—in particular they had available T cells that could be trained to react to the tumor.

The researchers conducted an animal trial with the transgenic (CEA+) mice using four groups of animals in which murine D28 murine tumors (transfected to be CEA+) had been implanted. Groups included those receiving the vaccine itself, ^{90}Y-COL antibody or a combination of vaccination followed by TRT using the same amount (100 μCi) of ^{90}Y-COL antibody to CEA. A control cadre received only a buffer solution. Tumor size was then determined at 56 days after implantation of 3×10^5 transfected tumor cells.

TABLE 10.8 Vaccine Therapy Combined with TRT
in Colorectal Cancer in Mice

TUMOR SIZE AT 56 DAYS	THERAPY	SIGNIFICANCE VERSUS CONTROL
0.59 cm³	Vaccine + TRT (⁹⁰Y-Mab)	0.001
1.28	TRT alone	0.001
2.65	Vaccine alone	0.051
3.31	Control	N/A

Source: Chakraborty, M. et al., *Can. Immun. Immunother.*, 57, 8, 2008. With kind permission of the American Association for Cancer Research.

Results of the experiments are given in Table 10.8. It was found that the tumor volume was minimal in the population that received both the vaccine and the TRT. Compared with the control group, lesion size was reduced approximately sixfold. All three experimental groups showed significant decrease in lesion size compared with buffer with the *p*-value being marginal in the case of pure vaccine treatment ($p = .051$). In the case of the combination therapy, the value was significant at the .001 level of confidence.

The authors presented evidence in their report that T cells had been activated in the vaccinated mice. Such cells attack the tumor *in situ* as part of the general immune system response. As a further encouragement, other T cells were detected that were reactive against other possible tumor antigens including the fas protein.

10.7 SUMMARY

While clearly of use in treating B-cell lymphomas, TRT in solid tumors has been generally ineffective due to the lack of a significant radiation dose. While comparable to the absorbed doses seen in treating NHL, solid-tumor absorbed dose results are not sufficient, of themselves, to treat disease sites. To improve the achievable dose values, one can try to increase the amount of target material in the tumor or to enhance the given dose by using a radiation sensitizer. Enhancement in the amount of local antigen can also, in principle, lead to commensurate increases in the dose since the uptake will depend directly on the amount of antigen present.

A number of traditional therapy treatment modalities have been invoked to enhance TRT in the solid tumor context. Using the radionuclide therapy postsurgery is one of the more promising of these strategies. Researchers have treated residual tissues including breast and brain sites with direct injections of the radioactive drugs. While the breast study, done in Italy, was promoted as a logistical advantage to external beam irradiation, there is manifest hope that glioma patients may directly benefit from such postsurgery injections to clear up residual disease in the surgical cavity. Such a need follows from the neurologist's general attempt to spare normal brain as much as possible during the surgical procedure. Both ¹³¹I (beta ray) and ²¹¹At (alpha ray) emissions have been applied in this context of presumably small lesions or even small numbers of residual tumor cells. Recalling the results of Chapter 3, we can justify the therapy of smaller lesions or cell clusters since the uptake (%ID/g) would presumably be higher for these structures than for large tumors. In other words, the oncologist would prefer to treat minimal residual disease. We should also recall the inversion of this strategy as already indicated, where investigators used TRT to reduce the size of hepatic lesions prior to surgery. It would seem that surgeons have the same preference as the medical oncologist—treat the smallest lesions possible.

A second and almost equally traditional method of treating cancer is tissue temperature increases. Here, the therapist attempts to achieve local tumor temperatures of approximately 42 to 43°C. Hyperthermia can, in principle, be invoked during the targeted radionuclide therapy with a commensurate tumor-cell killing enhancement of a given Gy dose. In cellular studies, the amount of ionizing radiation dose (D_0) required to kill 63% of the cancer cells can be reduced by almost an order of magnitude using concurrent hyperthermia. Such treatments are difficult to manage since maintaining—and measuring—temperature elevation is more problematic than even the corresponding exercises in ionizing radiation. Hyperthermia is particularly liable to be nonuniform due to blood flow variations within the tumor and adjacent normal tissues. Direct measurement of the temperature increase can be a difficult as well as a possibly morbid complication of this combination procedure.

Probably the most exciting area of combination therapies involving TRT is use of simultaneous treatment with various traditional chemotherapeutics. In many of the latter examples, the chemotherapy agents are already given as combinations themselves due to the historical evolution of solid-tumor therapy. Individually, these agents are of two types: those that are simply sensitizers to ionizing radiation and those that are also toxic in their own right. Preclinical (animal) and early clinical studies have shown that these chemical agents generally act in a linear fashion and are not synergistic in their results. One improvement that seems valid in such studies is the reduction of human immune response to the antibody aspects of the therapy. The importance of this outcome is that subsequent therapies are made possible in those patients whose initial therapeutic result is either not optimal or is a complete response initially with a recurrence some time later.

A final combinatory tactic is the use of immune modulators in conjunction with the TRT. We include in this category up-regulation of the antigen of interest in the tumor. An example was given of using γ-Interferon to increase the density of CEA molecules in human colorectal tumors. In nude mouse studies, the increase in u_T was approximately a factor of two if the interferon was given over a 6-day interval including some 4 days of TRT using a specific antibody to CEA (ZCE025). No such enhancement was seen with a nonspecific antibody (ZME018).

The two most common TRT procedures now being done—Zevalin and Bexxar treatment of NHL—both use prior injection of unlabeled, anti-CD20 antibodies. The justification for this somewhat surprising philosophy is that the cold material enables a more effective subsequent biodistribution of the radioactive agents. Large organs such as the liver and spleen are less likely to receive the majority of the radioactive anti-CD20 material with a commensurate enhancement in the possibly multiple lesion sites. It is expected that such strategies may be continued to expand in number in the future—perhaps involving solid tumors as well as lymphomas.

REFERENCES

Chakraborty, M., A. Gelbard, et al. 2008. Use of radiolabeled monoclonal antibody to enhance vaccine-mediated antitumor effects. *Cancer Immunol Immunother* **57**(8): 1173–83.

Davies Cde, L., E. K. Rofstad, et al. 1985. Hyperthermia-induced changes in antigen expression on human FME melanoma cells. *Cancer Res* **45**(9): 4109–14.

Gold, D. V., D. E. Modrak, et al. 2004. Combined 90Yttrium-DOTA-labeled PAM4 antibody radioimmunotherapy and gemcitabine radiosensitization for the treatment of a human pancreatic cancer xenograft. *Int J Cancer* **109**(4): 618–26.

Kaminski, M. S., A. D. Zelenetz, et al. 2001. Pivotal study of iodine I 131 tositumomab for chemotherapy-refractory low-grade or transformed low-grade B-cell non-Hodgkin's lymphomas. *J Clin Oncol* **19**(19): 3918–28.

Kievit, E., H. M. Pinedo, et al. 1997. Addition of cisplatin improves efficacy of 131I-labeled monoclonal antibody 323/A3 in experimental human ovarian cancer. *Int J Radiat Oncol Biol Phys* **38**(2): 419–28.

Kuhn, J. A., B. G. Beatty, et al. 1991. Interferon enhancement of radioimmunotherapy for colon carcinoma. *Cancer Res* **51**(9): 2335–9.

Maxon III, H. R., E. E. Englaro, et al. 1992. Radioiodine-131 therapy for well-differentiated thyroid cancer--a quantitative radiation dosimetric approach: outcome and validation in 85 patients. *J Nucl Med* **33**(6): 1132–6.

Mittal, B. B., A. M. Zimmer, et al. 1992. Effects of hyperthermia and iodine-131-labeled anti-carcinoembryonic antigen monoclonal antibody on human tumor xenografts in nude mice. *Cancer* **70**(12): 2785–91.

Mittal, B. B., M. A. Zimmer, et al. 1996. Phase I/II trial of combined 131I anti-CEA monoclonal antibody and hyperthermia in patients with advanced colorectal adenocarcinoma. *Cancer* **78**(9): 1861–70.

Msirikale, J. S., J. L. Klein, et al. 1987. Radiation enhancement of radiolabelled antibody deposition in tumors. *Int J Radiat Oncol Biol Phys* **13**(12): 1839–44.

O'Donnell, R. T., S. J. DeNardo, et al. 2002. Combined modality radioimmunotherapy for human prostate cancer xenografts with taxanes and 90yttrium-DOTA-peptide-ChL6. *Prostate* **50**(1): 27–37.

Paganelli, G., C. De Cicco, et al. 2009. Intraoperative avidination for radionuclide treatment as a radiotherapy boost in breast cancer: results of a phase II study with (90)Y-labeled biotin. *Eur J Nucl Med Mol Imaging* **37**(2): 203–11.

Pennington, K., A.J. Guarino, A.N. Serafini, C. Roche-Lima, K. Suppiah, C.J. Schneider, D.V. Gold, R.M. Sharkey, W.A. Wegener, and D. M. Goldenberg. 2009. Multicenter study of radiosensitizing gemcitabine combined with fractionated radioimmunothrapy for repearted treatement cycles in advanced pancreatic cancer. *Journal of Clinical Oncology* **27**(155): 4620.

Reardon, D. A., G. Akabani, et al. 2002. Phase II trial of murine (131)I-labeled antitenascin monoclonal antibody 81C6 administered into surgically created resection cavities of patients with newly diagnosed malignant gliomas. *J Clin Oncol* **20**(5): 1389–97.

Sakurai, H., N. Mitsuhashi, et al. 1998. Cytotoxic enhancement of low dose-rate irradiation in human lung cancer cells by mild hyperthermia. *Anticancer Res* **18**(4A): 2525–8.

Sakurai, H., N. Mitsuhasi, et al. 1996. Enhanced cytotoxicity in combination of low dose-rate irradiation with hyperthermia in vitro. *Int J Hyperthermia* **12**(3): 355–66.

Seldin, D. W. 2002. Techniques for using Bexxar for the treatment of non-Hodgkin's lymphoma. *J Nucl Med Technol* **30**(3): 109–14.

Supiot, S., F. Thillays, et al. 2007. Gemcitabine radiosensitizes multiple myeloma cells to low let, but not high let, irradiation. *Radiother Oncol* **83**(1): 97–101.

Williams, L. E., G. L. DeNardo, et al. 2008. Targeted radionuclide therapy. *Med Phys* **35**(7): 3062–8.

Witzig, T. E., L. I. Gordon, et al. 2002. Randomized controlled trial of yttrium-90-labeled ibri-tumomab tiuxetan radioimmunotherapy versus rituximab immunotherapy for patients with relapsed or refractory low-grade, follicular, or transformed B-cell non-Hodgkin's lymphoma. *J Clin Oncol* **20**(10): 2453–63.

Wong, J. Y., S. Shibata, et al. 2003. A Phase I trial of 90Y-anti-carcinoembryonic antigen chimeric T84.66 radioimmunotherapy with 5-fluorouracil in patients with metastatic colorectal cancer. *Clin Cancer Res* **9**(16 Pt 1): 5842–52.

Wong, J. Y., L. E. Williams, et al. 1989. The effects of tumor mass, tumor age, and external beam radiation on tumor-specific antibody uptake. *Int J Radiat Oncol Biol Phys* **16**(3): 715–20.

Zalutsky, M. R., D. A. Reardon, et al. 2008. Clinical experience with alpha-particle emitting 211At: treatment of recurrent brain tumor patients with 211At-labeled chimeric antitenascin monoclonal antibody 81C6. *J Nucl Med* **49**(1): 30–8.

Zeng, Z. C., Z. Y. Tang, et al. 1998. Improved long-term survival for unresectable hepatocellular carcinoma (HCC) with a combination of surgery and intrahepatic arterial infusion of 131I-anti-HCC mAb. Phase I/II clinical trials. *J Cancer Res Clin Oncol* **124**(5): 275–80.

Allometry (Of Mice and Men)

11

11.1 INTRODUCTION

Two issues present themselves when using animals to justify and subsequently to predict clinical trial outcomes. First is the question of whether the test species' result has any clear relationship to the eventual patient study. In a majority of trials, there is little effort devoted to making the logical connection between the two results. Researchers compartmentalize their thinking and often keep these data sets literally in separate files with no crossover comparisons ever being made. It would seem important to "close the circle" (Williams, Lopatin, et al. 2008) and to make direct comparisons of parameters seen in the two sequential studies. Additionally, the preclinical study has taken considerable time and associated cost. If the animal evaluation does not have predictive value for the clinical trial, there should be no further use of the test species with that particular agent. While an alternative animal model may eventually be discovered, it may well be that the appropriate test animal should be *homo sapiens* from the beginning.

Associated with the scientific question of animal–human correspondence is the ethical issue of using animals to test novel pharmaceuticals or radiopharmaceuticals (RPs). Unless there can be shown a clear relationship between the biodistributions of the two (or more) species, there is little reason scientifically to recommend the nonhuman testing. Many animal activist groups have complained about the manifest lack of this logical connection (Miller 2009). Ideally, the correspondence will be a mathematical one that gives both volumetric and kinetic parameters of the particular agent or radiopharmaceutical in vivo.

The precise form of the relationship between animal and human results is often unclear even if the two results are statistically correlated. An analyst might imagine a large number of different mathematical forms that could be used to predict human results given a sequence of animal data. These relationships might be of a linear, exponential, power-law, or other format. As we will see herein, investigators have used as their dominant concept the power-law representation to go from animal to human data sets. Yet there may be other, as yet little used, relationships that may prove more effective in this context. By more effective, we mean possibly having fewer free parameters as well as better clinical predictions.

One can argue that the proof of a correlation between the two species can come only after the animal and clinical trials are completed. Thus, the question of the appropriateness of the animal work may appear to be moot. Yet what is hoped for is that the physiological correspondence, of whatever form, becomes known as a function of the type of agent being tested. For example, the question can be raised as to how intact antibodies are handled by mice and men. If gross disparities are found, then subsequent murine trials may be forgone for that specific class of agent. We have argued earlier that the investigator should at least be required to provide comparisons between the original animal data and the eventual clinical parameters. In this way, the predictive value of animal trials can be tested and the form of the correspondence established or disproved. One would also wish to see such relationships—or lack of same—published in the public literature. In this way, disparities between species could be taken into account more readily by other investigators who may be naïve in their knowledge of the behavior of the agent they are about to test. We have suggested that such data be required as part of the U.S. Food and Drug Administration (FDA) approval process (Williams et al. 2008).

It should be made clear that the biodistribution data to be compared are those corrected for radiodecay. In other words, the cross-species analyst is interested in the pharmacokinetics of the hypothetical (unlabeled) pharmaceutical agent. The usual assumptions regarding label being permanently attached (Chapter 2) are made in these comparisons. This chapter includes a number of agents that are not labeled; thus, some of the biodistributions are of ordinary pharmaceuticals. In such cases, there are only blood data available since the other organs cannot be assayed by external counting using a gamma camera or positron emission tomography (PET) scanner.

In absorbed dose calculations (Chapters 4 and 8), any kinetic differences seen in the pharmaceutical behavior between animal species will necessarily be mitigated due to the presence of radiodecay. As the physical half-life becomes very short, such as in some of the commonly used positron emitters, almost any species variability would be essentially immaterial to dose estimates since the most rapid rate in the computation would be that of physical decay. Yet we still need to know the kinetic variation of the agent to make absorbed dose predictions in general, for example, in the case of a possible future label that may have a more extended physical half-life. Thus, the fundamental issues are distribution volumes and pharmacokinetics of the unlabeled material.

11.2 ALLOMETRY IN NATURE

One method to compare biological data across two (or probably more) species is allometry. It is a termed coined by Julian Huxley (1932) to describe differences in tissue masses and rate constants in going from one species to another. These comparisons may be extended beyond mammalian systems to birds and even cold-blooded animals. Some investigators have attempted to extend the analysis from intact animals down to the cellular and even subcellular levels of biological organisms. Probably the best known of these allometric studies is a comparison of animal major organ system masses.

Most allometry effort has involved mammals and major organ mass changes going from the shrew to the blue whale. It has been established that organ sizes vary in a power-law format as described in the modeling chapter (Chapter 6). In effect, one finds that the major mammalian organs show a variation of mass (m) given by

$$m = m_0 M^b \qquad (11.1)$$

where m_0 and b are fitted constants, and M is the total body mass of the mammal. At the present geological time, the range of mammalian M is probably the greatest it has ever been. Total body mass goes from a few grams in the shrew to hundreds of thousands of kilograms in the blue whale, the largest mammal ever known to have lived.

As noted in Chapter 6, power-law equalities have great intellectual appeal to the mathematically minded researcher. All the animals are handled in the same analytic format with only two parameters being required once the organ or metabolic process has been identified. By looking at a log-log graph of any such variable, the observer cannot identify the mass region of the plot. A reader could be looking at the small rodent comparison, the canine segment, or the elephant sector of the power law. All of these intervals will look alike much as in the power-law picture of organ retention of activity as given in Chapter 6. This is another aspect of the esthetic appeal of the power-law representation.

In Chapter 6, we emphasized that power-law models are applicable only over a limited range. In particular, such representations must fail at the lower limit (very low or zero mass) and at very large M values. The lower limit is no problem in zoology since there can be no animal that has literally zero mass. Even single cells, if we were to include them as our logically simplest living system, would still have a finite size. The upper limit of $M \sim 10^8$ g is still within the scope of the relationship. The blue whale, however, is allowed in nature only due to its living in a water medium. If we restricted our species to land-dwelling mammals, the upper value of M would be reduced by more than one order of magnitude.

Table 11.1 gives some best-fit results (Lindstedt and Schaeffer 2002) for the masses of a number of major organs. In the compilation, the total animal mass (M) is in kg and the organ or tissue mass (m) is in g or ml. One important factor in using a power law is determining exactly which physical dimensions are appropriate for both the m_0 and the b parameters. In some published tabulations, the units are not clear, and an example should be done by the reader to find out what has been implicitly assumed by the authors.

Let us do a sample calculation of human liver mass to show how the formula of Equation 11.1 should be used. From Table 11.1, we substitute $m_0 = 39.4$ and $b = 0.917$ into the equality with the result

$$m(\text{human liver}) = 39.4 \text{ g } (70 \text{ kg})^{0.917} = 39.4 * 49.19 = 1{,}938 \text{ g}$$

This value is numerically reasonable and indicates that our assumptions of kg for M and g for organ mass (m) are correct.

TABLE 11.1   Mass Law Exponents for Mammalian Organ Sizes

ORGAN IN MAMMALS	COEFFICIENT OF THE TOTAL BODY MASS (m_0)	EXPONENT OF TOTAL BODY MASS (b)
Size (g)		
Brain	8.16	0.716
Liver	39.4	0.917
Kidneys	8.35	0.853
Muscle	383	1.00
Heart	5.68	1.00
Skeleton	93	1.16
Lungs	9.27	1.00
Blood volume	71.5	1.01
Metabolic Rates (ml/s)		
Cardiac output	3.72	0.750
Ventilation	7.72	0.745

Source: Lindstedt, S. L. & Schaeffer, P. J., *Lab Anim*, 36, 2002. With permission.
Note: In the table, the adult animal's body mass is in kg and the organ or tissue mass is in g or ml.

While describing power-law relationships, we should also discuss what happens if there is a change in the dimensions (i.e., the physical units) being used to measure the quantities of interest. Remember that all entities in Equation 11.1 must carry units since they are based on measurements. If one assumes the equality, natural logarithms may be taken of both sides of the equation with the result that

$$\ln(m) = \ln(m_0) + b \ln(M) \tag{11.2a}$$

While we are using natural logarithms, base-10 or other formats are equally acceptable. In our first case, we let the dimensions of the quantity on the left-hand side (LHS) change so that m is replaced by λm, where λ is a constant. For example, if m were in grams originally, and we decide to replace it with units of kg, then λ is 1×10^{-3}. As a result, we find that Equation 11.2a becomes

$$\ln(\lambda m) = \ln(\lambda) + \ln(m) = \ln(m_0) + b \ln(M) \tag{11.2b}$$

The result is seen to be the same as the original Equation 11.2a with the exception that there is an additional term, $\ln(\lambda)$, on the LHS. Thus, there is no fundamental change in the nature of the power-law format. In particular, the exponent (b) is unaltered. A similar variation may be considered if we allow the units of M to change whereby we replace M by ηM. The argument goes as before, and we now alter the right-hand side (RHS) of Equation 11.2a with the result

$$\ln(m) = \ln(m_0) + b[\ln(\eta) + \ln(M)] = \ln(m_0) + b \ln(\eta) + b \ln(M) \tag{11.2c}$$

This last relationship is similar to Equation 11.2a with the exception of an additional term on the RHS: $b\ln(\eta)$. This is a constant such that the basic form of the power law is not altered and b remains the same in this revised format. Our conclusion is that the exponent is unchanged if the units of either m or M or both are altered. The only variation comes in the constant term (m_0) in the power-law analysis. The interested reader can test the invariant form of the original equation if the logarithm type is altered (e.g., to go to a base-10 logarithm).

Probably the most important result of these power-law, least-square fits is that the exponent (b) in Equation 11.1 lies between 0.8 to unity for the size (mass) of almost every major mammalian organ system. This is an extremely narrow numerical range. The only system requiring a marginally higher exponent is the bone mass of the skeleton. Skeletal mass increasing faster than the total body size presumably follows from mechanical structural constraints on Earth-bound mammals, which are a majority of the examples in the least-squares fitting. The proportionality coefficient m_0, on the other hand, shows much greater variability—approximately two orders of magnitude—over the range of known mammals.

Lindstedt and Schaeffer (2002) pointed out that there is a tendency for certain so-called control organs to have an exponent significantly less than unity—near the value of 0.8. A summary of their results is included in Table 11.1. Brain, kidneys, and liver are examples of such tissues whose sizes are not directly proportional to the total body mass. By control, the authors imply that two of these organs control the blood chemistry (kidneys and liver) and one of them operation of the body in general (the brain). If we take values of these organ masses and divide by M, the resultant ratio is the fraction of the total mass devoted to system oversight. This ratio, for the three organs indicated, decreases with increasing total body mass. Thus, the larger mammals devote significantly less of their size to control than do the smaller mammals. We will return to this aspect of the analysis later in this chapter when we consider what might be termed second-order corrections to the fitting of power-law results. We will find that some authors make more elaborate correlations using products of terms on the left-hand side of equalities like Equation 11.1.

From Table 11.1, the total blood volume is one of the tissues with an exponent of essentially unity. If we imagine a given hematocrit of approximately 50%, we would expect that plasma volume in mammals is also proportional to total body mass to the first power. Thus, if we consider the injection of a RP limited to the plasma space, the human dilution factor would be 3,500 compared with the mouse. Here we assume a 20 g mouse and a 70 kg patient. Hospital trials then have, as their "natural" uptake unit, %ID/kg (Chapter 2) rather than %ID/g as in the 20 g mouse. Similar arguments hold for the uptake in other organs such as muscle and lung. For control organs, the variation in going from mouse to man would be reduced even further since the smaller mammal would have correspondingly greater fractional uptake in these special tissues. In either event, the dilution factor has led some investigators to be concerned that the amount in the plasma, and hence in the tumor, would be greatly reduced in a clinical trial. This is a false anxiety. As compensation, the investigator must increases the amount of injected activity by the same dilution factor. In this way, the kBq activity injection of the murine experiment becomes MBq for a clinical trial of the same material.

In the lower segment of Table 11.1 are included some metabolic power-law results. It is seen that these exponents are also significantly less than 1. For example, the cardiac output has $m_0 = 3.72$ and slope (b) of 0.750. Other rates are governed by b values near 0.75. Some have argued that this is a general result, but the conjecture remains controversial. We will next describe some recent analysis of kinetic variables in terms of a fixed exponent equal to 0.75.

Reasons for the power-law relationships and the appearance of values near 0.75 seen in metabolism in mammals are often argued in heuristic ways. West and Brown (2004), for example, used a branching network strategy to predict a metabolic exponent of precisely 3/4. According to these authors, the result holds experimentally over 27 orders of magnitude—down to parts of the cellular structure itself. In their mathematical analysis, the particular value of 3/4 is attributable to living in three dimensions. More generally, the metabolic rate scaling should be $d/(d + 1)$ in a system with d dimensions. Other observers justify using a somewhat different exponent on the basis of heat loss through the surface area of an animal. If we assume a 3-D solid spherical object as our "animal," the total volume would be proportional to M with surface area proportion to $M^{2/3}$. To maintain the animal's internal temperature, the metabolism (e.g., measured in watts) would have to be proportional to the surface area so that the species does not become cold or overheat. Precisely the same argument was used in Chapter 3 to derive the power-law dependence of tumor uptake upon tumor mass for spherical and cylindrical lesions.

Metabolic rates are one example of kinetic parameter variation seen in the biological world. Their variation is clearly very large if we assume even the $M^{2/3}$ power-law value. By going from a creature of approximately 5 g to one of 10^8 g, the relative metabolic rate change would be on the order of 3×10^5-fold. Comparing the shrew with the blue whale is similar to comparing a single home with a relatively large city in terms of energy consumption rates. One is perhaps misled by this argument to believe that kinetic parameters, in general, would show variations of a similar magnitude in going from the smallest to the largest mammal. In particular, if we imagine going from mouse to man, a change of mass of approximately 3.5×10^3-fold, the metabolic rate variation predicted would be $(3.5 \times 10^3)^{0.75}$ or a factor of 455. Would this apparently large value be representative of the variation in kinetic rate constants (the k values of a typical model from Chapter 6) in going from mouse to man in pharmacokinetic parameters? This large difference is, in fact, not the case, and the two species, although quite disparate in size, have relatively similar kinetic variables. The latter parameters have been measured by using various RP agents that have been tested in murine species and then have gone on to clinical evaluation. For the sake of a more general discussion of rate variations, we should also describe some other time variations seen in mammalian experience.

11.3 HISTORICAL TEMPORAL AND KINETIC CORRESPONDENCES

Because of historical veterinarian records that document situations such as family pets and zoo animals over extended periods, there are now lifetime measurements in a number of animal species. It is conventionally said that 1 human year is approximately 7 dog years; that is, a dog ages seven times faster than its master. Likewise, mice under optimal lab conditions survive for perhaps 1 to 2 years, so 1 human year is 35 to 70 mouse years. If we were to continue this heuristic argument, we would expect that the mouse clearance of a given pharmaceutical would be comparable to this enhanced aging factor; that is, mice should lose agents from the blood at rates that are 30 to 70 times faster than that seen clinically. Again, this argument is simply incorrect. As in the case of the $b = 3/4$ rule, the predicted clearance difference between mouse and man is much too large. The rate constant disparity based on lifetimes is not as great as predicted using Equation 11.1 with the assumption that $b = 0.75$. Such a relative lifetime hypothesis is too large by about two orders of magnitude. Let us now introduce kinetic data analysis for two or more species to see what the variations actually are.

First, it is important to review some of the available comparison measurements made of various agents by pharmacologists following the movement of therapeutic drugs in the blood system. These early data did not involve radiolabels and were made using chemical assay on animal and patient blood samples taken over a time course between a few days up to several weeks. Restriction of the assay to blood follows from a lack of radiolabel on the pharmaceutical. While other organ data would be useful, they were not available in these measurements.

One of the most fundamental results coming from the early work is the concept of relative time exemplified by analysis of methotrexate in a number of mammals (Dedrick, Bischoff, and Zaharko 1970). Here, the authors were able to shown that the plasma clearance of the chemotherapeutic could be made comparable between species by scaling clock time for each animal by a factor proportional to $M^{-1/4}$. The results are shown in Figure 11.1 for the five species: mouse, rat, monkey, dog, and human. Note that the ordinate is the plasma concentration divided by the dose per gram of body mass (M). When the curve shown is differentiated, the resultant slope is given by

$$d \ln(C)/dt = -k/M^{1/4} \tag{11.3}$$

where k is a constant. This result implies that the invariant time (i.e., the time that can be compared directly between species) is clock time divided by $M^{1/4}$. If we take this exponent into our mouse and man comparison, we would predict that their kinetic differences would be in the ratio of $(3,500)^{1/4}$ or approximately 7.7. This result is considerably less than the ratio of 455 given previously if we assume an exponent of 0.75. Although not directly comparable with these analyses, the history of physics can contribute to the idea that time may pass at apparently different rates as seen in different systems of coordinates. One example is from the decay of radioactive particles.

Subjective (or "local") time is a familiar concept for physicists who are accustomed to its appearance in the context of radioactivity and special relativity. Observers, moving at constant velocity with respect to each other, see objective decay events unfold at apparently different rates. As physicists measure cosmic ray muons passing through the earth's atmosphere at relatively high speed, the particle's apparent half-life appears extended compared with the half-life when the muon is at rest relative to the laboratory. In this case, the true half-life given in the literature is the one measured when moving alongside the particle—that is, when the muon is at rest to the laboratory system. Time

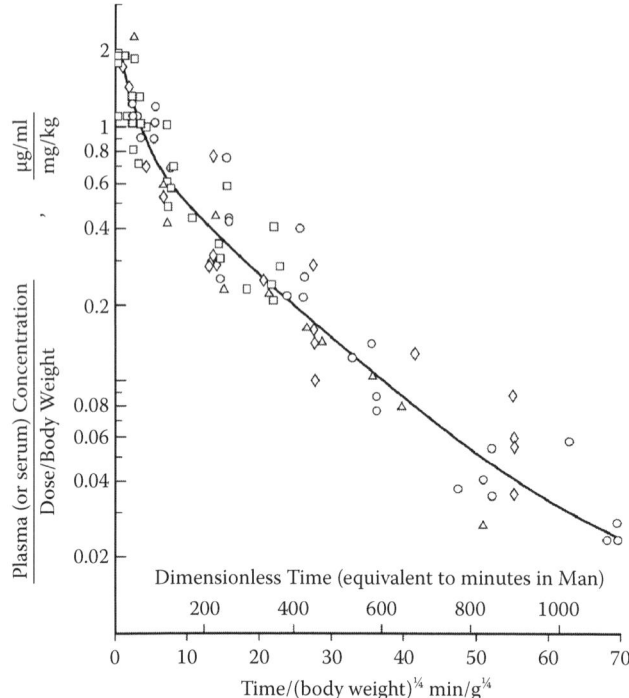

FIGURE 11.1   Plot (Dedrick 1970) of methotrexate blood clearance in several species. Diamond = mice. Circles = rats. Triangles = monkeys or dogs. Squares = patients (male). The various data sets can be made to overlie each other if the time is adjusted with an $M^{1/4}$ variation, where M is animal's total body mass. This quarter-power invariance may not occur with other agents. (Modified from Dedrick, R. L., Bischoff, K. B., and Zaharko, D. S., *Cancer Chemotherapy Reports*, 54, 1970.)

measured in this system is called the *proper time* in physics literature. In allometry, there is no "proper" set of kinetic parameters, only variants on a theme driven by $M^{1/4}$ or perhaps some other time–dilation rule. If we are arrogant, we could define the proper value as that measured in the human.

11.3.1 Measured Protein Kinetic Parameters Using Simple Allometry

Allometric comparisons are still relatively rare in radiopharmaceutical research. Today, we typically have to rely more on unlabeled pharmaceutical studies that are necessarily restricted purely to blood curve evaluations. Most of these data have been generated in the past 20 years using a variety of proteins—mostly human-based antibodies. Because of historical reasons, such results are usually described by parameters characteristic of chemotherapeutic agent analyses. As described in Chapter 6, such variables generally include CL, V_i, and V_{ss}. Some investigators, however, also include rate constants (k) and particular serum or plasma half-times in the analyses. Recall that a half-life is $\ln(2)/k$.

Mordenti, Chen, et al. (1991) investigated blood clearance of five engineered proteins having masses between 6 kDa and 98 kDa. There is interest in each of these molecules as they may have important applications in a variety of human conditions and diseases. Relaxin, the lowest-mass molecule of the group at 6 kDa, is proposed for use in uterine thinning and softening to promote mammalian birth processes. Human growth hormone (hGH) is a recombinant human form used to increase stature in young patients. Recombinant CD4 (rCD4) is a cell-surface glycoprotein found on mature T cells and has a molecular weight (MW) of 50 kDa. It is the receptor for HIV. This agent blocks HIV-1 infections of T cells in vitro. Tissue-plasmogen activation factor (rt-PA) is appropriate for treatment of acute myocardial infarctions and has a MW of 59 kDa. Finally, CD4-IgG (MW = 98 kDa) is a constructed protein that has an extended serum half-life compared with rCD4. All of these molecular weights are below that of the intact IgG molecules' value of approximately 150 kDa.

All five proteins were injected intravenously into five mammalian species: mouse, rat, rabbit, Rhesus monkey, and human. The three previously listed pharmacokinetic (PK) parameters were then measured using the plasma curve for each species. Finally, these values were compared across the species by use of Equation 11.1 to select a most probable exponent (b) as well as the coefficient m_0 by least-squares methods. Results of these analyses are shown in Table 11.2.

Exponents for the two volumes—that of the initial distribution (V_i) and the steady-state volume (V_{ss})—were both close to unity. Values of b varied between 0.83 and 1.05 with a mean value of 0.94 for these two cases. Clearance rates (CLs) had exponents between 0.65 and 0.84 with a mean value of 0.75 for the five proteins. One may conclude that the volume of distribution is slightly less than proportional to the total mass of the animal. No obvious expansion of this volume was seen in going to the lowest mass proteins such as Relaxin. Clearance showed more variation but had a mean that was comparable to the West and Brown (2004) ideal value of 0.75 as observed in metabolism.

Other protein analysts (e.g., Mordenti et al. 1991), at Genentech Inc. subsequently did comparable work but with only a single protein at a time in the experiment. Kelley, Gelzleichter, et al. (2006) described a humanized anti-CD40 monoclonal antibody (Mab) in three mammals: mouse, rat, and cynomolgus monkey. Species average masses

TABLE 11.2 Blood Kinetic Exponents (b) for Five Protein Constructs

PROTEIN	MW (kDa)	b(CL)	b(V_i)	b(V_ss)
Relaxin	6	0.80	0.87	0.92
hGH	20	0.71	0.83	0.84
rCD4	50	0.65	1.02	1.02
Rt-PA	59	0.84	0.93	0.84
CD4-IgG	98	0.74	1.05	1.01

Source: Mordenti, J. et al., *Pharm. Res.*, 8, 1991. With kind permission from Springer Science and Business Media.

were 0.031, 0.239, and 1.68 kg, respectively. CD40 is a protein on the surface of B cells and has some evidence of being up-regulated in lymphoma. In this 150 kDa example, the exponent of CL was found to be 1.01. One important result of this study was that some of the monkeys, upon repeat injections, developed antihuman antibodies to the anti-CD40 protein. This led to much increased clearance in those individuals due to the formation of antigen–antibody complexes that deposit in the liver. This topic is described further herein.

In an analogous study, Lin, Nguyen, et al. (1999) described the behavior of a recombinant humanized anti-VEGF (vascular endothelial growth factor) antibody rhuMab VEGF. This antibody targets the growth factor receptor in the vascular system. Activation of the receptor allows the tumor to increase in size by expanding its vascular network. The same animal species were included as in the work by Kelley, Gelzleichter, et al. (2006). Among the parameters measured were the traditional three of pharmacology (CL, V_i, and V_{ss}) as well as the terminal half-life ($T_{1/2late}$). Exponent values for CL, V_i, and V_{ss} ranged over the interval 0.79 to 0.92 and were comparable to those measured by Mordenti et al. (1991). The late $T_{1/2}$ was an interesting case where the exponent was relatively large at 1.05. Using animal data, the predicted human value of 12 days was seen to be roughly comparable to $17d$ value found in a related study on hospital patients. While this difference may appear large, the impression is based on our usual linear view of the world. If we select a logarithmic viewpoint, the difference is only 15%. Such is the benefit of a power-law approach as well as a cause of some cynicism among the nonbelievers.

Khor, McCarthy, et al. (2000) more recently performed a study on the rPSGL recombinant antibody, which binds the P-selectin molecule. The selectin molecule family mediates the binding of leukocytes to the surface of endothelial cells. By reducing this binding, inflammatory processes may be reduced in magnitude. Diseases to be treated might include deep vein thrombosis (DVT) and acute infarctions. This study followed the outline of Mordenti et al. (1991) in its conceptual outline except that the rabbit species was replaced with a pig and no human data were included. Results were obtained for the power-law parameters for CL, V_c, V_{ss}, and $T_{1/2}$ terminal-phase half-life. Again, CL and the volumetric variables were found to have exponents close to unity. However, the $T_{1/2}$ late phase had an exponent of only 0.159. Notice that as the exponent approaches zero, there is no dependence on species' size for the associated parameter. Following, we will see other protein examples where the overall clearance has little dependence on total body mass M.

Another single-protein study was done by Richter, Gallati, et al. (1999) at Hoffmann-La Roche Ltd. on a fusion construct featuring union of the tumor necrosis factor (TNF) receptor and segments of human immunoglobulin. Named lenercept, this 120 kDa entity binds TNF to decrease the severity of septic shock. For this allometry trial, rats, rabbits, cynomologus monkeys, and dogs were the four test species. Exponents for CL and V_{ss} were found to be 1.06 and 0.97, respectively. In the analysis, $T_{1/2late}$ exponent was on the order of zero: −0.0789. Because of the negative exponents, the human half-life values were predicted to be lower than those seen in the animal work. This is a rather unlikely result and shows that the magnitude of the uncertainty in the exponent is probably on the order of 0.08 in this example. In any event, late-half-time exponents were seen to be essentially zero so that the animal half-times were quite close to the measured clinical values (Haak-Frendscho, Marsters, et al. 1994).

As in the work by Lin et al (1999), Richter and colleagues (1999) predicted human results for their fusion protein. In the case of V_{ss}, the predicted human value and the measured were both 70 ml/kg. Agreements for the CL parameter were not quite as good: the predicted clearance rate was 0.66 ml/m, whereas the measured was only 0.33 ml/m. The anticipated $T_{1/2late}$ value for humans was 4.2 days, whereas the measured result was somewhat longer at 7.0 days.

A common feature of the previous sequence of allometry studies was the onset of immune response in a large subset of the animals after only a single injection. Thus, researchers must be careful when using an essentially human antibody repeatedly in these various animal species. At a certain time post the first intravenous (IV) injection, the animal will develop antihuman antibodies much as patients develop antimouse or antichimeric antibodies against repeated injections of murine-based proteins in clinical trials. This topic was described in Chapter 3.

11.3.2 Kinetic Variations Using a More Sophisticated Analysis

As mentioned in Chapter 4, there is a simple absorbed dose analysis and a more elaborate format called the biologically effective dose (BED). The latter is based on the former with added features that one or another investigator feels may be important to the final agreement between data and analysis. This stylistic elaboration of a simple concept is also becoming of interest to pharmacists—particularly one group at the FDA. These authors (Mahmood and Balian 1996) tried to incorporate at least two features previously mentioned into a more complicated allometric formulation. The new features include the idea that there is a dichotomy separating ordinary organs from control organs as cited already (i.e., two separate but comparable power-law representations). In addition, the FDA team used measured values of animal lifetime to augment simple allometry.

Mahmood and Balian (1996) described both of these elaborations in detail. With regard to the control aspect, the FDA group uses a brain–weight correction factor whereby each CL value of a drug is first multiplied by the animal's brain weight before the correlation is calculated with total body mass. Such brain masses are known so that there is no difficulty in making this multiplication with the original data set for the animals being studied. The authors, because of the two parameters in the power-law model, rightly insist on their being at least three animal species in the analytic process. This constraint is also followed in other manipulations of the clearance parameter including that involving the total lifetime of the animals.

To take into account the variations in lifetime among the different species, a two-step algorithm must be followed. The first step is to determine an abstract value, called the maximum life potential (MLP), for each species in question. A best-fit equation for this quantity has previously been derived by Sacher (Mahmood 2009):

$$MLP = 185.4 \left(m_{\text{brain}}^{0.636} \right) \big/ M^{0.225} \tag{11.4}$$

In Equation 11.4 the units are years for *MLP* and kg for the two masses. Note the appearance of the brain mass (m_{brain}) in the numerator of the equality. The second step of this algorithm is, in analogy to the previous brain mass multiplication, to multiply each *CL* value by *MLP* for the animal in which clearance was measured. The product of *CL* and *MLP* is then fed into the regression analysis versus the animal total body mass (*M*).

It is important to understand, in either of the manipulations already described, that the final regression form is one of an algebraic product versus the body mass of the animal. This is the fundamental difference from the simple form of Equation 11.1. In the first case, the product is the brain weight times the *CL* value; in the second, the analyst uses the product of *MLP* and *CL* vs total body mass. Therefore, to actually find the predicted *CL* value for an unmeasured species, the pharmacist must divide out the brain weight or *MLP,* respectively. Since both of these are known quantities for man or most other mammals, this last step is essentially trivial.

The FDA group emphasized in its use of these mathematical products that the actual formula to be applied is chosen via what is termed the *rule of exponents* (ROE). Using the ROE plan, an analyst has one of four choices to make in the prediction of clinical results given animal data. Initially, it is prescribed that one perform a simple computation as shown in Equation 11.1—that is, a calculation with no correction for either brain mass or MLP. If, in this first result, the resultant exponent (*b*) lies in the range $0.55 < b < 0.70$, then the process is halted and simple allometry is used to predict the human *CL* value given a sequence of animal clearance data. The authors recommend that the *MLP*CL* approach be followed if the initial exponent lies between 0.71 and 0.99. Finally, if *b* becomes larger than unity but less than 1.3, the *CL* value should be predicted using the product of *CL* and brain mass. Exponents outside the range are probably not going to help with human clearance predictions. For example, if the exponent is under 0.55, it is thought that the animal *CL* data will predict a human result that is incorrect and manifestly too small. The converse holds when *b* becomes large—in particular when $b > 1.3$. In these last two extreme cases, there is no direct application of allometry to the clinical situation and animal data cannot be used to predict the human result.

An example of this type of ROE analysis is given in Table 11.3. In the listing, Mahmood (2009) gives the clearance results (*ml/h*) for six antibodies studied

TABLE 11.3   Comparison of Methods to Find Human Antibody Clearance (CL)

ANTIBODY	EXPONENT IN SA	ERROR IN SA (%)	ERROR IN mbrain (%)	ERROR IN MLP (%)
Horse F(ab')$_2$	0.534	57	97	91
Digoxin Fab	0.667	18	94	95
RSHZ 19	0.732	45	60	53
VEGF	0.785	13	74	59
EGF/r3	1.020	500	6	82
Lenercept	1.052	109	42	6

Source: Mahmood, I., *Jour. Pharm. Sci.,* 98, 2009. With permission.
Notes: SA, simple allometry. mbrain, correction for brain size. MLP, correction for maximum lifetime potential.

in a number of small animals. It is seen that as the exponent of the simple analysis approaches or exceeds 1 the use of *MLP* or brain weight (m_{brain}) reduces the percent error in the estimate of human CL values. The converse is true when the *b* value is substantially below unity such as in the case of Horse $F(ab')_2$ antibody and VEGF. In these cases, the simple approach is considerably better than the augmented methods previously outlined.

11.4 COMPARISONS OF TUMOR UPTAKE AS A FUNCTION OF TUMOR MASS

One important area of comparison between animals and patients is assessment of the relationship of tumor uptake to tumor mass. Such correlations cannot easily be followed with ordinary pharmaceuticals and must be studied with labeled agents. This relationship, if valid, is a very important issue since absorbed dose rate is proportional to u_T when using a labeled RP. A power-law equality between u_T and *m* has been established for liposomes and intact antibodies in the mouse model as described in Chapter 3. One would hope that a similar relationship, albeit with a lower amplitude, would hold in the human case.

For clinical assessment of tumor uptake, surgical and pathology involvement is generally needed. In a majority of such cases, the RP investigators must restrict themselves to a single time point. Logistical difficulties in the operating room often lead to this single designated point becoming, instead, an extended time interval. This last slippage factor occurs as a necessity in most patient assays since the investigator cannot get the subject into surgery at the same precise time postinjection of the radiolabeled tracer. This variation alone may lead to an apparent lack of correlation between tumor (or other tissue) uptake and anatomical mass in a hospital study.

A second complication of using clinical data is that there are interpatient variations in the study cadre. Changes in tumor expression of the molecular marker as well as tumor perfusion may occur in any sequence of patients under study. The investigator cannot assume that all patients will show the same concentration of this marker and any correction for its presence in the tumor sample is difficult to make.

In an effort to test the power-law formula in patients, the author and colleagues from Aachen University in Germany (Williams, Bares, et al. 1993) were able to analyze 19 consecutive colorectal lesions (positive for carcinoembryonic antigen [CEA]) obtained from patients with recurrent disease. Here, the antibody was [131]I- BW 431/26, an anti-CEA IgG_1 given at 0.5 to 1.0 mg per injection. Sample masses were taken from lesions obtained at surgery and assayed in a well counter. Total tumor mass was determined using the three dimensions observed in the pathology specimen. As described above, tumor resection occurred at variable times postinjection.

Figure 11.2 shows the analysis with Equation 11.1 and all 19 patients. The resultant correlation ($\rho = -0.51$) was significant at the 2% level of confidence. Here, since we used all of the data, time of tumor acquisition varied between 4 and 14 days post-injection of the RP. If we further restricted the time interval to lie between 4 and 8 days, the correlation

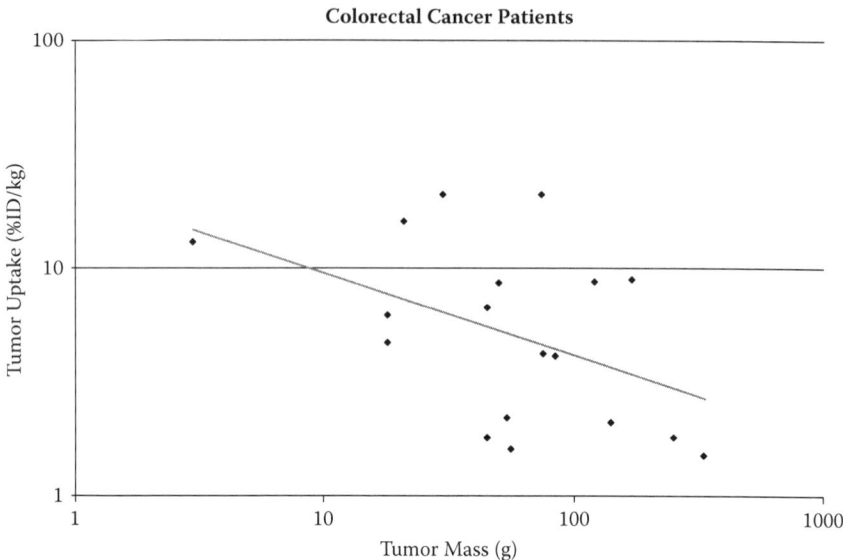

FIGURE 11.2 Variation of human colorectal tumor uptake (%ID/kg) versus tumor size as found in pathology specimens. An anti-CEA antibody was used in these studies involving 19 sequential patients at various times postinjection of the ^{131}I-labeled BW 431/26 anti-CEA protein. (Reprinted from Williams, L. E. et al., *Eur. Jour. Nuc. Med.*, 20, 1993. With permission of Springer-Verlag.)

TABLE 11.4 Correlation of Tumor Uptake and Tumor Mass in Recurrent Colon Cancer Patients

NUMBER OF PATIENTS	u_0 (%ID/kg)	b	ρ CORRELATION COEFFICIENT	SIGNIFICANCE (%)
19	21.9	−0.362	−0.510	2
14	25.0	−0.381	−0.623	1

Source: Williams, L. E. et al., *Eur. Jour. Nuc. Med.*, 20, 1993.

improved to a 1% level of confidence ($\rho = -0.623$) for the resultant 14 patients. A summary of the best-fit parameters as well as the correlation coefficients is given in Table 11.4.

We should caution that patients from Aachen University had to be studied with a third confounding factor: use of ^{131}I as the anti-CEA antibody radiolabel. As we have seen previously, radioactive iodine may be cleaved from the Mab due to dehalogenation enzymes and could be lost at different rates in different patients. This factor makes the assay of tumor uptake as a function of tumor mass more difficult when evaluating iodine-labeled agents.

We can conclude that, although there were a number of issues including variable time of sampling, interpatient variation, and a radioiodine label, there was a significant correlation between the tumor uptake (in %ID/kg) and the associate lesion mass (in g). Resultant exponents were on the order of −0.4, a value in the middle of the range cited in

Chapter 3 as being relevant to the murine experiments done originally at City of Hope. The magnitude (u_0) parameter was, as we would expect, on the order of 20 %ID/kg—a value that has the anticipated three-order-of-magnitude shift downward compared with a murine study. To follow the allometric argument to a logical conclusion, a more extensive comparison of the mouse and human amplitude values is warranted.

If we use the murine parameters (u_0 and b) from our original article (Williams, Duda, et al. 1988) on tumor uptake as a function of tumor mass, prediction of the corresponding a value for the patients may be made. Here, we assumed the same exponent as in the mouse studies but reduced the amplitude parameter by a whole-body mass scaling factor of 20 g/70 kg or 1/3500. The predicted human result was found to be parallel to but significantly lower (approximately fivefold) than that of the 19 Aachen patients. As a partial explanation of this discrepancy, our group at City of Hope (Williams, Beatty, et al. 1990) showed that this magnitude of difference is expected since the concentration of CEA in our original murine xenografts (LS174T) has been found to be essentially five times lower than a typical wild-type colon cancer seen in patients. If this factor is taken into account, the two curves (mouse and man) essentially were found to lie atop each other.

A Dutch study on head and neck cancer patients (HNSCC) is of interest as these investigators used ^{99m}Tc as the radiolabel on the E48 and U36 antibodies. Both Mabs were developed against squamous cell carcinoma at the Vrije Universiteit in Amsterdam (de Bree, Kuik, et al. 1998). In this clinical evaluation, due to the 6 h half-life of Technetium-99m, there was a strict guideline that each patient went to surgery at Day 2. A total of 17 patients were injected with E48 and 10 with U36 antibody. Tumor sizes were determined preoperatively using either magnetic resonance imaging (MRI) or computed tomography (CT) and integrating over slices in the images. Biopsy specimens obtained at the 48 h time period were counted and %ID/kg determined by correcting for decay. A pathologist confirmed that each surgical specimen was that of a malignancy. The group then analyzed the 27 results with a stepwise multivariate analysis of tumor uptake with regard to multiple clinical factors, including age, sex, tumor staging, antigen expression, antibody, tumor mass, and protein amount (mg). It was found that the dominant factor in the correlation between tumor uptake and outside variables was that of tumor mass. Explicitly, it was determined that u_T depended on the inverse of tumor mass. The authors did not explore mathematical specifics of this correlation, although they did show a graph of $u_T(m)$.

Table 11.5 gives a summary of the results I obtained using the three standard formats of linear regression introduced in Chapter 3: linear, exponential, and power law. It

TABLE 11.5 Correlation of Tumor Uptake and Tumor Mass in Squamous Cell Carcinoma of the Head and Neck (HNSCC) with Three Possible Unique Models

MODEL	AMPLITUDE (u_0)	SLOPE (b)	CORRELATION (ρ)	SIGNIFICANCE
(n = 27 lesions)				
Power law	42.9	−0.344	−0.516	0.01
Exponential	27.0	-1.7×10^{-2}	−0.487	0.01
Linear	31.3	−0.334	−0.329	0.10

was seen that the best correlation occurred with a power-law format and that the exponent of the mass was −0.34—a value in the range suggested in Chapter 3. A graph of u_T versus tumor mass for the 27 lesions is shown in Figure 11.3. There were certainly large variations seen in the uptake parameter for a given size tumor. This type of uncertainty arises out of the interpatient variability due to changes in the amount of antigen present as well as the degree of necrosis as already noted.

Chatal, Saccavini, et al. (1989) at Nantes University also commented on the variation of u_T with lesion size in a study of ovarian carcinoma. Here, the patient underwent abdominal therapy using an implanted catheter through which the ^{111}In-OC125 antibody (against CA 125 antigen present in serous ovarian carcinoma) was infused. Both intact and $F(ab')_2$ fragments were used in this study and average data reported for the latter antibody fragment's uptake in tumors versus lesion size in a cadre of 12 patients operated on within 3 days of intraperitoneal (IP) infusion. The surgeon removed a majority of the tumor at that time as well as malignant nodules and other suspicious structures in the peritoneal cavity.

In their report, the French investigators only gave average uptake and associated mass values for four smaller and nine relatively larger nodules recovered by the surgeon. In the former case, the uptake was 56.7 ± 11.1 %ID/kg at 21.5 ± 6 mg mass. With the larger nodules, the comparable values were 19.6 ±− 14.5 %ID/kg and 78 ± 47 mg mass. No individual mass and uptake results were quoted in the report. Again, it appears that there is an inverse correlation of uptake with lesion mass in this example of a radiometal-labeled antibody fragment.

It would be useful if investigators could eliminate the various outside influences on the patient data set with which the researchers at Aachen, Amsterdam, and Nantes had to contend. One clinical group was able to avoid most of these issues by following a single patient over several days using an antibody agent labeled with ^{111}In. Thus, there was no interpatient variation, the delay between injection and counting was controlled, and radioiodine was not used. In this investigation, Macey, DeNardo, et al. (1988) at the University of California–Davis (UC–Davis) were able to count the activity in a cluster of four melanoma lesions on the inguinal skin of a single person. This patient had more than 20 total lesions, but only 4 could be counted directly. Activity data [$A(t)$] were determined with a gamma camera after the patient had been injected with the ^{111}In-labeled antibody ZME-018 (antimelanoma). Lesion lateral dimensions were measured with a set of calipers. Since each tumor was present at the skin line, the investigators were also able to determine the thickness with the caliper technique. Gamma imaging was done at five time points: Days 1, 3, 6, 7, and 9. At each time, I used the mass law of Equation 11.2 to represent the data. Results of the best-fit parameters are given in Table 11.6.

From the least-squares fitting, it was seen that the amplitude factor decrease monotonically with time after 3 days. The slope (b) variable, however, was relatively constant and on the order of −0.6 for each day of lesion counting. This value, while quite stable over the 9 days of the UC–Davis experiments, was somewhat larger in absolute magnitude than the corresponding parameter found in the original murine assay (Williams et al. 1988) or the Aachen clinical results (Williams et al. 1993). At the present time, we cannot offer a conceptual explanation for this variation and must consider it to be a fact of melanoma uptake. Notice that the mass-law exponent is more negative than even the theoretical result for a cylindrical lesion (−0.50).

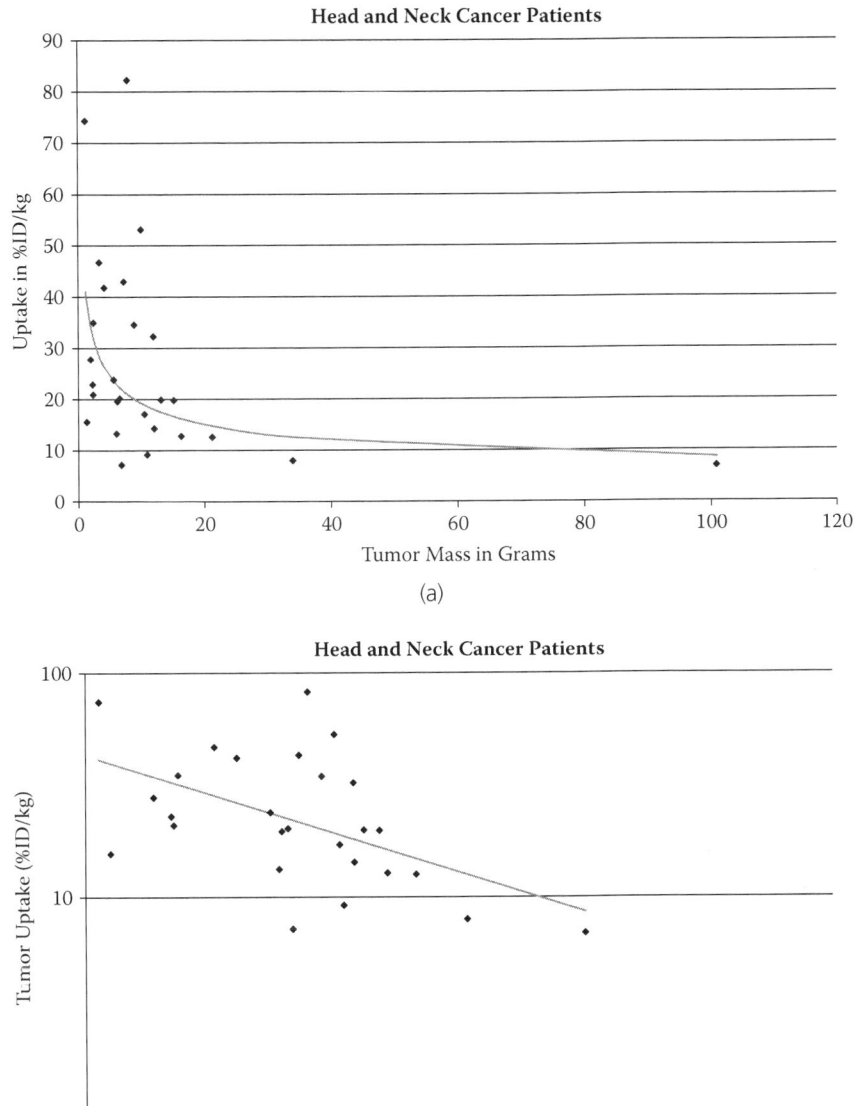

FIGURE 11.3   Uptake of antibody E48 or U36 against HNSCC in a sequence of 27 patients. Surgery was performed at 48 h postinjection of the ^{99m}Tc-labeled protein (de Bree, Kuik et al. 1998). (a) a linear plot. (b) a logarithmic plot with a power-law correlation.

TABLE 11.6 Correlation of Melanoma Uptake (%ID/kg) with Tumor Mass (g) in a Single Patient Having Four Measurable Lesions

TIME POSTINJECTION (d)	u_0 (%ID/kg)	b	CORRELATION (ρ)	SIGNIFICANCE
1	42.4	−0.546	−0.867	n.s.
3	72.1	−0.544	−0.974	0.05
6	42.8	−0.633	−0.991	0.05
7	29.4	−0.609	−0.985	0.05
9	27.3	−0.647	0.994	0.05

Notes: The equality assumed is $u_T = u_0 m^b$. This is a two-parameter model with four (tumor) data points at each time point.

Having described several examples of lesions in patients, we should point out that there are certainly experimental situations where there is little evidence of a power-law relationship between u_T and tumor mass. One particular case of interest is that of multiple, small lesions grown in the liver of nude rats and nude mice. This scenario is derived from a single injection into the animal's spleen of a large number of human colorectal tumor cells. These cells than migrate, via the portal vein, into the hepatic space and grow there as separate structures. In the rat model, for example, Sundin, Ahlstrom, et al. (1993) were able to obtain pathology samples of the various ($n \geq 20$) lesions and found that they were uniformly perfused and that their uptake was independent of their mass. Typical lesion sizes are less than 100 mg. A similar result ($b = 0$) has been obtained for the same implantation method using a nude mouse model (Yoshida, Rivoire, et al. 1990). As in the case of the rat, uptake in the multiple lesions was essentially independent of their size. Thus, the previously described mass law was not valid. We would anticipate that the uptake of radioactivity, although not measured, was probably uniform within these multiple, small hepatic tumors in the nude mouse.

Although we have seen two similar examples where a type of lesion implantation does lead to uniform uptake, we would emphasize that clinical situations are generally not of the type previously described for the nude rodent models. The patient will often have one—or perhaps a few—large lesions often with necrotic centers. If an imaging agent is given, the gamma camera images usually demonstrate a hot rim and a relatively cold interior. Thus, we would predict that if such structures could be excised intact and assayed for activity the picture of Chapter 3 would be valid and uptake would probably be a decreasing function of lesion mass.

11.5 SINGLE-PARAMETER COMPARISONS OF MOUSE AND HUMAN KINETICS

Until now, we have emphasized relatively simple models such as the single indicators (V_i, CL, or V_{ss}) arising from traditional analysis of pharmaceuticals or the power-law

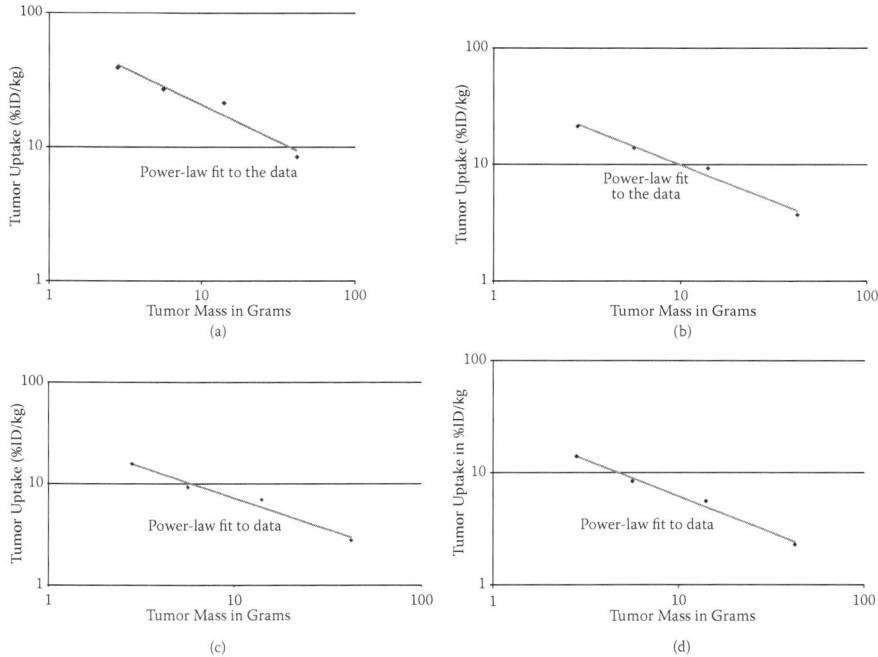

FIGURE 11.4 Variation of melanoma tumor uptake (%ID/kg) in a single patient at (a) 3, (b) 6, (c) 7, and (d) 9 days after antibody injection. Activity values were obtained using a camera system that was calibrated to a planar ^{111}In source (Macey, Denardo, et al. 1988). The camera was fitted with a medium-energy parallel-hole collimator. Power-law correlations are given at each of 4 days during the course of the study.

format with only a pair (u_0, b) of fitted parameters. In a typical modeling analysis, however, there are a much larger number of resultant parameters determined as an outcome of a single fitting episode in either an animal or patient. Probably the lowest order result would be the four-plex (A_1, A_2, k_1, k_2) resulting from a biexponential fit to a blood or tumor curve. An obvious question is how to compare two such sets from disparate species—or from the same species for two different radiopharmaceuticals.

First, we acknowledge that to quote only the rate constants k_1 and k_2 is not sufficient for any comparison. This is because, even if the two species had identical values for these two parameters—or for the corresponding half-lives—that agreement alone would not prove direct congruence between species. That is because there may be unequal amplitudes; for example, one animal may have a relatively larger A_1 value such that the first rate constant is favored for that species. Thus, as is mentioned in Chapter 6, the analyst must consider the entire set of fitted parameters and needs to treat them as a single bundled entity. This association of the fitted parameters cannot be broken and is the fundamental result of the compartmental or noncompartmental modeling of the RP movement in a species.

In an attempt to simplify this issue of comparison, our group has developed the concept of temporal moment analysis of blood or other tissue biodistribution curves. Here, one defines a first moment as

TABLE 11.7   First Moment Blood Comparisons (<t>)

AGENT	MW (kDa)	MOUSE <t> (h)	MAN <t> (h)	RATIO (HUMAN/MOUSE)
Intact Mab	160	97	129	1.33
F(ab')$_2$	120	8.8	12.7	1.4
Minibody	80	7.2	30.2	4.2

Source: Williams, L. E. et al., *Cancer Biother. Radiopharm.*, 23, 6, 2008.

$$\langle t \rangle = \int_0^\infty tu(t)dt \bigg/ \int_0^\infty u(t)dt \tag{11.5}$$

where $u(t)$ is an uptake curve (%ID/g) taken from the biodistribution data set for a given animal or patient. In this context, the term *first moment* means that the exponent of time is unity. If we increase that exponent in the numerator of Equation 11.5, other higher moments may also be computed. Notice that this first moment involves integration over the entire curve to give one effective time for that uptake function. Recall that $u(t)$ is corrected for decay so that we are presenting a first moment that represents the pharmacokinetic curve and not the radiodecay-driven curve.

Table 11.7 gives comparisons between nude mice and patients for three cT84.66 anti-CEA antibodies developed at City of Hope (Wu and Yazaki 2000). It is found that the intact antibody (MW = 160 kDa) has almost the same <t> value in humans as the mouse, although the first moment is slightly greater in the larger species. A similar result holds for the F(ab')$_2$ fragment with mass of 120 kDa. Yet, when we investigated the Minibody (80 kDa), there was considerable difference between murine and patient first moments for the blood. In this case, the blood curve of these intermediate sized antibodies is considerably faster than that of the average patient. If we calculate the ratio of the human first moment to that of the mouse, a quotient of approximately 4.2 is found. Thus, the nude mouse clears the intermediate mass chimeric antibodies approximately four times as fast as the typical patient.

11.6 COMPARING THE RATE CONSTANTS IN A COMPARTMENTAL MODEL: HUMAN VERSUS MOUSE

My colleagues and I at City of Hope attempted to make a direct comparison between murine and patient results using average activity data $a(t)$ and $A(t)$ for both species (Williams et al. 2008). Figure 11.5 demonstrates the mammillary compartmental model whereby the sampled tissues are connected directly to the blood. In this case we had an intact chimeric antibody (cT84.66) labeled with ^{131}I for the animal trial and

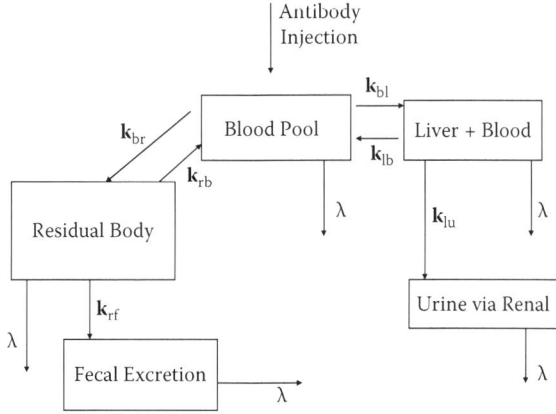

FIGURE 11.5 Compartmental model for the cT84.66 intact anti-CEA antibody. Among the adjustable parameters (Williams, Lopatin, et al. 2008) are the total volume (V), the various rate constants (k), and the fraction of blood in the liver (f_{bl}). This last variable is needed since the animal liver is taken without bleeding the animal and the patient is imaging with blood in the liver. No other organ was assigned an analogous variable for simplicity. (From Williams, L. E. et al., *Cancer Biother. Radiopharm.*, 23, 6, 2008. With permission.)

[111]In for the clinical data set. This labeling disparity was due to the need of ^{111}In in the clinical situation to act as a surrogate for ^{90}Y in the ensuing therapy component of the trial. Generally, we used radioiodine isotopes for animal work in the development phase of a new antibody due to the relatively smaller amounts of protein needed for direct labeling with iodine rather than indirectly labeled (via bifunctional chelators) with radiometals.

Best-fit parameters for the compartmental model are listed in Table 11.8. Notice that the error bars are smaller relative to the size of the parameters for the mice compared with the patients. This relatively reduced value for murine covariance presumably reflects the similarity among the animals in the study. The much larger covariance values among the colorectal cancer patients indicates their variation in terms of age, mass, and sex—parameters that could not be controlled in the clinical study.

TABLE 11.8 Pharmacokinetic Parameters in a Compartmental Model

KINETIC PARAMETER	MOUSE VALUE	PATIENT VALUE
V (Volume of distribution)	2.3 ± 0.20 ml	$5.36 \pm 0.52 \times 10^3$ ml
F_{bl} (Fraction of blood in liver)	0.12 ± 0.01	0.17 ± 0.01
K_{bl}	$1.4 \pm 3.0 \ 10^{-3}$ h^{-1}	$17 \pm 4 \ 10^{-3}$ h^{-1}
K_{lb}	37 ± 61	13 ± 9
K_{br}	87 ± 21	27 ± 10
K_{rb}	0.60 ± 0.15	28 ± 24
K_{rf}	12 ± 14	7 ± 2
k_{lu}	150 ± 750	5.4 ± 0.5

Source: Williams, L. E. et al., *Cancer Biother. Radiopharm.*, 23, 6, 2008. With permission.

Volumes were, as expected, different by about three orders of magnitude in going from mouse to man. In both species, the volume accessible to the intact Mab was approximately the total blood volume. This parameter was necessary since blood was sampled on a per ml basis [$a(t)$] such that the total amount of activity in the blood had to be estimated via the model. The fraction of the blood volume in the liver was on the order of 15% if we average over both animals and patients.

Generally, kinetic parameters were in approximate agreement when we consider the size of the uncertainties—particularly among the patients. There was one exception to this set of comparable results. A 20-fold increase was seen in the excretion parameter k_{lu} in going from man to mouse. This disparity can be explained by the use of an iodine label in the murine study. Urination of the radioactive metabolic residuals out of the body was much more pronounced in the case of iodine compared with indium-111. Our conclusion at the end of this analysis was that the two species handled the intact Mab in approximately the same way. The only exceptions were the total volume of distribution and the excretion rate constant from the kidneys.

11.7 SUMMARY

There are two issues at the heart of animal experimentation. One is the scientific question of the relevance of animal data with regard to any future human applications. In the present environment, such correlations are not required and are not of interest to many investigators. Instead, the animal results are seen only as a necessary step in obtaining regulatory approval of the clinical pharmaceutical trial. I have argued that, instead, the applicant must eventually provide evidence for the correspondence between the murine (or other animal) and human biodistribution data sets. Such closure should be part of the review process and the correlation—or lack thereof—would then be available as a public document. In other words, the applicants would be required to investigate this issue and subsequently report in an explicit forum. The long history of pharmaceutical company secrecy would be broken by this methodology. Yet there would be compensating advantages in that, for particular types of agents, animal trials could simply be avoided such that the human becomes the "test animal" *a priori*.

A second issue is the use of animals in general. Unless the investigators demonstrate an explicit correspondence between animal and human data, there is a basic question of the justification of the animal trial. Ethics extends beyond the human species. With a novel type of RP or unlabeled agent, such assurances cannot be given and must, of course, be made part of the scientific process. After the method of closure cited herein is instituted, however, the number of irrelevant animal protocols can be reduced. For classes of agents whose correspondence has been shown to be invalid, the animal study should not be permitted. Eventually, there would then be a tabulated set of comparisons of various types of agent and the correlation between mouse (or other animal) and man in each case.

This chapter shows that there are relatively few allometric relationships of either a volumetric or kinetic type in the RP literature. Some knowledge of protein uptake and clearance parameters has been established over a range of MW and molecular species. Generally, the uptake parameter goes as the inverse of the total body mass (M), whereas the late half-life is often found to be a slowly varying function of the animal's size. The latter correspondence is encouraging as it permits absorbed dose to be reliably estimated before the start of a clinical protocol.

Because of the fundamental issue of absorbed dose and cancer therapy, one primary interest in allometry is whether there is evidence that the uptake of an RP by human tumors shows a mass dependence similar to that found in murine studies. In the latter (Chapter 3), an inverse variation with tumor mass is seen with an exponent between -0.33 and -0.50. These correlations occurred in the cases of liposomes and intact antibodies. Several clinical studies using intact antibodies have indeed shown that the human lesions do, in general, follow this rule. Work at UC–Davis, using a single patient with multiple melanoma sites, demonstrated a variation of $u_T(t)$ with tumor mass that corresponded to an exponent of approximately -0.60. A comparable study, done at Aachen University with a group of 19 patients, also showed a power-law correlation of $u_T(t)$ with tumor mass. In that situation, the exponent was closer to -0.40 for the patient cadre. In the Aachen data set, there were confounding factors of patient variation, iodination, and time delays, but the overall result appeared to correspond to earlier murine studies done at City of Hope. In fact, if we accounted for the average difference between CEA concentration in the murine xenografts and the wild-type (patient) lesions, the curves of tumor uptake versus tumor mass were essentially similar for both species.

Analytic complexities, due to multiple kinetic variables, have bedeviled the comparisons between animal and human since the beginning of allometry. Our group has attempted to derive a single parameter method to compare one blood uptake curve with another. The first moment of the curve is given here as an example of this type of analysis. Using antibody data from City of Hope studies, it is found that this moment is indeed comparable between species for intact antibodies but begins to differ if the MW goes far below 150 kDa. As we approach lower MW engineered proteins, there is approximately a factor of threefold difference in magnitude between mouse and human blood first moments with the former being smaller than the latter. This disparity might be traditionally rationalized as due to faster circulation times in the smaller species, yet this same argument would not hold if carried to the intact form of the protein. Thus, the allometric relationship must be given as an empirical function of protein MW.

REFERENCES

Chatal, J. F., J. C. Saccavini, et al. 1989. Biodistribution of indium-111-labeled OC 125 monoclonal antibody intraperitoneally injected into patients operated on for ovarian carcinomas. *Cancer Res* **49**(11): 3087–94.

de Bree, R., D. J. Kuik, et al. 1998. The impact of tumour volume and other characteristics on uptake of radiolabelled monoclonal antibodies in tumour tissue of head and neck cancer patients. *Eur J Nucl Med* **25**(11): 1562–5.

Dedrick, R. L., K.B. Bischoff and D.S. Zaharko. 1970. Interspecies correlation of plasma concentration history of methotrexate (NSC-740). *Cancer Chemotherapy Reports* **54**: 95–101.

Haak-Frendscho, M., S. A. Marsters, et al. 1994. Inhibition of TNF by a TNF receptor immunoadhesin. Comparison to an anti-TNF monoclonal antibody. *J Immunol* **152**(3): 1347–53.

Huxley, J. S. 1932. *Problems of Relative Growth*. New York, Dial Press.

Kelley, S. K., T. Gelzleichter, et al. 2006. Preclinical pharmacokinetics, pharmacodynamics, and activity of a humanized anti-CD40 antibody (SGN-40) in rodents and non-human primates. *Br J Pharmacol* **148**(8): 1116–23.

Khor, S. P., K. McCarthy, et al. 2000. Pharmacokinetics, pharmacodynamics, allometry, and dose selection of rPSGL-Ig for phase I trial. *J Pharmacol Exp Ther* **293**(2): 618–24.

Lin, Y. S., C. Nguyen, et al. 1999. Preclinical pharmacokinetics, interspecies scaling, and tissue distribution of a humanized monoclonal antibody against vascular endothelial growth factor. *J Pharmacol Exp Ther* **288**(1): 371–8.

Lindstedt, S. L., and P. J. Schaeffer. 2002. Use of allometry in predicting anatomical and physiological parameters of mammals. *Laboratory Animals* 36: 1–19.

Macey, D. J., S. J. Denardo, et al. 1988. Uptake of indium-111-labeled monoclonal antibody ZME-018 as a function of tumor size in a patient with melanoma. *Am J Physiol Imaging* **3**(1): 1–6.

Mahmood, I. 2009. Pharmacokinetic allometric scaling of antibodies: application to the first-in-human dose estimation. *J Pharm Sci* **98**(10): 3850–61.

Mahmood, I. and J. D. Balian. 1996. Interspecies scaling: predicting clearance of drugs in humans. Three different approaches. *Xenobiotica* **26**(9): 887–95.

Miller, G. 2009. Science and society. Hundreds gather for rally to defend animal research. *Science* **324**(5927): 574.

Mordenti, J., S. A. Chen, et al. 1991. Interspecies scaling of clearance and volume of distribution data for five therapeutic proteins. *Pharm Res* **8**(11): 1351–9.

Richter, W. F., H. Gallati, et al. 1999. Animal pharmacokinetics of the tumor necrosis factor receptor-immunoglobulin fusion protein lenercept and their extrapolation to humans. *Drug Metab Dispos* **27**(1): 21–5.

Sundin, A., H. Ahlstrom, et al. 1993. Radioimmunolocalization of hepatic metastases and subcutaneous xenografts from a human colonic cancer in the nude rat. Aspects of tumour implantation site and mode of antibody administration. *Acta Oncol* **32**(7–8): 877–85.

West, G. B. and Brown, J.H. 2004. *Life's universal scaling laws*.

Williams, L. E., R. B. Bares, et al. 1993. Uptake of radiolabeled anti-CEA antibodies in human colorectal primary tumors as a function of tumor mass. *Eur J Nucl Med* **20**(4): 345–7.

Williams, L. E., B. G. Beatty, et al. 1990. Estimation of monoclonal antibody-associated 90Y activity needed to achieve certain tumor radiation doses in colorectal cancer patients. *Cancer Res* **50**(3 Suppl): 1029s–1030s.

Williams, L. E., R. B. Duda, et al. 1988. Tumor uptake as a function of tumor mass: a mathematic model. *J Nucl Med* **29**(1): 103–9.

Williams, L. E., G. Lopatin, et al. 2008. Update on selection of optimal radiopharmaceuticals for clinical trials. *Cancer Biother Radiopharm*.

Wu, A. M. and P. J. Yazaki. 2000. Designer genes: recombinant antibody fragments for biological imaging. *Q J Nucl Med* **44**(3): 268–83.

Yoshida, K., M. Rivoire, et al. 1990. Radiolocalization of monoclonal antibodies in hepatic metastases from human colon cancer in congenitally athymic mice. *Cancer Res* **50**(3 Suppl): 862s–865s.

Summary of Radiopharm- aceuticals and Dose Estimation

12

12.1 INTRODUCTION (CHAPTER 1)

Among many possible radiopharmaceutical (RP) agents, contemporary molecular imaging and therapy studies have been involved with liposomes, antibodies, segments of RNA and DNA, selective high-affinity ligands (SHALs), and a variety of nanoparticles. The number of reports has now grown to thousands of published articles on these targeting structures. The entries are not mutually exclusive as combinations are possible. More generic types will certainly be developed, so the question of which agent is best for a given role is unclear today. Historically, figures of merit based on ratios of uptake or of ratios of areas under the curve (AUCs) have been generated as an *ad hoc* quality indicator for imaging and therapy, respectively. As has been shown, such historical figures of merit (FOMs) are not acceptable as true descriptors because of their ratio formulations. Ratios lead to a cancellation of absolute factors, so the actual amount in the lesion or target site is lost in the analysis. In other words, whereas the ratio may be large, the numerator may still be so miniscule as to make imaging or therapy impossible in any finite time or with any realistic amount of activity, respectively.

The essence of the nuclear medicine drug development situation is that there are many possible RP agents, yet relatively few patients to try them on. In the proteins alone, literally astronomical numbers of different engineered antibodies might be constructed. To preselect as many of the agents as possible, animal studies must be compared to find a possible optimal agent for the clinical trial. Among these, multiple engineering and other techniques allow optimization. An additional aspect of these agents is the possibility that radiation absorbed dose might be provided by the RP. This dose is important in diagnostic work but is absolutely vital in therapy practice. Thus, unlike typical pharmaceuticals, the RP may demonstrate an additional aspect in its clinical performance. Yet direct methods to perform absorbed dose accountings are difficult to implement and may be dangerous to the patient. As a result, indirect and necessarily complicated calculations are required to estimate the associated radiation doses.

The primary application of the various agents in this text has been in the targeting to tumors distributed in the patient's body. Such situations are generally untreatable at the present time. As mentioned already, it is anticipated that some fraction of the lesions will not be observable with any nuclear or anatomic imaging modality. This problem arises because the resolution of the system is comparable to or larger than the intrinsic size of the object within the patient's body.

In gamma camera and positron emission tomography (PET) scanners, this imaging difficulty is termed the count recovery issue. For small enough lesions, any nuclear imager will detect only a fraction of the photons emitted by the tumor source. For the gamma camera the spatial resolution value monotonically increases with depth into the patient. At the patient's surface, the value is approximately 1.5 cm. For PET, the resolution is essentially fixed as the source moves inside the body and is on the order of 0.5 cm. As depth increases, the gamma camera (and its associated collimator) will show considerably worsening resolution with results approaching several cm at 10 cm inside a tissue-equivalent phantom. As the tumor or other small-size source approaches or becomes less than these resolution numbers, its image becomes confused with the blood or other background activities in the animal or patient.

We should note that computed tomography (CT) and magnetic resonance imaging (MRI) provide sub-mm resolution throughout the phantom or patient anatomy. The numbers shown in the resultant images are not determined by targeting of a molecular marker but rather because of intrinsic properties of the imaged object. In the case of CT, it is the linear attenuation coefficient, in Hounsfield numbers, that is tabulated in the various anatomical sections. For MRI, proton density moderated by the decay times of the proton signal are tabulated in such a spatial registry. Neither anatomic imager however can show a quantitative spatial distribution of particular molecules in the patient. Thus, these sharp-eyed anatomic imagers cannot, at the present time at least, be used in lieu of the fuzzy-eyed nuclear images based on receptor density and other physiological parameters. In recent times, the anatomic imagers have been brought into action as a complementary device installed on the same imaging bed as the nuclear to provide the single-photon emission computed tomography (SPECT)–CT or PET–CT technology. Future directions will involve SPECT–MRI and PET–MRI.

12.2 ANIMAL RESULTS (CHAPTER 2)

Biodistributions are a requirement of RP investigations. Tumor uptake results, generally given in terms of the decay-corrected $u_T(t)$ uptake variable, are conventionally quoted in %ID/g in an animal study. In most instances, the model is an appropriate mouse. For antibodies, the standard animal has been the nude rodent (mouse or rat), which lacks T cells and thereby allows human tumor implantation. Proof of targeting is discerned when u_T exceeds a uniform distribution situation. In the murine model, this is 4 to 5 %ID/g if the agent is free to move throughout the anatomy and is not rapidly sequestered or excreted. Corresponding human values are approximately three orders of magnitude smaller.

Unlike conventional pharmaceuticals, which are usually followed only in the bloodstream, RP agents permit data to be acquired on a per-organ basis via counting gamma ray emissions. Statistically, the best method is to image the same test group of animals over a period of time. Miniature scanners (both gamma camera and PET) are important in this work. At the end of the imaging, some animals may then be sacrificed to calibrate the system so that absolute values of activity per gram of tissue in each organ as a function of time [$a(t)$ in units of %IA/g] are made available. If we integrate over the organs, the total activity [$A(t)$] in each is also available. Notice that these last two parameters are not corrected for decay.

To make biodistribution data clearest to other researchers, RP inventors correct their measured activity [$a(t)$] results for radiodecay. Hence, any published biodistribution shown [$u(t)$] will be for a hypothetical material that supposedly exists in the animal or patient. We refer to these results as those of the pharmaceutical. It may be, of course, that the radioactivity has become detached from the original, intact test material and the imager is following the track of only the radiolabel, or a labeled metabolite, and not anything more. For the physicist, this nuance is usually of no concern since the entity of interest is the activity—not the location of the chemical moiety. For the developer of novel agents, the difference between location of activity and location of moiety is much more important. In the case of iodine labels these differences can become vary large over extended periods in vivo.

12.3 FIGURES OF MERIT FOR CLINICAL TRIALS (CHAPTER 3)

Multiple possible agents and their required preclinical evaluations lead to large numbers of biodistributions. Using engineering practice as our guide, the most important concept in comparing various uptake data sets is the use of one or more FOMs. Though fundamental in an aircraft or ship design context, such practical evaluations are not yet common in biological engineering. Unlike ordinary pharmaceutical studies, RP evaluations involve both imaging and therapy of the tumor or other target entity. Thus, two figures of merit need to be applied. Historically, investigators have used a simple ratio (r) of tumor to blood uptake (u_T/u_B) as a necessary and sufficient FOM for tumor imaging. Similarly, the traditional therapy figure of merit has been the corresponding ratio of integrals of activity: AUC(tumor)/AUC(blood). Here, blood is being used as a surrogate for the red marrow (RM) or, more generally, for the body background. In either imaging or therapy, other denominators may be used in lieu of the blood. Sensitive organs such as the liver or kidneys may be substituted, and corresponding simple ratio figures of merit can be computed. Yet simple ratios such as these are limited and can lead to paradoxical results.

Because of these restrictions, a new imaging figure of merit (IFOM) has been derived (Williams, Liu, et al. 1995) based on the statistical likelihood of finding a lesion in a blood or other tissue background. Explicitly

$$\text{IFOM} = 1/\Delta t = CV_0(z)^2 \varepsilon u_T \exp(-\lambda t)A_0 Vol[1 - 1/r]^2/[1 + 1/r] \qquad (12.1)$$

Details of this derivation as well as definition of the various symbols are given in Chapter 3. This IFOM has been shown to correlate with the clarity of the tumor image in both a limited set of animal data as well as in Monte Carlo camera–collimator simulations. Such results are important since they cannot be shown to be true for the r index. Likewise, the IFOM clearly shows the importance of the half-life of the radioactive label as well as its photon energies. Notice that this indicator can be applied to novel situations such as using a radiolabel that has not been evaluated in an experiment—or to a new camera–collimator design that is to be built in the future.

Advantages of the IFOM over the traditional value r are several-fold. First, one may use the new figure of merit to predict the optimal time for imaging a given RP that is labeled with a particular radionuclide. This prediction is made incorrectly using the r index as is shown in the Monte Carlo simulations of two constructs in Chapter 3. One may, theoretically, change the radioactive tag on a given agent to see if the imaging will be improved at a particular time for any particular detector system. Also, one may compare radioiodinated and radiometal-labeled versions of the same pharmaceutical. In the latter case, these comparisons have been generalized to PET imaging. Here, we make the assumption that a 3-D system uses a set of 2-D projections to generate the tomographic sections.

As was shown in Chapter 3, the r value for each of the five iodinated cognate anti-carcinoembryonic antigen (anti-CEA) antibodies approached infinity as time increased. This behavior was not true for the IFOM. In fact, IFOM decreased as both time and r go to infinity due to the exponential decay factor in its numerator as seen in Equation 12.1. Again, the use of a ratio to predict optimal imaging times was inherently incorrect, and the IFOM was a manifestly superior indicator of image quality.

A novel therapy figure of merit (TFOM) is also now available to replace the traditional (R) value in the literature (Williams et al. 1995). As in the case of the imaging figure of merit, the newer concept is generated to evaluate the statistical difference of absorbed doses between tumor and blood. Explicitly, we have

$$\text{TFOM} = \frac{k[1-1/R]^2 \text{AUC}(u_T \exp(-\lambda t))}{1+1/R} \qquad (12.2)$$

More details on this derivation and definition of the symbols in Equation 12.2 are given in Chapter 3. This novel TFOM has several advantages over R. Primary among these is that the ratio R may become arbitrarily large due to a vanishingly small denominator (blood curve). Yet such behavior may not indicate a useful therapeutic agent since the numerator, the tumor dose, may also be very small. Likewise, due to use of a ratio in R, the energy of the emitter is lost in the analysis. Herein we will vary the k parameter, which includes the mean beta energy, in an analysis of antibody fragments.

TABLE 12.1 Variation of Blood Kinetics via Changing the Fc
Segment of an Engineered scFv–Fc Antibody to CEA

ANTIBODY	A_α (%ID/g)	$T_{1/2\alpha}$ (h)	A_β (%ID/g)	$T_{1/2\beta}$ (h)
Wild Form	23.3	4.22	22.4	289
H435R	19.5	1.87	26.6	83.4
H435Q	20.9	1.53	28.0	52.6
I253A	28.9	2.31	18.8	51.7
H310A	32.6	1.66	20.2	27.2
H310A/H435Q	25.2	0.86	32.3	8.0

Source: Kenanova, V. et al., *Cancer Res.*, 65, 2, 2005. With permission of
the American Association of Cancer Research.

Several recent reports from Anna Wu and her colleagues at the University of California–Los Angeles (UCLA; Kenanova, Olafsen, et al. 2005) demonstrated the use of molecular engineering to affect the kinetics of anti-CEA proteins in the blood of animals. In this work, the authors concentrated on the Brambell receptor (BR) (Brambell, Halliday, et al. 1958). It is generally considered that the Brambell receptor is responsible for the transference of immunity from mother to fetus. Relevant to protein engineering, the BR is also held to moderate the circulation times in the blood of an antibody (or antibody-like structure) that contains an intact Fc region. By referring to Chapter 3, it may be seen that a number of the five-member anti-CEA cognate set developed on the original murine T84.66 antibody do not contain the Fc region. Two examples are the diabody and minibody. By reducing the binding of the Fc fragment of the designed antibody to the Brambell receptor, one can greatly reduce both the $T_{1/2\alpha}$ and $T_{1/2\beta}$ of the blood curve of the construct. Table 12.1 contains a summary of five variants on the single-chain variant (scFv)–Fc construct as well as the wild type or natural form of this structure. It was seen that the $T_{1/2\alpha}$ decreased by a factor of approximately 5-fold and the $T_{1/2\beta}$ by a factor of almost 40-fold by going to the double-mutated H310A/H435Q form of the engineered protein. If we use the single parameter (first moment of the blood curve; cf. Chapter 3), the ratio of first moments between the wild-type scFv–Fc and the double mutation is 411 h/11 h, or 37.3. It was noteworthy that this effect was produced by only two amino acid changes on the wild-type protein.

Following this initial work, biodistributions were obtained in nude mice with implanted CEA-positive tumors. The three fastest-clearing scFv–Fc agents were tested in these animals: I253A, H310A, and the double mutant H310A/H435Q. Lesion size was constrained to be approximately 100 mg in these studies using both ^{125}I and ^{111}In labels on the injected fragments (Kenanova, Olafsen, et al. 2007). Results for the TFOM—including the mean energy of the beta emissions for possible ^{131}I or ^{90}Y radionuclides—are given in Figure 12.1. It was found that the optimal agent was H310A with a relative value of approximately 5,000 with the yttrium label. Both the I253A and double mutant had TFOM (^{90}Y) values near 4,000. Iodine-131-labeled fragments had TFOM values that were almost an order of magnitude smaller than those seen with yttrium-90. This analysis assumed that all the emitted energy was absorbed in the lesions.

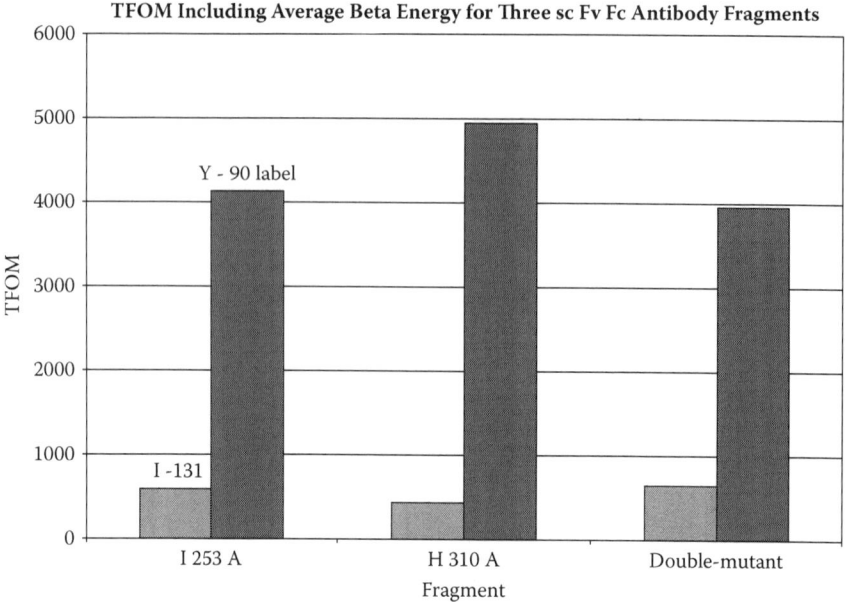

FIGURE 12.1 Therapy figure of merit (TFOM) for three scFv-Fc mutant antibodies using a nude mouse model xenografted with LS174T human colorectal carcinoma. In this case I have included the mean beta energy in the TFOM as well as the different biodistributions of the engineered proteins (Data adapted from Kenanova, V. et al., *Cancer Res.*, 67, 2, 2007.

12.4 ABSORBED DOSE ESTIMATION (CHAPTER 4)

Radiopharmaceuticals have an important and unique aspect that is roughly analogous to one found in traditional chemotherapeutics. An RP provides ionizing radiation and hence absorbed dose at locations corresponding to target molecules or structures. If the analyst integrates the time–activity curve in a given organ, the toxic effects of ionizing radiation produced locally by the RP may be estimated. In addition, there is the possibility that ionizing radiation may come from a number of other tissues or organs in the same body. Usually this contribution is due to photons coming from such sources, but abutting tissues may also send ^{90}Y beta radiation over distances of 10 mm. Thus, the total absorbed dose depends on activity locally as well as that deposited elsewhere in the animal or patient. Overall, the toxic effect is quantitated, to lowest order, using the concept of absorbed dose; that is, the ratio of energy is deposited in a tissue of interest divided by the mass of that tissue. The standard unit is the gray, which is defined as 1 joule per kilogram. Imaging absorbed doses are typically 1 to 5 cGy per protocol.

Therapy, on the other hand, requires lesion doses beyond 50 Gy for most solid tumors. Malignancies in B cells are much more sensitive to ionizing radiation dose (5–10 Gy may be effective); this result has led to a number of important clinical applications in non-Hodgkin's lymphoma (NHL) that are becoming a standard part of the therapy of B-cell diseases (Vose, Wahl, et al. 2000; Wiseman, Leigh, et al. 2003).

Almost all dose estimates done in nuclear medicine are based on a standard form of computation that divides the activity aspects from the energy and spatial parameters. In the canonical form of the Medical Internal Radiation Dose (MIRD) Committee, the average organ dose (D) is given by the matrix product

$$D = \mathbf{S} * \tilde{A} \tag{12.3}$$

where $\mathbf{S}$ is an array that includes the probability of energy deposition at the organ of interest, and $\tilde{A}$ is the vector set of areas under the curve for the source organ activities $A(t)$. More information on the makeup of $\mathbf{S}$ is included in Chapters 4 and 7. Generally, phantoms are used to generate $\mathbf{S}$ so that the geometry of source and target is well defined spatially. Such results are available at the RADAR site (Stabin 2003) for a number of phantoms and various emitters commonly used in nuclear medicine.

In principle, the result shown in Equation 12.3 can be generalized to any geometry. For example, Auger emitters within the cytoplasm or alpha emitters on the cell or nuclear membrane may be evaluated as to their dose estimates using the matrix product approach but with different structures (and geometries) being involved. In the case of such short-range charged particles, subcellular geometric entities would be appropriate for estimation of microscopic absorbed dose values. Obtaining biodistributions at such small scales is only possible using a biopsy approach due to the finite resolution of nuclear imaging systems described earlier.

In Chapter 4, it is shown that absorbed dose is probably only a crude, first approximation to the actual effect of the ionizing radiation in any living tissue. Associated (usually enhancing) factors include parameters such as the type of ionizing particle, the spatial distribution of dose within the tissue, and the rate of delivery of the ionizing energy. The last of this triad is studied in relation to the cellular division rate in both normal and tumor tissues. The final outcome of these analytic aspects has become the biologically effective dose (BED).

One unfortunate term that has been allowed, perhaps even encouraged in the dose assessment world, is to describe any and all of these estimated absorbed dose values as *dosimetry*. Compounding that turn of phrase is a general lack of probable errors in these estimates. One sees doses given in the literature that have no ± values associated with them. Only actual dose measurements using calibrated dosimeters should ever be called dosimetry. Chapter 4 gives one example of this type of dose measurement using a humanoid phantom. Clearly, there will be limitations that restrict the use of physical dosimeters on living animals or patients due to both scientific as well as ethical considerations. It is hoped that such measurements will be carried out more extensively in the future so that the dosimetry literature can grow and then be compared with the calculations generally required in nuclear medicine practice.

12.5 DETERMINING ACTIVITY AT DEPTH IN THE PATIENT (CHAPTER 5)

A classical problem of nuclear medicine has been the inability of its practitioners to find the absolute amount of activity at a given location in the body of the subject. A typical nuclear radiologist reads the study out as a set of relative accumulations in the tissues under review. One aspect of this issue is the ambiguity in planar imaging. A weak source at a shallow depth may give a similar signal as a strong source at a greater distance from the collimator face. Astronomers would have similar problems if they did not use the red shift of the emitted photons to determine absolute distance from source to observer. Also, the astronomer usually contends only with relatively small sources of radiation and not ones that subtend large angles as seen from the detector face. Because of the size of these sources, there is inevitable overlap with consequential confusion of identifying what anatomic source is responsible for which detected photon.

As noted in Chapter 5, there is still no optimal method for quantitation of activity after more than 50 years of nuclear medicine technical and mathematical evolution. A set of six different techniques is listed in Chapter 5. Generally, several of these methods are customarily applied simultaneously to a given animal or patient study. One question in any of the strategies is the amount of time required to perform the quantitative assay of the amount of radioactive emitter in a particular tissue or lesion. A second item of interest is the accuracy of such assays.

In planar imaging, the ambiguity can be resolved, at least in principle, by rotating the gamma camera around the sources. Yet such SPECT studies take extensive time and require that the patient remain immobile for yet more imaging. Several simpler methods have been proposed to resolve the situation of determining the amount of activity at depth. The oldest is use of the geometric mean (GM) of two counts obtained at opposite sides of the patient. Typically, the anterior and posterior projections are the best for this purpose because a human is thinnest in this direction. It may be, of course, that a given source can be seen only from a single direction to make the GM method impossible to apply. Even if two opposite views are available, the method is difficult to use in a situation where more than one emitting source lies along the direction (mathematical ray) of interest. If these two circumstances can be circumvented, absolute quantitation is on the order of ± 30% for the GM method.

Use of PET imaging resolves the ambiguity of activity measurements by having, due to the presence of back-to-back photons, a physical ray passing through the emitting body. Thus, every coincidence event defines this axis. Sensitivity is known since the imaging device has a calibrated attenuation correction for the total distance that the photon pair has traversed during its passage from source decay to the simultaneous pair of detection events.

Yet PET is not expected to become a panacea for absolute activity detection. Due to the limited number of positron emitters that exist in nature (cf. Table 1.1), a clinical practitioner may be hard pressed to employ PET technology in the case of a novel radiopharmaceutical. If a cyclotron is at hand, ^{11}C, ^{13}N, and ^{15}O may be produced and quickly labeled onto the agent of choice. For those a bit farther away from the accelerator, ^{18}F can still be generated and taken by courier to the clinical site. Longer-lived examples such as ^{64}Cu and ^{124}I allow airfreight to ship the label to the user. Even in these cases, the probability of positron emission is strictly limited due to the competition of electron capture by the radioactive nucleus. For ^{124}I, an interesting possible label due to the ease with which a halogen may be attached to biological molecules, the probability of positron emission is only 23%, so large amounts (5x) of the radionuclide must be purchased to provide the same number of decays per unit time as with a given activity of ^{18}F. This isotope of iodine also has multiple high-energy photons that make its practical use difficult due to interference with the 511 keV photons from its positron decay.

To bring quantitative uptake into common clinical practice, two novel ideas are undergoing current development. Both methods use coregistered CT (or MRI) anatomic data sets to allow better estimation of activity at depth, such as in overlapping organs as seen in a planar view. An example of a first-order strategy is CT-assisted matrix inversion (CAMI), which may be implemented on a planar imaging device (Liu et al. 1996). Since the algorithm has the geometric data built into the analysis, this technique allows the user to assess activity values from whole-body images, for example, with accuracies on the order of ± 7%. Scatter radiation near strong sources is one limitation to the method. A more common technique is quantitative single-photon emission computed tomography (QSPECT) (Koral, Dewaraja, et al. 2003), which uses calibration sources and knowledge of the attenuation, scatter, and collimator construction to permit accuracies approaching ± 5% in humanoid phantoms.

Both of these camera-based methods have difficulty if the size of the source (emitter) is ≤ twice the resolution of the imaging system. Here, as in PET imaging at this size level, the reconstructed activity value is systematically less than that in the source, and a recovery coefficient must be invoked to correct for this undercounting. Such coefficients are essentially *ad hoc* and must be determined on site with a given imaging system using a set of spherical phantom lesions that can be filled with activity of various types and concentrations. In such cases, a background level of activity is also necessary to simulate the clinical situation.

The net result of imaging is a set of activity values $A(t)$ for various tissues as a function of time. In addition, the observer usually has blood activity concentration values (in units of MBq/cc). This additional datum is essentially $a(t)$ for the blood. The next step in dose assessment is to find the area under the curve for each of these tissues and also for the blood. Such interpolation and extrapolation imply a need for a mathematical model.

12.6 MODELING OF BIODISTRIBUTIONS AND OTHER DATA (CHAPTER 6)

Modeling, in some sense of the word, is a necessity in the handling of biodistribution data. A mathematical formula to represent blood data has a long history in unlabeled pharmaceutical research. In the case of an RP, a mathematical representation is needed in integrating activity information to go from $A(t)$ to $\tilde{A}$. Yet there are other aspects. One is the possibility of finding errors in a biodistribution data set. Interpolation of measured data is another useful result from the modeler's perspective. Finally, there is the possibility that one data set can be compared with another using the modeled parameters rather than the raw information. Such comparisons can even extend, if we use the concept of allometry, from one species to another. This last correlation is one of the most important in the evaluation of a novel RP.

It is pointed out in Chapter 6 that modeling may be done on data as collected [i.e., $a(t)$] or data corrected for decay [i.e., $u(t)$]. Because of the possibility of certain nonlinear terms in the model's equations, the former data are mathematically required in the analysis. In addition, uncorrected data have statistical validity that is more transparent since they are not inflated by correcting for decay. This limitation of the data type prior to analysis is particularly important for radiolabels with short half-lives such as those used in PET.

Two fundamental styles of models occur. Compartmental modeling, which resembles a chemical engineer's view of reality, considers all the tissues communicating to each other via a set of reactions that are generally first order. An unknown rate constant (represented as an "arrow") then governs motion from one compartment ("box") to another; generally, a second constant describes the reverse motion. This simultaneous assay of all the data is sometimes difficult to implement due to the complexity of the resultant differential equations. It is no surprise, therefore, that noncompartmental or "open" models are probably the most frequently seen in the RP literature. Here, each organ is separately represented by a functional form that has some intellectual or statistical appeal to the investigator. The most common format is a set of two or more exponential functions fitted to each time–activity curve. Such exponential models, it should be noted, arise naturally out of the mathematics of the set of coupled differential equations that describe the multiple organs or tissues in a compartmental or "closed" system. In practice, it is found that more and more exponentials have to be added to a given organ's analysis as observations are extended to longer and longer times. In the days prior to computers, this phenomenon led to the concept of curve stripping whereby the slowest rate constant (equivalent to the longest half-life) was determined first, the next slowest rate constant second, and so forth. By looking at the various regions of the multiexponential curves, one can find various phases of uptake and clearance of the RP.

One important alternative to a multiexponential model is the power law. It can be shown that the power-law function is an integral over a set of exponentials with a weighting function that is the rate constant considered as a continuous variable. Power laws are an attractive result for the mathematically minded analyst since the organ curve now becomes a function whereby all temporal intervals look the same; that is, we have an equivalent of fractal behavior—albeit in time rather than space. By looking at a power-law representation, there is no "early phase" or "late phase" result, only a single curve given by two parameters. In a log-log plot, this curve becomes a straight line. We should add that only clearance curves are described via the power-law format—and only over a limited range of time. These limitations are described more explicitly herein.

One interesting power-law result outside the temporal realm was the demonstration that, at a given time, tumor uptake u_T (%ID/g) is given by a power-law function of tumor mass. Over a range probably extending down to sizes on the order of diffusion distances and out to dimensions of centimeters, a number of both animal and human tumors have been documented to have such variations. Originally done in the liposome and intact antibody contexts (Williams, Duda, et al. 1988), a similar argument should hold for most cases. A heuristic explanation, essentially assuming that tumor uptake is determined by the ratio of lesion vascularized area divided by lesion volume, was presented in Chapter 3. A caveat was also introduced in that not all tumors are of this format; in particular, those that are uniformly perfused as they develop will show uptake that is independent of mass. Tumors that have been implanted into nude rodent livers (via injection into the spleen) are an example whereby perfusion, and the u_T values, are independent of lesion size.

One impact of the power-law result is that radiation absorbed dose rate is highest in the smallest lesions. If we assume a variation of the form

$$u_T = u_0 m_T^{-1/3} \tag{12.4}$$

the resultant increase in absorbed dose rate is approximately a factor of 2.5-fold for a 10-fold reduction in tumor size. Thus, the smaller lesions are expected to be the most susceptible to radiation therapy via targeted RP agents. Because of the limitations of nuclear imaging of small objects as already noted, such optimal lesions may not even be clearly detectable in the patients. An ironic implication is that an unobservable tumor may be the one most susceptible to ionizing radiation therapy. Adjuvant therapies may therefore become the best application of tumor-targeting agents. One can conjecture that a similar argument may hold for chemotherapeutic agents due to similar restrictions on their access to the tumor volume. This conjecture is significant and should be tested using radiolabeling of known chemotherapeutic molecules.

Power-law applications in other areas of modeling lead us to the concept of fractal behavior. As mentioned, a power law in the time realm implies that the system behaves the same whether one observes its operation at the earliest or the latest phase of RP clearance. Yet power laws must break down at both ends of their useful (and limited) range. At very short times we cannot use the relationship due to inadequate mixing of the tracer into the blood and other tissues. Also, at short times, there is an uptake phase of the RP in organs such that a monotonic decreasing function as Equation 12.4 cannot be used. Note that we would replace m_T in Equation 12.4 with time (t) to represent clearance of a tumor or normal organ. Finally, at sufficiently long times, no signal can be detected above background to make a power law difficult to apply. It is important to find the limits of such log-log correlations and to determine new relationships that may occur at these endpoints. Similar arguments may be made for the limited mass range for the $u_T(m)$ law of Equation 12.4.

Convolution is another form of temporal modeling whereby the tumor (or organ) uptake curve is represented as a convolution of the measured blood curve with a second, but unknown, function. This last variable is called the impulse–response function [$h(t)$], and it may be used, as with IFOM, to compare various RP agents. By using impulse–response functions, the analyst can compare all relevant agents on a theoretically equal footing. For example, it may occur that certain low-mass proteins are so rapidly excreted via the kidneys that there is little opportunity for targeting to the intended tumor sites.

In Chapter 3, the cT84.66 diabody was shown to be this type of agent. Yet if one calculated the impulse response function, the diabody is found to be the optimal molecular form among these five anti-CEA cT84.66 cognates. By this we mean that the $h(t)$ function is the broadest in time and highest for the diabody among the five molecular forms for cT84.66. Thus, if the diabody could be kept from being excreted, it would have the highest tumor uptake and the maximum IFOM value. One method, becoming popular in the literature, is to use stealth technology such that the circulating protein is enclosed within a layer of polyethylene glycol (PEG) to make renal excretion less probable. An example of this type of stealth diabody has been published by our group (Li, Yazaki, et al. 2006).

From a physics perspective, one important aspect of mathematical representations is to predict behavior of a system for conditions not used to develop the model. In other words, does the model do more than simply represent a set of data? In this vein, several groups have considered the case of multiple tumors in a single laboratory animal. Because of the complexity of the problem, it is most easily handled if we solve only for the activities in the multiple lesions in the early phase of the study (i.e., out to 48 h in the case of intact antibodies). At the City of Hope, we studied only the forward movement of the tracer (Williams, Beatty, et al. 2001) and neglected completely reflux back from the tumor and normal organ targets. This limited form of analysis is analogous to the Patlak strategy (Patlak, Blasberg, et al. 1983) used in the determination of tumor uptake in patients receiving ^{18}FDG.

In this limited analytic approach, we first fitted the model parameters to the single-lesion case using nude mice with LS174T human colorectal cancer xenografts. Each animal in the initial study had only one implanted tumor, and the assay time was fixed at 48 h after the injection of the RP. Some of the animals, however, were also pretreated with relatively large amounts of the same unlabeled protein (200 μg of T84.66 antibody to CEA). The idea was to see if this pretreatment regimen reduced the amount of available CEA at the hepatic site to permit more labeled antibody, given subsequently, to target to the single lesion. Both untreated and specific antibody pretreatment (SAP) single-tumor groups were fitted to a set of forward-only modeling equations

$$dA_T(t)/dt = k_0 m^\alpha b(t)$$

$$dA_L(t)/dt = [k_{L0} + \varepsilon m^\beta] b(t)$$

$$dA_R(t)/dt = k_R b(t)$$

$$db(t)/dt = -[k_0 m^\alpha + k_{L0} + \varepsilon m^\beta + k_R] b(t)$$

(12.5)

where subscripts T, L, and R refer to tumors, liver, and residual body of the nude mouse, respectively, and $b(t)$ is the blood curve. Notice that we are using the activity in the various tissues (A) and not the uptake in the analysis. This allows us to directly invoke conservation of total activity in the analytic process. The equations are of the compartmental form where, for mathematical simplicity, we explicitly neglected any return of activity from the various tissues back into the blood. We were thus restricting our analysis to include only a mammillary picture of activity coming from blood to tumors, liver, and residual body.

A single lesion (m) was assumed in Equation 12.5. If more than one tumor were present, the set of equalities would have one additional equality, analogous to that of line one, for each added tumor. This increase in complexity is the reason the problem rapidly becomes intractable if one were to use a complete compartment description. The exponent α, since we have multiplied uptake (u) through by mass to convert to activity (A), is now positive and is expected to be approximately 0.67 by the arguments of Chapter 3. The possibility of tumor mass affecting liver activity is included in the epsilon term in the second equality of Equation 12.5 since CEA, the target antigen, is known to be secreted by the tumors into the blood and is rapidly sequestered (within a few minutes) by the liver. It was possible to integrate the four equations in closed form, and they were then fitted to the two data sets: untreated and SAP animals. Table 12.2 gives the resultant best-fit values for the various parameters of the model applied to a single-tumor implant. The parameters were found to be similar with the exception of the epsilon term, which decreased from $2.25 \times 10^{-2}\,h^{-1}$ in the control group to only $0.06 \times 10^{-2}\,h^{-1}$ in the pretreated cadre. Presumably, this change reflected the blocking of CEA sites in the liver by the cold antibody given as an injection prior to the radiolabeled antibody. The change was approximately a 40-fold reduction.

Figures 12.2 and 12.3 demonstrate the modeling results for the phosphate buffered saline (PBS) and SAP single-tumor examples, respectively. In these 48 h biodistribution results, the PBS case led to essentially a saturation curve with a limiting value of 15 %ID. As the tumor mass went much beyond 1.0 g, there was no further increase in the total accumulation in the lesion. On the other hand, if the animal was pretreated with 200 µg of cold anti-CEA protein, the accumulation showed an essentially linear increase with tumor mass well beyond 1.6 g.

We then used the results of Table 12.2 to predict 48 h results for both control and SAP animals with four to five tumors in each animal. It was found that the single-tumor modeling results, without adjustment, could be used to predict multiple-tumor results with relatively high accuracy. In general, it was seen that multiple lesions lead to greatly

TABLE 12.2 Fitted Parameters of the Single Tumor Model of Equation 12.5

PARAMETER	CONTROL ($n = 52$)	SAP TREATMENT ($n = 86$)
K_0	$1.02 \times 10^{-2}\,h^{-1}$	$0.81 \times 10^{-2}\,h^{-1}$
α	0.561	1.10
β	0.744	0.523
ε	$2.25 \times 10^{-2}\,h^{-1}$	$0.06 \times 10^{-2}\,h^{-1}$
k_{L0}	$3.0 \times 10^{-2}\,h^{-1}$	$3.1 \times 10^{-2}\,h^{-1}$
k_R	$3.9 \times 10^{-2}\,h^{-1}$	$2.8 \times 10^{-2}\,h^{-1}$

Source: Williams, L. E. et al., *Cancer Biother. Radiopharm.*, 16, 2001. With permission.
Note: The number of lesions (n) is indicated.

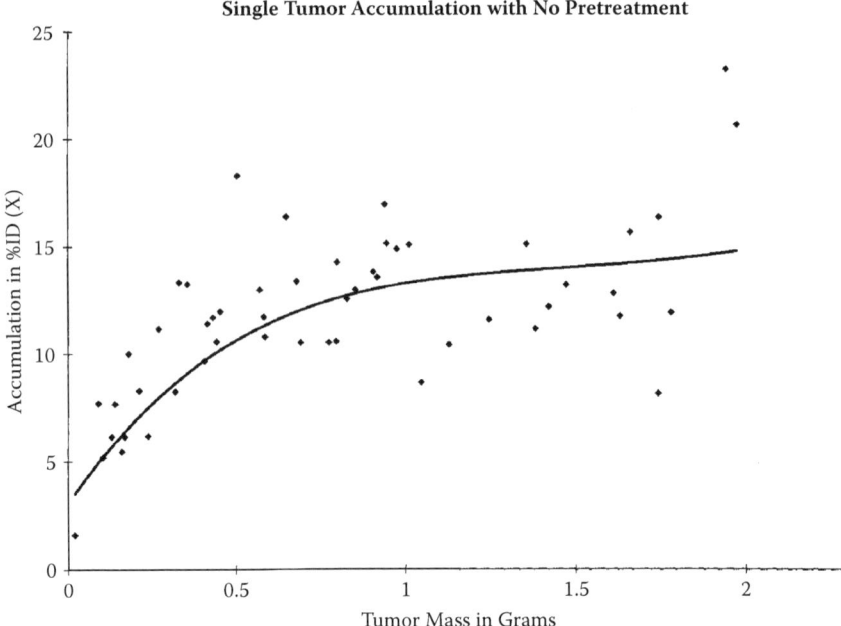

FIGURE 12.2 Predicted and measured representations of single tumor data in the nude mouse model. No specific antibody pretreatment (SAP) was used in this example. (From Williams, L. E. et al., *Cancer Biother. Radiopharm.*, 16, 2001, published by Mary Ann Liebert, Inc. With permission.)

reduced activity in any given size tumor mass. Reductions are on the order of threefold or higher in animals where large lesions, on the order of 1 to 2 grams, take up most of the injected activity. A representative example is shown in Figure 12.4 for an animal with five lesions of sizes ranging from 0.4 to 0.85 g in size each. Recall that total body mass for the mouse is approximately 20 g.

Results in the SAP case were not as striking. Again, the accumulation was reduced for a given size lesion, but not as much as in the PBS case. Figure 12.5 shows a single pretreated animal with five lesions; the decrease from the single-lesion accumulation value was only 50% due, presumably, to the reduced hepatic accumulation.

Application of these results to multitumor patient data would be applicable if two conditions were satisfied. First, the antigen must be tumor secreted (as is CEA) and taken up by the liver or some other normal tissue. Second, the dominant human lesions would have to be a comparable fraction of the total body mass (i.e., approached kilogram quantities). This is an important result and is indicative of the difficulty of treating patients with large (perhaps) untreated primaries existing in conjunction with metastatic sites. A dominant site will essential starve the smaller sites of the injected therapeutic tracer and may leave them untreated. This result is then an argument for taking the patient to surgery initially to reduce the tumor burden prior to beginning internal emitter therapy with radionuclides. A similar logic is possible for chemotherapy. These strategies are discussed more extensively in Chapter 10.

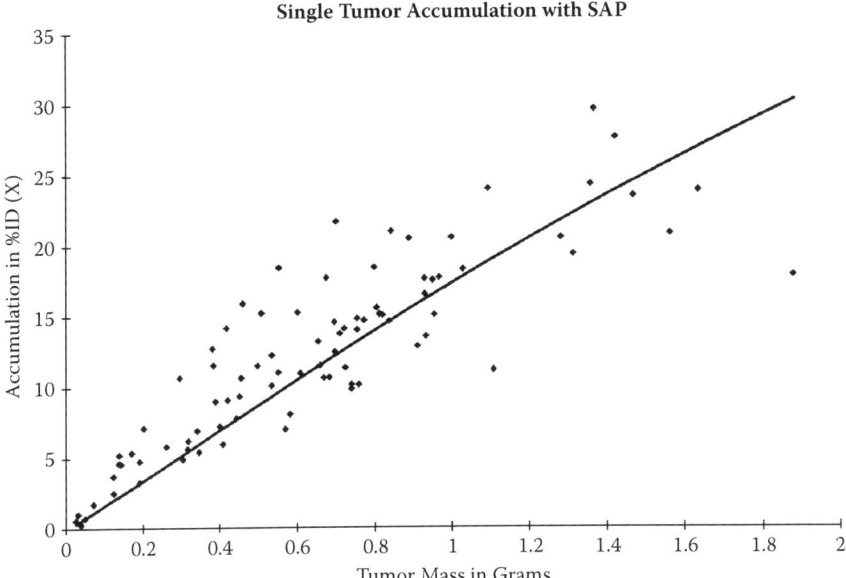

FIGURE 12.3 As in Figure 12.2 except SAP (200 μg) was applied to the nude mice prior to their receiving the radiolabeled anti-CEA antibody. (From Williams, L. E. et al., *Cancer Biother. Radiopharm.*, 16, 2001, published by Mary Ann Liebert, Inc. With permission.)

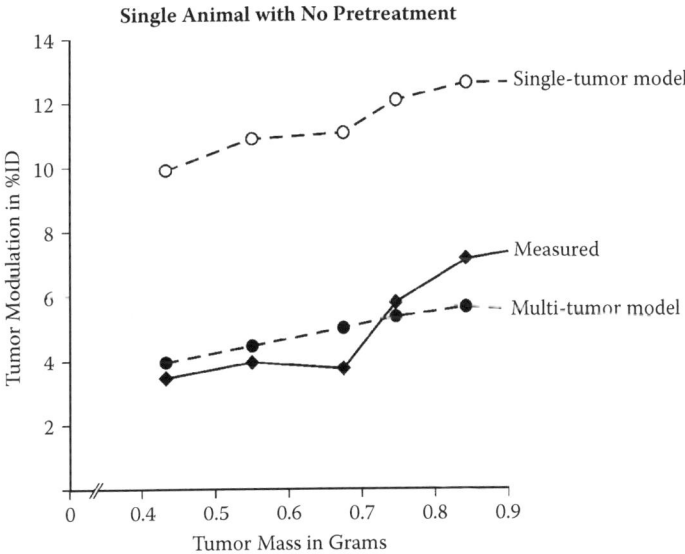

FIGURE 12.4 Single mouse with five lesions and no pretreatment. Note the disparity between the actual accumulation (%ID) and that predicted using the single-tumor parameters. With the multiple-tumor model, the agreement is much improved. (From Williams, L. E. et al., *Cancer Biother. Radiopharm.*, 16, 2001, published by Mary Ann Liebert, Inc. With permission.)

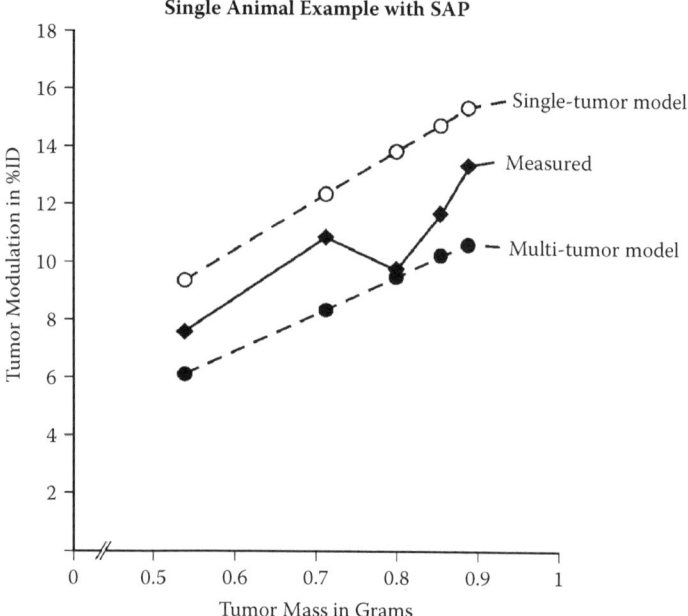

FIGURE 12.5   Single mouse with five lesions and SAP. As in the case of the control animals (cf. Figure 12.4), there is a decrease in the accumulation at 48 h compared with the prediction of the single-tumor model. The disparity is not as marked as in the control animals due to reduction of hepatic uptake of the labeled tracer. (From Williams, L. E. et al., *Cancer Biother. Radiopharm.*, 16, 2001. With permission.)

12.7 NUMERICAL VALUES OF S AND OTHER DOSE ESTIMATION FUNCTIONS (CHAPTER 7)

Given the integral of activity at a point, in a voxel or in an entire organ, the estimator faces the next challenge in absorbed dose calculations. This is the problem of finding a dose function (DF), which allows the target tissue (or voxel) energy deposition to be determined. We have shown three types of functions in this regard. First is the point-source function (PSF), which gives the absorbed dose at a point in a uniform tissue-equivalent medium that extends to infinity. Several such functions are known, and others may be made using Monte Carlo methods. This computation is a point-to-point DF. A question immediately arises as to the knowledge of point-by-point activity distribution in the sources. Generally, such information cannot be available due to finite resolution of any imaging system. Chapter 4 demonstrates that theoretical estimates may be made using point-source functions for simple geometric situations that arise in dosimetry.

A second level of dose function is the voxel source kernel. Here the dose estimator admits that point-source information is not obtainable and goes to the next higher level of spatial information. Voxel data are determined by either camera or PET means, and a 3-D table of dose functions (voxel source kernels) is determined by Monte Carlo methods. Notice that this is a 3-D calculation and not a matrix manipulation. Thus, the voxel source kernel gives dose in all relevant adjacent voxels. For pure beta emitters, this may be up to 1 cm or so from the source. As in the case of point-source functions, the medium inside the phantom or patient is assumed to be uniform. Likewise, the universe is assumed to be infinitely large (no edge effects) in the generation of the voxel source kernel.

The most common dose function is the **S** matrix of the MIRD Committee. This is a rectangular array, which permits organ-to-organ absorbed dose estimates. The calculation is done at the highest geometric level with an entire organ being the source and an entire organ being the target. In the usual tables of **S** matrices, the source organ is assumed to have uniform activity distribution, and the absorbed dose is the average given to the target. By far the majority of dose estimates made in nuclear medicine use this canonical form of the **S** matrix. In these computations, there will be one or another phantom assumed so that the Monte Carlo calculation of **S** may be carried out.

Some have considered the idea that the average patient, by the law of large numbers, will converge to the set of organ mass values assumed in one or another standard phantom. Therefore, the average of a large group of individuals should be the phantom value. This simple idea has two limitations. First, we must treat an individual patient so that the average is irrelevant to the entire therapy process. Admittedly, the average might be appropriate for diagnostic absorbed dose estimates, but those values are generally not of primary importance in nuclear medicine. Second, the patient will still demonstrate specific activity values and kinetics, which must be incorporated into the absorbed dose estimation process. In the final analysis, both activity data and organ size data are unique to the patient and must be handled as such. Echoing the disclaimer in detective novels, any resemblance of real patients or animals to a particular phantom is purely coincidental.

Multiple **S** matrices are known in the literature, and segments of several of these are tabulated in Chapter 7. One general feature of these matrices is that they have approximate symmetry about the diagonal values. Secondly, for nonpenetrating (alpha or beta) radiation, the diagonal elements are dependent on the inverse of the source (= target) tissue mass. In such cases, **S** is a diagonal matrix since the cross-organ terms vanish. Here, we explicitly do not include brake radiation in our analysis.

In the text, we have emphasized that two generic types of human dose estimates are generally of clinical interest. More common (Type I) is the calculation of the absorbed dose to one or another humanoid phantom given animal biodistribution data. In this case, the animal $\tilde{A}$ values are modulated on a per-organ basis using Equation 4.8. The phantom is chosen to be consistent with the objectives of the study; for example, a breast cancer agent would require an adult female humanoid **S** value. This sort of estimate is required in obtaining an investigational new drug (IND) approval from the U.S. Food and Drug Administration (FDA) since no direct human information is probably available for the initial application.

A second type of dose estimate (Type II) is one required for an individual patient therapy. Here, the $\tilde{A}$ value of that individual is used as obtained, but a novel **S** value set must be generated for the person in question. If the emitter is of particulate form, individual diagonal **S** values may be modulated by the ratio of the mass ratio of the organ of interest

$$S_{pt} = S_{phantom} \, (m_{phantom}/m_{pt})$$ (12.6)

where $m_{phantom}$ is the phantom's mass for that organ, and m_{pt} is the patient's organ mass. The latter is assumed to be best obtained via CT or MRI anatomic imaging as the diffuse nature of nuclear images would not be as adequate for assay.

Many calculations in the literature, including some done by this author, are not consistent with either of these two fundamental types. Instead, due to lack of anatomic data, the organ mass value of the phantom is used instead of true mass for the tissue of interest. In such cases, errors in the final dose may be factors of twofold or larger due to normal and disease-induced changes in the tissue mass value. A particularly important example of the latter is the enlargement of the splenic mass due to lymphoma. Another practical limitation is the lack of a mass value for the patient's red marrow. No direct means is available for such latter estimates, although it is known that the marrow mass decreases monotonically with age and with a variety of chemical therapies.

12.8 ABSORBED DOSE ESTIMATES WITHOUT CORRELATIONS (CHAPTER 8)

Chapter 8 described a number of calculations of absorbed dose for situations where there was no correlation with clinical outcome or no correlation attempted with such results. Such situations are very commonly the case in the dose estimation process. Nonetheless, one may compare one dose estimate with another if similar patient populations are being studied. As an example, let us consider a University of Alabama report on a multistep targeting regimen for treating gastrointestinal (GI) malignancies using the biotin–avidin system (Shen, Forero, et al. 2005). In this nine-patient study, a first injection of the fusion product CC49-avidin was given as a cold protein. A clearing agent was then injected at 42 or 72 h later, followed, in turn, by labeled biotin 24 h after the clearing step. The radiolabel would be [111]In (5 mCi) for imaging (and dose estimation) and ^{90}Y (10 mCi/m^2) for therapy of the various metastatic sites. Because of the small molecular weight (MW) of the labeled biotin, the radioactive material cleared relatively quickly via the renal system. Organ sizes were corrected using patient CT scans, so this was a Type II estimation of absorbed doses. Tumors were visible in some of these colorectal cancer patients, and their dose values were also estimated. An example is shown in Figure 12.6.

Absorbed dose estimates are given in Table 12.3 for the University of Alabama results as well as an earlier trial using CC49 as an intact antibody. In the prior example, Leichner, Akabani, et al. (1997) also corrected for patient organ size, so the two results are comparable as Type II calculations. The manifest improvement in the normal organ

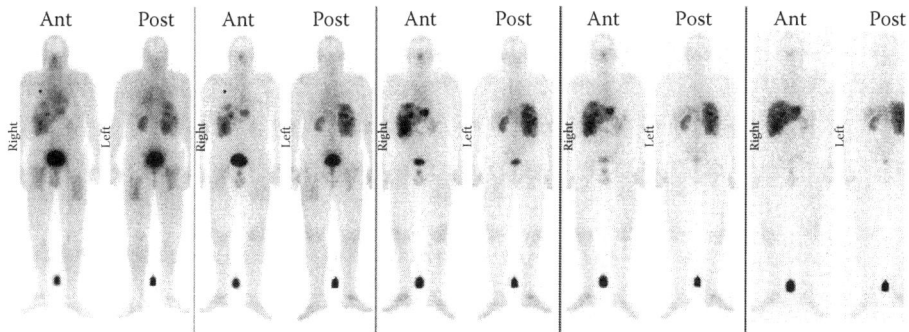

FIGURE 12.6 Three-step targeting of a patient with colorectal metastatic disease in the liver. Note the rapid clearance of the ^{111}In-Biotin out of the body with some indication of both kidneys and bladder. Times are 0, 3h, 1d, 2d, and 4d, from left to right. (From Shen, S. et al., *J. Nucl. Med.*, 46, 4, 2005. Reprinted with permission from the Society of Nuclear Medicine.)

TABLE 12.3 Patient-Specific Mean Absorbed Dose Estimates in GI Cancer Patients

ORGAN OR LESION	INTACT CC49 ANTIBODY[a]	PRETARGETED BIOTIN[b]
Liver	10.1 (4.6–22) mGy/MBq	1.01 (0.17–1.9) mGy/MBq
Red marrow	0.68 (0.43–1.1)	0.060 (0.032–0.92)
Spleen	8.1 (3.2–17)	0.63 (0.16–1.2)
Tumors	9.3 (1.9–22)	7.8 (1.1–33)

Source: Shen, S. et al., *J. Nucl. Med.*, 46, 4, 2005. With permission.
Note: Values in parentheses are the range of the data sets.
[a] From Leichner, P. K. et al., *J. Nucl. Med.*, 38, 4, 1997.
[b] Shen, S. et al. (2005).

dose estimates using the three-step process is essentially one order of magnitude for liver, spleen, and red marrow. At the same time, the tumor dose estimates were lower only by about 20% compared with the intact antibody of the same type. There was a clear advantage in the use of a small MW compound as the one carrying the therapeutic radionuclide into the patient. We should add that there was no specific comparison to the Leichner et al. results for kidney dose estimates as no renal values were available from that earlier data set. The Alabama group did report a mean value of 1.9 mGy/MBq for the renal system. While larger than the estimates for liver, RM, and spleen, the kidney dose per activity is still well under that delivered to the metastatic colorectal lesions in the three-step process. Of particular interest was the very low specific absorbed dose estimate for the red marrow using labeled biotin.

Use of avidin in the context of patient therapy has one significant risk factor: development of patient antiavidin antibodies. Thus, there may only be a single therapy possible with a given patient in a multistep process involving avidin. It is for this reason that other moieties have been used to provide the binding necessary to target multiple tumor sites during the labeled step of the process.

12.9 ABSORBED DOSE CORRELATIONS WITH BIOLOGICAL EFFECTS (CHAPTER 9)

In published reports, there are relatively few instances wherein the tumor or normal organ dose estimates correlate with clinical results. Several examples have been cited in Chapter 9 including the measurement of chromosome translocations. The lymphoma study from the University of Michigan was also featured as it represented an example at the statistical edge of significance for tumor regression versus absorbed dose estimates. In a similar vein, a recent report from Switzerland (Delaloye, Antonescu, et al. 2009) contains a summary of a 13-nation Phase III trial involving treating indolent lymphoma with ^{90}Y ibritumomab tiuxetan (Zevalin). As described in Chapter 8, unlabeled antibody (Rituxan) was given at 250 mg/m^2 just prior to the targeted radionuclide therapy (TRT). The investigators made dose estimates for 57 patients of 70 accrued to the trial. MIRDOSE3 (Stabin 1996) phantom data were used to estimate **S** values for the patients, so this was a mixed type of dose estimation (neither Type I or type II) for the normal organs. Whole-body and red marrow estimated doses did not differ significantly between the partial recovery (PR) and complete recovery (CR) patients. Yet both the whole-body and RM doses did correlate positively with progression-free survival (PFS). In the former case, the *p*-value was .0006, and in the RM example it was $p = .03$. The patients were followed for a median time of 3.5 years in this study. Consistent with most other clinical trials, neither absorbed dose estimate correlated with hematological toxicity.

The Swiss report noted that, due to the earlier chemotherapy of these patients, there was minimal disease at the outset of their TRT trial. Hence, no tumor sites could be adequately visualized during the ^{111}In-ibritumomab tiuxetan aspect of the protocol. Therefore, no direct comparison could be made to the earlier Michigan results regarding tumor dose estimates versus tumor size regression. It is important to notice that the therapy in the 13-nation report is presumably due to treatment of minimal residual disease—that is, very small lesions that could not be visualized using nuclear medicine techniques. In this case, wherein the mass law of Chapter 3 may be valid, one might expect that the minimally sized lesions are most amenable to radiation treatment by internal emitters. This is a topic that has possible future importance in the follow-up care of patients after an initial treatment using surgery, external beam irradiation, chemotherapy, or other interventions.

12.10 COMBINATIONS OF RADIATION AND OTHER THERAPIES (CHAPTER 10)

Multiple therapies have been applied in cancer treatment. Among the modalities are surgery, hyperthermia, chemotherapy, external beam therapy, and immune intervention. While any of these may be effective alone, a growing number of radiation oncologists have invoked the concept of using more than one modality at the same time in the patient. In the case of antibodies, one strategy of this type is giving the animal (not yet tried in patients) a predosing of an unlabeled agent that increases the expression of the antigen of interest.

A recent report by Sharkey, Press, et al. (2009) emphasized the importance of having the appropriate amount of unlabeled (cold) antibody being given prior to the TRT. One agent that has had this aspect investigated clinically has been Zevalin (ibritumomab tiuxetan). Effects of the unlabeled pretreatment include slower clearance from the blood as well as reduced splenic uptake of the ^{111}In form of ibritumomab tiuxetan. In the latter case, a fourfold reduction of the amount seen in the spleen was recorded (Knox, Goris, et al. 1996). There is contrary evidence that large amounts of cold antibody against the same molecular target as the labeled agent may reduce the uptake in lesions. Several animal studies have shown that ibritumomab tiuxetan uptake in xenografts can be substantially reduced when rituximab is given prior to the radioactive material (Gopal, Press, et al. 2008). Sharkey et al. argued that one solution to this problem is to use unlabeled agents that react with a different antigen than the radiolabeled agent. In this way, both the immune response of the mammal and the radiation dose of the TRT can act in concert.

Use of the sigmoidal functions in describing multimodality therapies may become a primary form of future analysis. One can, using a saturation function, determine which fraction of a therapy is assigned to each of two treatment types: for example, radiation versus antibody as exemplified by the University of Michigan studies of the tositumomab antibody labeled with [131]I. In that case, the pretreatment with cold B1 antibody was seen to be a major contribution to the treatment of NHL patients. Similar sigmoidal analyses would be an attractive tool to evaluate other multiple therapies (e.g., chemotherapy at the same time as TRT).

12.11 ALLOMETRY (CHAPTER 11)

An issue of concern in the development of novel RPs is the relationship (if any) between animal data sets and the eventual clinical results. In public literature, there has been a conspicuous lack of follow-up to the original targeting proof-of-concept biodistributions typically obtained with nude mice. Allometry permits the comparisons to be made directly or based on a mathematical model. Probably the most popular of the latter is the power-law format. One example is determination of an organ's mass given the total mass of the animal. Explicitly

$$m = m_0 M^b \tag{12.7}$$

where m is the mass of a major mammalian organ, and M is the animal's whole-body mass value. This relationship, as we noted in Chapter 11, must fail outside the endpoints of the organ distribution. In other words, as the mass of a given organ becomes sufficiently small, it cannot function in the mammal. At or below this size, the power law cannot hold for that tissue.

It is now hypothesized that mammalian tissues fall into two camps: the standard organ and the control organ. Standard organs such as lung, heart, and blood show a whole-body mass exponent (b in Equation 12.7) of close to unity. Control tissues such as brain, kidneys, and liver have b approximately $= 0.75$. Reasons for this dichotomy may be argued, but the essence of the situation is one of pragmatism. By having the power-law relationship, the observer can estimate tissue sizes for mammals in general—even if that animal has become extinct without being assayed or has never been captured for evaluation. Such relationships have had wide interest in the biophysics community but little usage in pharmaceutical development. The Rule of Exponents developed by workers at the FDA (Mahmood 2009) is one example of how to use this variation of exponents to better represent the data from ordinary and control organs.

Such mass relationships lead to pharmaceutical dilution factors that are very large when comparing mouse with man—the ratio being on the order of 3,500. Thus, the amount of mg of agent or MBq of activity must be increased by this factor when scaling up from the mouse experiments. It also leads to the natural unit of human uptake: %ID/kg. The natural unit of patient injected radioactivity is then the GBq rather than the MBq of a small rodent.

Variation of kinetic parameters from mouse to man is a more subtle feature than organ mass or distribution volume. A number of authors have shown that the clearance rates and half-times show only slight changes in going from the smaller to the larger of the two species. This variation is on the order of $M^{-1/4}$ or less, so the patient study is slower than the mouse—but not that much slower. In the assay of various cognate proteins, the variation in blood clearance is more marked in the lower MW compounds but tends to vanish for intact antibodies. In an attempt to simplify this issue, we have suggested use of a single parameter to represent the blood curve. The lowest-order form would be first moment of the blood biodistribution. When this parameter is computed for intact antibodies in mice and patients, the results are found to be similar.

12.12 SUMMARY

The development of novel radiopharmaceuticals is made difficult by the large number of possible agents and the multiple ways each may be used in targeting to disease sites or to a given molecule. Molecular engineering, as predicted by Feynman in his APS Pasadena talk in 1965, indeed does have "plenty of room at the bottom"—perhaps too much room. Yet there are relatively few available and willing patients with a given disease at any institution—or even within an entire country. Thus, obtaining a group of test subjects for a new RP is difficult. However, from statistical considerations, larger groups of possible patients are desirable due to variations inherent in any population being studied.

The lack of patients implies that agents must be preselected for any RP trial. In this text, I propose two figures of merit to expedite the RP selection process. The IFOM is one of these indicators. It is validated in two instances and, more importantly, is found to be superior to the traditional indicator (r), which is the ratio of tumor to blood uptake.

Generalization of IFOM to other background tissues such as nonblood organs is also included in the application of this new indicator. Of particular interest are the liver and kidneys for such comparisons, as these two organs may be dominant uptake tissues for any novel RP. For therapy, the analogous indicator is TFOM, which also has superior features to the traditional ratio of tumor to blood AUCs.

Along with the biodistribution data are the resultant absorbed doses to the various tissues in the body—including the tumor. Such quantities can almost never be measured, so dose estimation becomes our method of choice. Three strategies are present in the literature for such calculations. The most common form is the canonical equality relating dose to a matrix product of S and $\tilde{A}$. One or another of a certain set of phantoms is used to represent the anatomy of the patient during the calculation of the absorbed dose.

Both S and $\tilde{A}$ present problems to the estimator of dose. The former quantity is usually based on phantom analysis using Monte Carlo calculations, and the latter has no single optimal method of being determined in clinical practice. Thus, the uncertainty of their matrix product for a given patient may be quite high. Variation of patient mass is the primary problem with the use of tabulated S values in the literature. In our hands, these variations could be factors of twofold to threefold due to biological variation or disease state.

Modeling of the biodistribution data leads to multiexponential forms as well as power-law format. Integration is then possible in closed forms. Comparison of data sets is another result of the use of modeling. The last has application to the allometric studies of one species versus another with regard to clearance or other RP parameters.

Many estimates have been made of animal and patient absorbed doses. In the former case, uniformity of the test animal in terms of size, shape, and radiation response leads to relative good correlation between absorbed dose and effects. In particular, the absorbed dose is a better indicator than activity, activity/body surface area, or activity/body mass. Usually, however, there is little correspondence of absorbed dose estimates with either tumor regression or normal organ toxicity in the clinic. A few TRT trials have given good correlation of dose to outcomes. In such results, it is typically the case that the experimenter has used accurate tissue masses as well as quantitative activity levels in vivo. There has also been a growing use of BED to improve the relationship.

There are completed and ongoing animal and clinical trials using combinations of one type of therapy and another for cancer treatment. Many strategies have been applied; most are concerned with radiation adjuvants, which increase the effect of a given Gy dose to the tissue in question. Timing of the administration of these nonradioactive moieties relative to the internal emitter treatment is a key issue. Similar arguments can be made for physical interventions such as hyperthermia or external beam radiation.

Finally, we have the ultimate question of allometry. Do animal data, in some sense, predict a clinical result? This topic is, at the present time, under intense study. While organ mass sizes can be correlated across species using power laws, the kinetics of pharmaceuticals has not been examined nearly as extensively. We have shown that the clearance of the cT84.66 cognates from the blood of mice and men are reasonably similar. In fact, the intact (parent) antibody has mean residence times that are almost the same for the two species. If the mass of the protein decreases to values approaching 50 kDa, on the other hand, the mean time is found to be appreciably faster in the mouse; the ratio between the two species is on the order of fourfold. These results are examples of discovering what relationships exist between two mammals of very different sizes handling the same RP. One would hope that similar correspondences might be determined in future work in this field. We have argued that pharmacology experimenters should be required to publicly give such results for their agents, so we have greater confidence between applications in one animal and another—and between any given animal and humans.

REFERENCES

Brambell, F. W., R. Halliday, et al. 1958. Interference by human and bovine serum and serum protein fractions with the absorption of antibodies by suckling rats and mice. *Proc R Soc Lond B Biol Sci* **149**(934): 1–11.

Delaloye, A. B., C. Antonescu, et al. 2009. Dosimetry of 90Y-ibritumomab tiuxetan as consolidation of first remission in advanced-stage follicular lymphoma: results from the international phase 3 first-line indolent trial. *J Nucl Med* **50**(11): 1837–43.

Gopal, A. K., O. W. Press, et al. 2008. Rituximab blocks binding of radiolabeled anti-CD20 antibodies (Ab) but not radiolabeled anti-CD45 Ab. *Blood* **112**(3): 830–5.

Kenanova, V., T. Olafsen, et al. 2005. Tailoring the pharmacokinetics and positron emission tomography imaging properties of anti-carcinoembryonic antigen single-chain Fv-Fc antibody fragments. *Cancer Res* **65**(2): 622–31.

Kenanova, V., T. Olafsen, et al. 2007. Radioiodinated versus radiometal-labeled anti-carcinoembryonic antigen single-chain Fv-Fc antibody fragments: optimal pharmacokinetics for therapy. *Cancer Res* **67**(2): 718–26.

Knox, S. J., M. L. Goris, et al. 1996. Yttrium-90-labeled anti-CD20 monoclonal antibody therapy of recurrent B-cell lymphoma. *Clin Cancer Res* **2**(3): 457–70.

Koral, K. F., Y. Dewaraja, et al. 2003. Update on hybrid conjugate-view SPECT tumor dosimetry and response in 131I-tositumomab therapy of previously untreated lymphoma patients. *J Nucl Med* **44**(3): 457–64.

Leichner, P. K., G. Akabani, et al. 1997. Patient-specific dosimetry of indium-111- and yttrium-90-labeled monoclonal antibody CC49. *J Nucl Med* **38**(4): 512–6.

Li, L., P. J. Yazaki, et al. 2006. Improved biodistribution and radioimmunoimaging with poly(ethylene glycol)-DOTA-conjugated anti-CEA diabody. *Bioconjug Chem* **17**(1): 68–76.

Liu, A., L. E. Williams, et al. 1996. A CT assisted method for absolute quantitation of internal radioactivity. *Med Phys* **23**(11): 1919–28.

Mahmood, I. 2009. Pharmacokinetic allometric scaling of antibodies: application to the first-in-human dose estimation. *J Pharm Sci* **98**(10): 3850–61.

Patlak, C. S., R. G. Blasberg, et al. 1983. Graphical evaluation of blood-to-brain transfer constants from multiple-time uptake data. *J Cereb Blood Flow Metab* **3**(1): 1–7.

Sharkey, R. M., O. W. Press, et al. 2009. A re-examination of radioimmunotherapy in the treatment of non-Hodgkin lymphoma: prospects for dual-targeted antibody/radioantibody therapy. *Blood* **113**(17): 3891–5.

Shen, S., A. Forero, et al. 2005. Patient-specific dosimetry of pretargeted radioimmunotherapy using CC49 fusion protein in patients with gastrointestinal malignancies. *J Nucl Med* **46**(4): 642–51.

Stabin, M. G. 1996. MIRDOSE: personal computer software for internal dose assessment in nuclear medicine. *J Nucl Med* **37**(3): 538–46.

Stabin, M. G. 2003. Radiotherapy with internal emitters: what can dosimetrists offer? *Cancer Biother Radiopharm* **18**(4): 611–7.

Vose, J. M., R. L. Wahl, et al. 2000. Multicenter phase II study of iodine-131 tositumomab for chemotherapy-relapsed/refractory low-grade and transformed low-grade B-cell non-Hodgkin's lymphomas. *J Clin Oncol* **18**(6): 1316–23.

Williams, L. E., B. G. Beatty, et al. 2001. Accumulation of radiolabeled anti-CEA antibody (mT84.66) in the case of multiple LS174T tumors in a nude mouse model. *Cancer Biother Radiopharm* **16**(2): 147–57.

Williams, L. E., R. B. Duda, et al. 1988. Tumor uptake as a function of tumor mass: a mathematic model. *J Nucl Med* **29**(1): 103–9.

Williams, L. E., A. Liu, et al. 1995. Figures of merit (FOMs) for imaging and therapy using monoclonal antibodies. *Med Phys* **22**(12): 2025–7.

Wiseman, G. A., B. R. Leigh, et al. 2003. Radiation dosimetry results from a Phase II trial of ibritumomab tiuxetan (Zevalin) radioimmunotherapy for patients with non-Hodgkin's lymphoma and mild thrombocytopenia. *Cancer Biother Radiopharm* **18**(2): 165–78.

Index